Periodic Table of the Elements with the Gmelin System Numbers

1 H 2																	2 He 1
3 Li 20	4 Be 26										5 B 13	6 C 14	7 N 4	8 O 3	9 F 5	10 Ne 1
11 Na 21	12 Mg 27										13 Al 35	14 Si 15	15 P 16	16 S 9	17 Cl 6	18 Ar 1
19 * K 22	20 Ca 28	21 Sc 39	22 Ti 41	23 V 48	24 Cr 52	25 Mn 56	26 Fe 59	27 Co 58	28 Ni 57	29 Cu 60	30 Zn 32	31 Ga 36	32 Ge 45	33 As 17	34 Se 10	35 Br 7	36 Kr 1
37 Rb 24	38 Sr 29	39 Y 39	40 Zr 42	41 Nb 49	42 Mo 53	43 Tc 69	44 Ru 63	45 Rh 64	46 Pd 65	47 Ag 61	48 Cd 33	49 In 37	50 Sn 46	51 Sb 18	52 Te 11	53 I 8	54 Xe 1
55 Cs 25	56 Ba 30	57** La 39	72 Hf 43	73 Ta 50	74 W 54	75 Re 70	76 Os 66	77 Ir 67	78 Pt 68	79 Au 62	80 Hg 34	81 Tl 38	82 Pb 47	83 Bi 19	84 Po 12	85 At 8a	86 Rn 1
87 Fr 25a	88 Ra 31	89*** Ac 40	104 71	105 71													

* NH₄ 23

**Lanthanides 39	58 Ce 39	59 Pr	60 Nd	61 Pm	62 Sm	63 Eu	64 Gd	65 Tb	66 Dy	67 Ho	68 Er	69 Tm	70 Yb	71 Lu
***Actinides	90 Th 44	91 Pa 51	92 U 55	93 Np 71	94 Pu 71	95 Am 71	96 Cm 71	97 Bk 71	98 Cf 71	99 Es 71	100 Fm 71	101 Md 71	102 No 71	103 Lr 71

A Key to the Gmelin System is given on the Inside Back Cover

Gmelin Handbook of Inorganic Chemistry

8th Edition

Index 2, Vol. 3

Gmelin Handbook
of Inorganic Chemistry

8th Edition

Gmelin Handbuch der Anorganischen Chemie

Achte, völlig neu bearbeitete Auflage

Prepared
and issued by

Gmelin-Institut für Anorganische Chemie
der Max-Planck-Gesellschaft
zur Förderung der Wissenschaften

Director: Ekkehard Fluck

Founded by | Leopold Gmelin

8th Edition | 8th Edition begun under the auspices of the
Deutsche Chemische Gesellschaft by R. J. Meyer

Continued by | E.H.E. Pietsch and A. Kotowski, and by
Margot Becke-Goehring

Springer-Verlag Berlin Heidelberg GmbH 1989

The following Gmelin Formula Index volumes have been published up to now:

Formula Index

Volume	1	Ac–Au
Volume	2	$B-Br_2$
Volume	3	Br_3-C_3
Volume	4	C_4-C_7
Volume	5	C_8-C_{12}
Volume	6	$C_{13}-C_{23}$
Volume	7	$C_{24}-Ca$
Volume	8	Cb–Cl
Volume	9	Cm–Fr
Volume	10	Ga–I
Volume	11	In–Ns
Volume	12	O–Zr
		Elements 104–132

Formula Index 1st Supplement

Volume	1	Ac–Au
Volume	2	$B-B_{1.9}$
Volume	3	B_2-B_{100}
Volume	4	$Ba-C_7$
Volume	5	C_8-C_{17}
Volume	6	$C_{18}-C_x$
Volume	7	Ca–I
Volume	8	In–Zr
		Elements 104–120

Formula Index 2nd Supplement

Volume	1	$Ac-B_{1.9}$
Volume	2	B_2-Br_x
Volume	3	$C-C_{6.9}$ (present volume)

Gmelin Handbook
of Inorganic Chemistry

8th Edition

INDEX

Formula Index

2nd Supplement Volume 3

$C-C_{6.9}$

AUTHORS   Rainer Bohrer, Helga Hartwig, Renate Jonuschat,
Bernd Kalbskopf, Renate Nohl, Uwe Nohl,
Hans–Jürgen Richter–Ditten, Paul Velić,
Rudolf Warncke

SYSTEM SUPPORT   Gottfried Olbrich

CHIEF EDITOR   Helga Hartwig

Springer-Verlag Berlin Heidelberg GmbH 1989

THE VOLUMES OF THE GMELIN HANDBOOK ARE EVALUATED FROM 1980 THROUGH 1987

Library of Congress Catalog Card Number: Agr 25-1383

ISBN 978-3-662-06193-0   ISBN 978-3-662-06191-6 (eBook)
DOI 10.1007/978-3-662-06191-6

© by Springer-Verlag Berlin Heidelberg 1988
Originally published by Springer-Verlag, Berlin · Heidelberg · New York · London · Paris · Tokyo in 1988
Softcover reprint of the hardcover 8th edition 1988

Preface

The Gmelin Formula Index and its First Supplement covered those volumes of the Eighth Edition of the Gmelin Handbook which had been issued up to the end of 1979.

The present Second Supplement updates the Index by inclusion of the volumes which appeared up to the end of 1987. With this Second Supplement all compounds described in the Gmelin Handbook of Inorganic Chemistry in the period between 1924 and 1987 can be located. The basic structure of the Formula Index remains the same as in the previous editions.

Computer techniques were employed in the preparation and print of the Second Supplement. The data acquisition, sorting, and further data handling were performed with the aid of a series of computer programs developed by staff members of the former "Online Group" of the Gmelin Institute, now at Chemplex GmbH, and by the printer "Universitäts-druckerei H. Stürtz AG, Würzburg".

Whereas the Handbook itself will continue to appear in printed form, the present Second Supplement of the Index is intended to be the last one issued in print. The cumulated contents of the Index and its Supplements are contained in the Gmelin Formula Index (GFI) database which is available to the scientific community via STN. This database will be updated annually to include the published Handbook volumes.

Frankfurt am Main   Helga Hartwig
October 1988

Instructions for the Formula Index

First Column (Empirical Formula)

The empirical formulas are arranged by alphabetical order of the element symbols and by increasing values of the subscripts. Any indefinite subscripts are placed at the end. Ions always appear after the neutral species, positive ions precede the negative ones.

H_2O is included in the empirical formula only if it is an integral part of a complex, as written in the second column. Compounds which are isolated only as solvates are found under the empirical formulas both including the solvent molecule and excluding it. Multicomponent systems (solid solutions, melts, etc.) are listed under the empirical formulas of their respective components. However, solutions are found only under the solute. Polymers of type $(AB)_n$ are listed under AB.

Second Column (Conventional Formula)

The second column presents a structural formula as it is usually given in the handbook text. In many cases, however, another form is shown if additional structural features could be detailed. For the elements, the elemental names are given.

Entries with the same empirical formula but different structural formulas are arranged in the following order: elements or compounds, isotopic species, polymers, hydrates, and multicomponent systems.

For multicomponent systems the components are arranged in the sequence: inorganic components – organic components – water. The inorganic components are arranged alphabetically, and the organic components by the number of carbon atoms. If a component is a single element it is always represented by the unsubscripted symbol. The term "system" is used in a restricted sense in this index: it represents equilibrium mixtures described by phase diagrams or sometimes by, e.g., eutectic points.

Elements and compounds treated extensively in the handbook are subdivided by topics, e.g., physical properties, preparation, electrochemical behavior, or toxicity.

Third Column (Volume and Page Numbers)

The first symbol is the element which is treated in a given volume, followed by an abbreviated form of the type of volume, including the Part or Section. The page numbers are given after a hyphen. The following abbreviations are used for the type of volume:

MVol.	Main Volume (Hauptband)
SVol.	Supplement Volume (Ergänzungsband)
Org. Comp.	Organic Compounds
Org. Verb.	Organische Verbindungen
PFHOrg.	Perfluorohalogenoorganic Compounds of Main Group Elements
SVol.GD	Gmelin-Durrer, Metallurgy of Iron
Biol.Med.Ph.	Bor in Biologie, Medizin und Pharmazie

Volume descriptors like "1st Suppl. Vol. 2" are abbreviated as "SVol. 1/2". For instance, the entry "B: B Comp.SVol. 1/2-345" indicates that the information can be found on page 345 of the boron volume: Boron Compounds 1st Suppl. Vol. 2.

C – C$_{6.9}$

CClF₃O₂S .... CF₃SO₂Cl .... F: PFHOrg.SVol.3-158/9, 170/7, 180/1

CClF₃O₃S .... ClOSO₂CF₃ .... F: PFHOrg.SVol.3-55, 69, 82/3, 102/3

CClF₃S .... CF₂ClSF .... F: PFHOrg.SVol.2-116, 117

− .... CF₃SCl .... F: PFHOrg.SVol.2-117, 120/1, 124/31, 135/6, 138, 144/5, 148/63

CClF₃S₂ .... CF₃SSCl .... F: PFHOrg.SVol.2-173, 184, 196, 199/200

CClF₃Se .... CF₃SeCl .... F: PFHOrg.SVol.3-215, 227, 242

CClF₄N .... CF₃NFCl .... F: PFHOrg.8-154/5, 176, 205, 206

− .... ClCF₂NF₂ .... F: PFHOrg.8-154/5, 177, 206

CClF₄NO₂S .... (CF₃)(SO₂F)NCl .... F: PFHOrg.8-162, 187, 205

CClF₄NS .... ClCF₂NSF₂ .... F: PFHOrg.9-2, 13

CClF₄P .... CF₃PFCl .... F: PFHOrg.SVol.1-107, 108, 118

CClF₅N₂ .... ClCF(NF₂)₂ .... F: PFHOrg.8-163/4, 188

CClF₅S .... CF₂ClSF₃ .... F: PFHOrg.SVol.3-10, 33, 44

CClF₅Si .... CF₃SiF₂Cl .... F: PFHOrg.SVol.1-43/8

CClF₆PS .... CF₂ClSPF₄ .... F: PFHOrg.SVol.2-117, 244/5, 249

− .... CF₃SPF₃Cl .... F: PFHOrg.SVol.2-244, 250

CClF₇S .... CF₃SF₄Cl .... F: PFHOrg.SVol.3-187/8, 195, 207

CClH₃I₂Sn .... CH₃SnI₂Cl .... Sn: Org.Comp.8-144

CClH₇Se₂ .... CH₃Cl · 2 H₂Se · 17 H₂O = 8 CH₃Cl · 16 H₂Se · 136 H₂O .... Se: SVol.B1-62

CClH₁₅N₅O₃Rh ..... [Rh(NH₃)₅Cl]CO₃ .... Rh: SVol.B2-141

CClH₁₅N₅O₇Rh ..... [Rh(NH₃)₅(CO₃)](ClO₄) · H₂O .... Rh: SVol.B2-136

CClNO₃S .... ClCN · SO₃ .... S: SVol.3-316

CClNO₆S₂ .... ClCN · 2 SO₃ .... S: SVol.3-316

CClN₃S₂ .... S₂N₃CCl .... S: S-N Comp.4-1/2

CClN₅O₃S₅ .... S₄N₄ · C(O)NSO₂Cl .... S: S-N Comp.2-224

CClORu .... Ru(CO)Cl .... Pt: SVol.A1-45

CCl₂CuO⁻ .... [(CO)CuCl₂]⁻ .... Cu: Org.Comp.3-191, 203

CCl₂FHO₃S .... CFCl₂SO₃H .... F: PFHOrg.SVol.3-49

CCl₂FN .... Cl₂CNF .... F: PFHOrg.8-206

CCl₂FNO .... FC(O)NCl₂ .... F: PFHOrg.8-159, 183

CCl₂FNOS .... FC(O)NSCl₂ .... F: PFHOrg.9-5/6, 20, 34

CCl₂FNO₂ .... Cl₂CFNO₂ .... F: PFHOrg.8-19

CCl₂F₂ .... CCl₂F₂
           Sorption on tungsten .... W: SVol.A7-206

CCl₂F₂S .... CF₂ClSCl .... F: PFHOrg.SVol.2-117, 121, 125/7, 131/2, 142/5, 148, 157

CCl₂F₂S₂ .... CF₂ClSSCl .... F: PFHOrg.SVol.2-173

CCl₂F₂Se .... CF₂ClSeCl .... F: PFHOrg.SVol.3-217, 227

CCl$_9$Mn$_5$NS	Mn$_5$(SCN)Cl$_9$ · 2 H$_2$O	Mn:	MVol.C7-237/8
CCoNiO$_4$Tc	NiCoCTcO$_4$	Tc:	SVol.2-76
CCo$_2$N$_4$OS$_4$	Co$_2$(CO)S$_4$N$_4$ = S$_4$N$_4$ · Co$_2$(CO)	S:	S-N Comp.2-191, 225
CCrF$_3$O$_3$S^{2+}	[CrOSO$_2$CF$_3$]$^{2+}$	F:	PFHOrg.SVol.3-61
CCr$_3$	Cr$_3$C solid solutions		
	(Cr,Mn)$_3$C	Mn:	MVol.C7-167
CCr$_3$Ni$_2$Si	Cr$_3$Ni$_2$SiC	Si:	SVol.B3-368
CCr$_5$Si$_3$	Cr$_5$Si$_3$C solid solutions		
	Cr$_5$Si$_3$C-Cr$_5$Si$_3$	Si:	SVol.B3-365, 366
CCsFH$_2$O$_6$U	Cs[UO$_2$F(CO$_3$)(H$_2$O)]	U:	SVol.C13-25
CCsF$_2$NO$_3$S	Cs[N(SO$_2$F)C(O)F]	F:	PFHOrg.9-9, 31, 33/4
CCsF$_3$O$_2$S	CsOS(O)CF$_3$	F:	PFHOrg.SVol.3-3
CCsF$_3$O$_3$S	CsOSO$_2$CF$_3$	F:	PFHOrg.SVol.3-60, 102/4
CCsF$_3$Se	CF$_3$SeCs	F:	PFHOrg.SVol.3-215/6, 241
CCuF$_2$O	(CO)CuF$_2$	Cu:	Org.Comp.3-191
CCuF$_3$	CF$_3$Cu	Cu:	Org.Comp.1-18/22, 26
CCuF$_3$O$_3$S	CuOSO$_2$CF$_3$	F:	PFHOrg.SVol.3-62, 106/7
CCuF$_3$S	CuSCF$_3$	F:	PFHOrg.SVol.2-251, 257, 258/62
CCuHN	HCNCu	Cu:	Org.Comp.3-242
−	NCHCu (radical)	Cu:	Org.Comp.3-242
CCuHO$_5$S	[(CO)Cu]HSO$_4$	Cu:	Org.Comp.3-190
CCuH$_3$	CH$_3$Cu	Cu:	Org.Comp.1-8/16
−	CH$_3$Cu · x AlCl$_3$	Cu:	Org.Comp.1-328/30
−	CH$_3$Cu · x B(C$_2$H$_5$)$_3$ · y LiI	Cu:	Org.Comp.1-335, 353/4
−	CH$_3$Cu · x B(C$_4$H$_9$-n)$_3$ · y LiI	Cu:	Org.Comp.1-335, 352/3
−	CH$_3$Cu · x BCl$_3$ · y LiI	Cu:	Org.Comp.1-335, 352
−	CH$_3$Cu · x BF$_3$	Cu:	Org.Comp.1-333
−	CH$_3$Cu · x BF$_3$ · y LiCN	Cu:	Org.Comp.1-335, 347
−	CH$_3$Cu · x BF$_3$ · y LiHal	Cu:	Org.Comp.1-335, 339, 341
−	CH$_3$Cu · x BF$_3$ · y LiI	Cu:	Org.Comp.1-335/8, 341/7, 350
−	CH$_3$Cu · x BF$_3$ · y MgBrI	Cu:	Org.Comp.1-335, 349
−	CH$_3$Cu · x B(OCH$_3$)$_3$ · y LiI	Cu:	Org.Comp.1-335, 352
−	CH$_3$Cu · x CH$_3$CN · y LiCN	Cu:	Org.Comp.1-385
−	CH$_3$Cu · x C$_2$H$_4$N$_2$(CH$_3$)$_4$ · y LiCN	Cu:	Org.Comp.1-385
−	CH$_3$Cu · x C$_2$H$_4$N$_2$(CH$_3$)$_4$ · y LiI	Cu:	Org.Comp.1-385
−	CH$_3$Cu · x C$_2$H$_4$N$_2$(CH$_3$)$_4$ · y LiOCHC$_6$H$_5$CHCH$_3$N(CH$_3$)$_2$	Cu:	Org.Comp.1-385
−	CH$_3$Cu · x C$_2$H$_4$N$_2$(CH$_3$)$_4$ · y MgBr$_2$	Cu:	Org.Comp.1-387
−	CH$_3$Cu · x C$_4$H$_8$NH · y LiI	Cu:	Org.Comp.1-386
−	CH$_3$Cu · x C$_5$H$_5$N · y LiCN	Cu:	Org.Comp.1-386
−	CH$_3$Cu · x C$_6$H$_5$CH$_2$NH$_2$	Cu:	Org.Comp.1-384
−	CH$_3$Cu · x C$_8$H$_{15}$BC$_8$H$_{14}$ · y LiI	Cu:	Org.Comp.1-333
−	CH$_3$Cu · x C$_{15}$H$_{26}$N$_2$ · y MgClI	Cu:	Org.Comp.1-386
−	CH$_3$Cu · x C$_{15}$H$_{26}$N$_2$ · y MgI$_2$	Cu:	Org.Comp.1-387
−	CH$_3$Cu · x LiBr	Cu:	Org.Comp.1-55, 95, 117

CCuH$_3$ . . . .　CH$_3$Cu · x LiCF$_3$COCHCOCF$_3$ . . . .　Cu: Org.Comp.1-51
－ . . . .　CH$_3$Cu · x LiCH$_3$COCHCOCH$_3$ . . . .　Cu: Org.Comp.1-51
－ . . . .　CH$_3$Cu · x LiCN . . . .　Cu: Org.Comp.1-125/31
－ . . . .　CH$_3$Cu · x LiCl . . . .　Cu: Org.Comp.1-55, 84
－ . . . .　CH$_3$Cu · x LiI . . . .　Cu: Org.Comp.1-55, 80, 84,
　　　　　　　　　　　　　　　　　　　　　　　　　　　　　　　　　 91/6, 103/11, 117/8
－ . . . .　CH$_3$Cu · x LiMgBrCl$_2$ . . . .　Cu: Org.Comp.1-299, 305
－ . . . .　CH$_3$Cu · x LiMgBr$_2$Cl . . . .　Cu: Org.Comp.1-299, 306/13
－ . . . .　CH$_3$Cu · x LiMgBr$_2$Hal . . . .　Cu: Org.Comp.1-299, 307/9,
　　　　　　　　　　　　　　　　　　　　　　　　　　　　　　　　　 312/3
－ . . . .　CH$_3$Cu · x LiMgBr$_2$I . . . .　Cu: Org.Comp.1-299, 304,
　　　　　　　　　　　　　　　　　　　　　　　　　　　　　　　　　 310
－ . . . .　CH$_3$Cu · x LiMgBr$_3$ . . . .　Cu: Org.Comp.1-299, 306
－ . . . .　CH$_3$Cu · x LiMg(Hal)$_3$ . . . .　Cu: Org.Comp.1-304
－ . . . .　CH$_3$Cu · x LiN(CHCH$_3$C$_6$H$_5$)CH$_2$C$_6$H$_4$OCH$_3$　Cu: Org.Comp.1-195, 197
－ . . . .　CH$_3$Cu · x LiN(CH$_3$)CH(CH$_3$)C$_6$H$_5$ . . . .　Cu: Org.Comp.1-195/6
－ . . . .　CH$_3$Cu · x LiN(CH$_3$)CO$_2$C$_3$H$_7$-i . . . .　Cu: Org.Comp.1-198
－ . . . .　CH$_3$Cu · x LiN(CH$_3$)C$_6$H$_5$ . . . .　Cu: Org.Comp.1-197
－ . . . .　CH$_3$Cu · x LiN(C$_6$H$_5$)CO$_2$CHCH$_3$
　　　　　　　　　CHCHC$_6$H$_2$(CH$_3$)$_3$ . . . .　Cu: Org.Comp.1-198
－ . . . .　CH$_3$Cu · x LiN(C$_6$H$_5$)CO$_2$CHCH$_3$CHCHC$_6$H$_5$
　　　　　　　　　　　　　　　　　　　　　　　　　　　Cu: Org.Comp.1-198
－ . . . .　CH$_3$Cu · x LiN(C$_6$H$_5$)CO$_2$C$_6$H$_8$CH$_3$ . . . . . .　Cu: Org.Comp.1-198
－ . . . .　CH$_3$Cu · x LiN(C$_6$H$_5$)CO$_2$C$_6$H$_9$ . . . .　Cu: Org.Comp.1-198
－ . . . .　CH$_3$Cu · x LiN(C$_6$H$_5$)CO$_2$C$_8$H$_{11}$ . . . .　Cu: Org.Comp.1-199
－ . . . .　CH$_3$Cu · x LiN(C$_6$H$_{11}$-c)$_2$ . . . .　Cu: Org.Comp.1-197
－ . . . .　CH$_3$Cu · x LiNHCH(CH$_3$)CH$_2$C$_6$H$_5$ . . . .　Cu: Org.Comp.1-195/6
－ . . . .　CH$_3$Cu · x LiNHCH(CH$_3$)C$_6$H$_5$ . . . .　Cu: Org.Comp.1-195/6
－ . . . .　CH$_3$Cu · x LiNHC$_{14}$H$_{14}$(CH$_3$)$_2$C$_3$H$_7$-i . . . . .　Cu: Org.Comp.1-195/6
－ . . . .　CH$_3$Cu · x LiOC(CH$_3$)(CHCH$_2$)C$_2$H$_4$CHC(CH$_3$)$_2$
　　　　　　　　　　　　　　　　　　　　　　　　　　　Cu: Org.Comp.1-150
－ . . . .　CH$_3$Cu · x LiOC(CH$_3$)(C$_6$H$_5$)CCC$_4$H$_9$-n . . . .　Cu: Org.Comp.1-151
－ . . . .　CH$_3$Cu · x LiOCH(CHCHC$_6$H$_5$)CH$_2$Si(CH$_3$)$_3$　Cu: Org.Comp.1-163
－ . . . .　CH$_3$Cu · x LiOCH(CHCH$_2$)CCC$_4$H$_9$-n . . . . .　Cu: Org.Comp.1-151
－ . . . .　CH$_3$Cu · x LiOCH(CHCH$_2$)CCC$_6$H$_5$ . . . .　Cu: Org.Comp.1-151
－ . . . .　CH$_3$Cu · x LiOCH(CH$_3$)CHCHC$_6$H$_5$ . . . .　Cu: Org.Comp.1-144, 149
－ . . . .　CH$_3$Cu · x LiOCH(C$_2$H$_5$)CCC$_6$H$_5$ . . . .　Cu: Org.Comp.1-150
－ . . . .　CH$_3$Cu · x LiOCH(C$_4$H$_9$-n)CCSi(CH$_3$)$_3$ . . . .　Cu: Org.Comp.1-151
－ . . . .　CH$_3$Cu · x LiOCH(C$_6$H$_5$)CHCHCH$_3$ . . . .　Cu: Org.Comp.1-144, 149
－ . . . .　CH$_3$Cu · x LiOCH(C$_6$H$_5$)CHCHSi(CH$_3$)$_3$ . . .　Cu: Org.Comp.1-163
－ . . . .　CH$_3$Cu · x LiOCH(C$_6$H$_5$)CH(CH$_3$)N(CH$_3$)CH$_2$C$_6$H$_5$
　　　　　　　　　　　　　　　　　　　　　　　　　　　Cu: Org.Comp.1-142, 148
－ . . . .　CH$_3$Cu · x LiOCH(C$_6$H$_5$)CH(CH$_3$)N(CH$_3$)$_2$ . .　Cu: Org.Comp.1-142, 147/8
－ . . . .　CH$_3$Cu · x LiOCH(C$_6$H$_5$)CH(CH$_3$)NHCH$_3$ . . .　Cu: Org.Comp.1-142, 147
－ . . . .　CH$_3$Cu · x LiOCH(C$_6$H$_5$)CCC$_6$H$_5$ . . . .　Cu: Org.Comp.1-151
－ . . . .　CH$_3$Cu · x LiOCH(C$_6$H$_5$)C$_6$H$_{11}$-c . . . .　Cu: Org.Comp.1-142/3, 147
－ . . . .　CH$_3$Cu · x LiOCH(C$_9$H$_6$N)C$_7$H$_{11}$NCHCH$_2$ . .　Cu: Org.Comp.1-142, 146
－ . . . .　CH$_3$Cu · x LiOCH$_2$CHCCH$_3$C$_2$H$_4$CHC(CH$_3$)$_2$
　　　　　　　　　　　　　　　　　　　　　　　　　　　Cu: Org.Comp.1-150
－ . . . .　CH$_3$Cu · x LiOCH$_2$CH(C$_2$H$_5$)NH$_2$ . . . .　Cu: Org.Comp.1-142, 145

CCuH₃	CH₃Cu · x LiOCH₂CH(C₃H₇-i)NCHC₆H₂(OCH₃)₃	
		Cu: Org.Comp.1-142, 145
—	CH₃Cu · x LiOCH₂CH(C₃H₇-i)NH₂	Cu: Org.Comp.1-142, 145
—	CH₃Cu · x LiOCH₂CH(C₆H₅)NCHC₆H₂(OCH₃)₃	
		Cu: Org.Comp.1-142, 146
—	CH₃Cu · x LiOCH₂CH(C₆H₅)NH₂	Cu: Org.Comp.1-142, 145
—	CH₃Cu · x LiOCH₂C₄H₇NCH₂C₄H₇NCH₃ . . .	Cu: Org.Comp.1-163
—	CH₃Cu · x LiOCH₂C₄H₇NH	Cu: Org.Comp.1-142, 148
—	CH₃Cu · x LiOCH₂C₆H₅	Cu: Org.Comp.1-145
—	CH₃Cu · x LiOC₄H₉-t	Cu: Org.Comp.1-145
—	CH₃Cu · x LiOC₆H₄CHNCHCH₃C₆H₅	Cu: Org.Comp.1-142, 146
—	CH₃Cu · x LiOC₆H₈(C(CH₃)₂)CH₃	Cu: Org.Comp.1-149
—	CH₃Cu · x LiOC₆H₈(CH₂)CH₃	Cu: Org.Comp.1-149
—	CH₃Cu · x LiOC₆H₈CH₃	Cu: Org.Comp.1-144, 149/50
—	CH₃Cu · x LiOC₆H₉CH₃C₃H₇-i	Cu: Org.Comp.1-143, 148
—	CH₃Cu · x LiOC₇H₅O₅(CH₃)₄	Cu: Org.Comp.1-149
—	CH₃Cu · x LiOC₉H₆N	Cu: Org.Comp.1-51
—	CH₃Cu · x LiP(C₆H₅)₂	Cu: Org.Comp.1-202
—	CH₃Cu · x LiSC₆H₅	Cu: Org.Comp.1-165/6, 168/72
—	CH₃Cu · x LiSC₆H₉CH₃C₃H₇-i	Cu: Org.Comp.1-167
—	CH₃Cu · x LiSi(CH₃)₂C₆H₅ · y LiCN	Cu: Org.Comp.1-51
—	CH₃Cu · x MgBrCl	Cu: Org.Comp.1-208, 258, 274, 285, 290
—	CH₃Cu · x MgBrI	Cu: Org.Comp.1-209, 277, 285/6, 289
—	CH₃Cu · x MgBrOCH₂CH(C₂H₅)NCHC₆H₂ (OCH₃)₃	Cu: Org.Comp.1-317, 322
—	CH₃Cu · x MgBrOCH₂CH(C₃H₇-i) NCHC₆H₂(OCH₃)₃	Cu: Org.Comp.1-318, 322
—	CH₃Cu · x MgBrOCH₂CH(C₆H₅)NCHC₆H₂ (OCH₃)₃	Cu: Org.Comp.1-318, 322
—	CH₃Cu · x MgBrOCH₂C₄H₇NCH₃	Cu: Org.Comp.1-316/7
—	CH₃Cu · x MgBrOCH₂C₄H₇NCO₂C₄H₉-t . . .	Cu: Org.Comp.1-317
—	CH₃Cu · x MgBrOCH₂C₄H₇NH	Cu: Org.Comp.1-316
—	CH₃Cu · x MgBrOC₂H₄C₄H₇NCH₃	Cu: Org.Comp.1-317
—	CH₃Cu · x MgBrOC₂H₄C₄H₇NCO₂C₄H₉-t . .	Cu: Org.Comp.1-317
—	CH₃Cu · x MgBrOC₄H₉-t	Cu: Org.Comp.1-316
—	CH₃Cu · x MgBr₂	Cu: Org.Comp.1-208, 271/3
—	CH₃Cu · x MgClI	Cu: Org.Comp.1-209, 274, 285
—	CH₃Cu · x MgClOC₄H₉-t	Cu: Org.Comp.1-316
—	CH₃Cu · x MgCl₂	Cu: Org.Comp.1-209
—	CH₃Cu · x MgHal₂	Cu: Org.Comp.1-209, 275
—	CH₃Cu · x MgIOCH₂C₄H₇NCH₃	Cu: Org.Comp.1-317
—	CH₃Cu · x MgI₂	Cu: Org.Comp.1-209, 273, 287, 290
—	CH₃Cu · x N(C₂H₅)₃ · y LiNHCHCH₃CH₂C₆H₅	
		Cu: Org.Comp.1-384
—	CH₃Cu · x N(C₂H₅)₃ · y LiOCH₂C₄H₇NH . .	Cu: Org.Comp.1-385

CEu	EuC solid solutions		
	(Eu,U)C$_{1-x}$	U:	SVol.C12-95, 151
CEuHO$_2$$^{2+}$	[Eu(HCOO)]$^{2+}$	Sc:	MVol.D5-4/5
CEuH$_3$O$_3$P$^+$	[Eu(CH$_3$PO$_3$)]$^+$	Sc:	MVol.D4-143
CEuNS^{2+}	Eu(NCS)$^{2+}$	Sc:	MVol.D4-340/3
CEuO$_3$$^+$	Eu(CO$_3$)$^+$	Sc:	MVol.D4-338/40
CFHN$_2$O$_4$	HCF(NO$_2$)$_2$.	F:	PFHOrg.8-33, 35, 42
CFH$_2$NS	FC(S)NH$_2$.	F:	PFHOrg.SVol.1-192
CFH$_4$N$_3$O$_4$	[NH$_4$][CF(NO$_2$)$_2$]	F:	PFHOrg.8-4/5
CFH$_6$NO$_3$S.	NH$_3$FCH$_3$SO$_3$.	F:	SVol.5-3/4
CFH$_7$N$_3$O$_6$PU.	CN$_3$H$_6$[(UO$_2$)PO$_4$HF] · H$_2$O	U:	SVol.C14-129
CFIN$_2$O$_4$	ICF(NO$_2$)$_2$	F:	PFHOrg.8-4, 19, 21, 34,
			35, 42
CFKN$_2$O$_4$.	KCF(NO$_2$)$_2$.	F:	PFHOrg.8-42
CFN	FCN	F:	PFHOrg.9-115/6
CFNO.	FC(O)N	F:	PFHOrg.9-125
CFNO$_3$S.	SFO$_2$NCO	F:	PFHOrg.9-33
CFN$_2$O$_4$	CF(NO$_2$)$_2$	F:	PFHOrg.8-33
CFN$_2$O$_4$$^-$	CF(NO$_2$)$_2$$^-$	F:	PFHOrg.8-42
CFN$_3$O	CF(O)N$_3$	F:	PFHOrg.8-61, 87
CFN$_3$O$_6$	CF(NO$_2$)$_3$	F:	PFHOrg.8-4/5, 19, 21,
			33/4, 35, 42/3, 47
CFN$_3$O$_6$$^-$	CF(NO$_2$)$_3$$^-$	F:	PFHOrg.8-33
CFN$_3$S$_2$	S$_2$N$_3$CF	S:	S-N Comp.4-1/2
CFN$_5$O$_3$S$_5$	S$_4$N$_4$ · C(O)NSO$_2$F	S:	S-N Comp.2-224
CFN$_5$O$_4$	CF(NO$_2$)$_2$N$_3$	F:	PFHOrg.8-4/5, 21, 34
CFP	FCP	F:	PFHOrg.SVol.1-149,
			151/2
CF$_2$GeH$_2$	F$_2$CGeH$_2$	F:	PFHOrg.SVol.1-50
CF$_2$HIO$_3$S	CF$_2$ISO$_3$H	F:	PFHOrg.SVol.3-49
CF$_2$HN	CF$_2$NH	F:	PFHOrg.9-115/6
CF$_2$HNO$_2$	CHF$_2$NO$_2$.	F:	PFHOrg.8-35
CF$_2$HP	F$_2$CPH	F:	PFHOrg.SVol.1-81/3
CF$_2$H$_2$INO$_2$S	CF$_2$ISO$_2$NH$_2$	F:	PFHOrg.SVol.3-138, 142
CF$_2$H$_2$Si.	F$_2$CSiH$_2$	F:	PFHOrg.SVol.1-41
CF$_2$H$_4$N$_2$O	NHF$_2$ · NH$_2$CHO.	F:	SVol.5-55
CF$_2$H$_5$NO	NHF$_2$ · CH$_3$OH .	F:	SVol.5-48/9, 55
CF$_2$INO$_2$	ICF$_2$NO$_2$	F:	PFHOrg.8-19
CF$_2$IN$_3$O$_2$S	CF$_2$ISO$_2$N$_3$	F:	PFHOrg.SVol.3-137, 142
CF$_2$KNO$_3$S.	K[N(SO$_2$F)C(O)F]	F:	PFHOrg.9-9, 31, 33
CF$_2$Li$_2$	CF$_2$Li$_2$	F:	PFHOrg.SVol.1-1
CF$_2$N	F$_2$CN	F:	PFHOrg.9-125
CF$_2$NNaO$_3$S.	Na[N(SO$_2$F)C(O)F]	F:	PFHOrg.9-9, 31, 33
CF$_2$N$_2$	CF$_2$N$_2$	F:	PFHOrg.8-57/8, 84, 104,
			205/6
CF$_2$N$_2$O$_4$	CF$_2$(NO$_2$)$_2$	F:	PFHOrg.8-4, 19, 20, 33,
			35, 42/3
CF$_2$N$_2$O$_4$$^-$	CF$_2$(NO$_2$)$_2$$^-$	F:	PFHOrg.8-33, 42/3
CF$_2$OS	F$_2$CSO	F:	PFHOrg.SVol.3-1, 13, 19,
			38

CF₃SeTl..........	CF₃SeTl..........................	F:	PFHOrg.SVol.3-215/6, 241
CF₄Ge...........	F₂CGeF₂.........................	F:	PFHOrg.SVol.1-50
CF₄HN...........	CF₃NFH..........................	F:	PFHOrg.8-152, 175
CF₄HNO₂S.......	CF₃NHSO₂F.......................	F:	PFHOrg.8-205
CF₄HN₅.........	F₂NCF(NHF)N₃....................	F:	PFHOrg.8-59/60
CF₄HPS₂.........	F(CF₃)P(S)SH.....................	F:	PFHOrg.SVol.1-139/41, 144
CF₄H₃NO₃S.......	NH₃FCF₃SO₃......................	F:	PFHOrg.SVol.3-59, 77, 102
		F:	SVol.5-3/4
–..............	ND₃FCF₃SO₃......................	F:	SVol.5-4
CF₄INO₃.........	CF₃I(F)NO₃.......................	F:	PFHOrg.SVol.3-255, 258, 265
CF₄KNO.........	NF₂CF₂OK........................	F:	PFHOrg.8-157/8, 203, 205/7
CF₄N...........	CF₃NF..........................	F:	PFHOrg.8-203
CF₄N₂..........	CF(NF)NF₂.......................	F:	PFHOrg.8-172, 197/8, 202/4, 206, 222/3
CF₄N₂O.........	CF₃N(O)NF.......................	F:	PFHOrg.8-55, 82, 102
–..............	NF₂C(O)NF₂......................	F:	PFHOrg.8-204
CF₄N₂O₂.........	NF₂CF₂NO₂.......................	F:	PFHOrg.8-2/3, 20, 34
CF₄N₄..........	F₂C(NF₂)N₃......................	F:	PFHOrg.8-59/60
CF₄N₄O₄S₆.......	[S₄N₃][N(SO₂CF₃)SO₂F]...........	S:	S-N Comp.2-123
CF₄OS..........	CF₃S(O)F........................	F:	PFHOrg.SVol.3-11, 18, 34, 42
CF₄O₂S.........	CF₃SO₂F.........................	F:	PFHOrg.SVol.3-152/3, 161, 170/80
CF₄O₃S.........	CF₃SO₃F.........................	S:	SVol.3-205
CF₄O₃SXe.......	XeF(OSO₂CF₃)....................	F:	PFHOrg.SVol.3-54/5, 83, 85, 101
CF₄O₄S.........	CF₃OOSO₂F......................	S:	SVol.3-314
CF₄O₆S₂.........	CF₃OSO₂OSO₂F...................	S:	SVol.3-205
CF₄PS₂⁻.........	F(CF₃)P(S)S⁻.....................	F:	PFHOrg.SVol.1-144
CF₄S...........	CF₃SF..........................	F:	PFHOrg.SVol.2-115/7
CF₄S₂..........	CF₃SSF.........................	F:	PFHOrg.SVol.2-173, 184, 196, 199
CF₄Si..........	F₂CSiF₂.........................	F:	PFHOrg.SVol.1-41
CF₅HN₂.........	NF₂CF₂NFH......................	F:	PFHOrg.8-152, 175, 203
CF₅H₂N₃........	(NF₂)₂CFNH₂.....................	F:	PFHOrg.8-152, 175
CF₅H₃OU........	UF₅(OCH₃).......................	U:	SVol.C13-76, 82
		U:	SVol.D1-128
CF₅I...........	CF₃IF₂..........................	F:	PFHOrg.SVol.3-255/6, 258, 262, 264, 266
CF₅IO..........	CF₃I(O)F₂.......................	F:	PFHOrg.SVol.3-253, 257, 265/6
CF₅IO₆S₂.......	CF₃I(OSO₂F)₂....................	F:	PFHOrg.SVol.3-255, 264
CF₅ISi.........	CF₃SiF₂I........................	F:	PFHOrg.SVol.1-41/8
CF₅KNO₃PS.......	KN(SO₂CF₃)P(O)F₂................	F:	PFHOrg.SVol.3-141, 146

CF$_8$O$_3$STa$^-$	TaF$_5$CF$_3$SO$_3$$^-$	F:	PFHOrg.SVol.3-94
CF$_8$S	CF$_3$SF$_5$	F:	PFHOrg.SVol.3-187, 194
CF$_8$S$_2$	CF$_2$(SF$_3$)$_2$	F:	PFHOrg.SVol.3-10, 16, 32
CF$_8$Si$_2$	CF$_3$SiF$_2$SiF$_3$	F:	PFHOrg.SVol.1-43, 46
—	SiF$_3$CF$_2$SiF$_3$	F:	PFHOrg.SVol.1-43, 46
CF$_9$NS	(CF$_3$)(SF$_5$)NF	F:	PFHOrg.8-162, 186
CF$_{10}$NP	CF$_3$NF$_2$ · PF$_5$	F:	PFHOrg.8-206
CF$_{12}$S$_2$	CF$_2$(SF$_5$)$_2$	F:	PFHOrg.SVol.3-190, 200
CF$_{13}$HO$_3$SSb$_2$	CF$_3$SO$_3$H · 2 SbF$_5$	F:	PFHOrg.SVol.3-89
CF$_{13}$HO$_3$STa$_2$	CF$_3$SO$_3$H · 2 TaF$_5$	F:	PFHOrg.SVol.3-89
CF$_{13}$O$_3$SU$_2$$^-$	U$_2$F$_{10}$CF$_3$SO$_3$$^-$	F:	PFHOrg.SVol.3-94
CFeN$_4$OS$_4$	Fe(CO)S$_4$N$_4$ = S$_4$N$_4$ · Fe(CO)	S:	S-N Comp.2-191/2, 225
CFe$_4$	Fe$_4$C (cluster)	Fe:	Org.Comp.C7-199
CFrHO$_2$	FrHCO$_2$	Fr:	MVol.-114
CFrHO$_3$	FrHCO$_3$	Fr:	MVol.-114
CFrN	FrCN	Fr:	MVol.-115
CFrNO	FrCNO	Fr:	MVol.-114, 115
CFrNS	FrCNS	Fr:	MVol.-114, 115
CFr$_2$O$_3$	Fr$_2$CO$_3$	Fr:	MVol.-115
CGaH$_3$I$_2$	Ga(CH$_3$)I$_2$	Ga:	Org.Comp.1-162/3
CGaH$_3$O	[Ga(CH$_3$)O]$_n$	Ga:	Org.Comp.1-218/9
CGaH$_5$	Ga(CH$_3$)D$_2$	Ga:	Org.Comp.1-124
CGaH$_6$Na	Na[Ga(CH$_3$)H$_3$]	Ga:	Org.Comp.1-322
CGaMn$_3$	Mn$_3$GaC	Mn:	MVol.C7-148/54
—	Mn$_3$GaC solid solutions		
	Mn$_{3(1-x)}$Cr$_{3x}$GaC$_{1-x}$N$_x$ (x = 0.05)	Mn:	MVol.C7-229
	Mn$_3$GaC$_{1-x}$B$_x$	Mn:	MVol.C7-245
	Mn$_3$GaC$_{1-x}$N$_x$	Mn:	MVol.C7-226/8
	Mn$_3$Ga$_{1-x}$Zn$_x$C	Mn:	MVol.C7-155/8
	Mn$_3$Ga$_{1-x}$Zn$_x$C$_{1-x}$N$_x$ (x = 0.1)	Mn:	MVol.C7-228
	Mn$_3$Ge$_{1-x}$Ga$_x$C	Mn:	MVol.C7-160
	Mn$_{3+x}$Ga$_{1-x}$C	Mn:	MVol.C7-154/5
CGa$_2$H$_3$I$_5$	Ga$_2$(CH$_3$)I$_5$	Ga:	Org.Comp.1-166
CGd	GdC solid solutions		
	(Gd,U)C$_{1-x}$	U:	SVol.C12-95, 151
CGdHO$_2$$^{2+}$	[Gd(HCOO)]$^{2+}$	Sc:	MVol.D5-4/5
CGdH$_2$NOS^{2+}	[Gd(ONCHSH)]$^{2+}$	Sc:	MVol.D4-69
CGdH$_3$O$_3$P$^+$	[Gd(CH$_3$PO$_3$)]$^+$	Sc:	MVol.D4-143
CGdNS^{2+}	Gd(NCS)$^{2+}$	Sc:	MVol.D4-340/3
CGdO	Gd(CO)	Sc:	MVol.D6-166
CGeMn$_3$	Mn$_3$GeC	Mn:	MVol.C7-159/60
—	Mn$_3$GeC solid solutions		
	Mn$_3$Ge$_{1-x}$Ga$_x$C	Mn:	MVol.C7-160
CGe$_{2.4}$Mn$_{6.6}$	Mn$_{6.6}$Ge$_{2.4}$C	Mn:	MVol.C7-159
CHHoO$_2$$^{2+}$	[Ho(HCOO)]$^{2+}$	Sc:	MVol.D5-4/5
CHLaO$_2$$^{2+}$	[La(HCOO)]$^{2+}$	Sc:	MVol.D5-4/5
CHLaO$_3$$^{2+}$	La(HCO$_3$)$^{2+}$	Sc:	MVol.D4-338/40
CHLuO$_2$$^{2+}$	[Lu(HCOO)]$^{2+}$	Sc:	MVol.D5-4/5
CHMnO$_2$$^+$	[Mn(HCOO)]$^+$	Mn:	MVol.D2-2/3

CSi. . . . . SiC

CSi. . . . . SiC

$C_2CaH_2O_4$	$Ca(HCOO)_2$ systems		
	$Ca(HCOO)_2-H_3BO_3-H_2O$	B:	B Comp.SVol.1/1–188
		B:	B Comp.SVol.3/2–105
C_2CaMnO_6	$CaMn(CO_3)_2$	Mn:	MVol.C7-222
C_2CaO_7W	$Ca[WO_3(C_2O_4)]$	W:	SVol.B4-104
C_2CaO_8U	$Ca[UO_2(CO_3)_2]$ · 10 H_2O	U:	SVol.C13-10/2
$C_2CdF_6O_6S_2$	$Cd(OSO_2CF_3)_2$	F:	PFHOrg.SVol.3-60, 78, 102/4
C_2Ce	CeC_2 solid solutions		
	$(Ce,U)C_2$	U:	SVol.C12-94, 151/2
$C_2CeClH_2O_2^{2+}$	$[Ce(CH_2ClCOO)]^{2+}$	Sc:	MVol.D5-53/6
C_2CeClO_4	$CeCl(C_2O_4)$ · 3 H_2O	Sc:	MVol.D5-146/7
$C_2CeCl_2HO_2^{2+}$	$[Ce(CHCl_2COO)]^{2+}$	Sc:	MVol.D5-54
$C_2CeCl_3H_3N$	$CeCl_3$ · CH_3CN	Sc:	MVol.D1-102
$C_2CeCl_3H_3O_2$	$Ce(CH_3COO)Cl_3$ · 3 H_2O	Sc:	MVol.D5-53
$C_2CeCl_3H_8N_4O_2$	$CeCl_3$ · 2 $OC(NH_2)_2$	Sc:	MVol.D2-220
$C_2CeCl_3H_{10}N_2$	$CeCl_3$ · 2 CH_3NH_2	Sc:	MVol.D1-16
$C_2CeCl_3O_2^{2+}$	$[Ce(CCl_3COO)]^{2+}$	Sc:	MVol.D5-54
$C_2CeF_2H_3O_2$	$CeF_2(CH_3COO)$	Sc:	MVol.D5-50/1
$C_2CeH_2IO_2^{2+}$	$[Ce(CH_2ICOO)]^{2+}$	Sc:	MVol.D5-53/6
$C_2CeH_2O_4^+$	$[Ce(HCOO)_2]^+$	Sc:	MVol.D5-4/5
$C_2CeH_3O_2^{2+}$	$[Ce(CH_3COO)]^{2+}$	Sc:	MVol.D5-23, 25/8
$C_2CeH_3O_2S^{2+}$	$[Ce(OOCCH_2SH)]^{2+}$	Sc:	MVol.D4-53/4
$C_2CeH_3O_3$	$CeO(CH_3COO)$	Sc:	MVol.D5-43
$C_2CeH_3O_3^{2+}$	$[Ce(HOCH_2COO)]^{2+}$	Sc:	MVol.D5-222/30
$C_2CeH_3O_3^{3+}$	$[Ce(HOCH_2COO)]^{3+}$	Sc:	MVol.D5-231
$C_2CeH_3O_4^{2+}$	$[Ce((HO)_2CHCOO)]^{2+}$	Sc:	MVol.D5-249/50
$C_2CeH_4NO_2^{2+}$	$Ce(NH_2CH_2COO)^{2+}$	Sc:	MVol.D1-103
$C_2CeH_4O_3^{4+}$	$[Ce(HOCH_2COOH)]^{4+}$	Sc:	MVol.D5-231
$C_2CeH_4O_7P_2^-$	$Ce[H(OC(PO_3)_2CH_3)]^-$	Sc:	MVol.D4-147
$C_2CeH_5OS^{2+}$	$Ce(HOC_2H_4S)^{2+}$	Sc:	MVol.D4 49
$C_2CeH_5O_2^{2+}$	$[Ce(OCH_2CH_2OH)]^{2+}$	Sc:	MVol.D3-31/2
$C_2CeH_5O_3^{3+}$	$CeOH(CH_3COOH)^{3+}$	Sc:	MVol.D5-52
$C_2CeH_5O_7P_2$	$Ce[H_2(OC(PO_3)_2CH_3)]$ · 5 H_2O	Sc:	MVol.D4-147
$C_2CeH_6O_6P_2^-$	$[Ce(CH_3PO_3)_2]^-$	Sc:	MVol.D4-143
$C_2CeH_8N_7O_{11}$	$Ce(NO_3)_3$ · 2 $OC(NH_2)_2$ · 2 H_2O	Sc:	MVol.D2-220
$C_2CeN_2S_2^+$	$[Ce(NCS)_2]^+$	Sc:	MVol.D4-340/3
$C_2CeO_4^+$	$[Ce(C_2O_4)]^+$	Sc:	MVol.D5-120/4
C_2CeO_5	$CeO(C_2O_4)$	Sc:	MVol.D5-146/7
$C_2CeO_{12}S_2^{2-}$	$[Ce(SO_4)_2(C_2O_4)]^{2-}$	Sc:	MVol.D5-148/9
$C_2Ce_2Cl_4O_4$	$Ce_2Cl_4(C_2O_4)$ · x H_2O	Sc:	MVol.D5-146/7
$C_2Ce_2H_8N_4O_{14}S_3$ ...	$Ce_2(SO_4)_3$ · 2 $OC(NH_2)_2$ · 5 H_2O	Sc:	MVol.D2-222/3
$C_2Ce_3H_4O_5^{6+}$	$Ce_3O_3(CH_3COOH)^{6+}$	Sc:	MVol.D5-52
C_2ClCu	$ClCCCu$ · x LiI	Cu:	Org.Comp.3-152, 155
C_2ClCuH_2	$(C_2H_2)CuCl$	Cu:	Org.Comp.4-44
C_2ClCuH_4	CH_2CH_2CuCl	Cu:	Org.Comp.4-5
$C_2ClCuH_4O_4$	$(CH_2CH_2Cu)ClO_4$	Cu:	Org.Comp.4-5/6, 16
C_2ClCuH_6Mg	$(CH_3)_2CuMgCl$	Cu:	Org.Comp.2-213, 217/23
$C_2ClCuH_8O_6$	$CH_2CH_2Cu(H_2O)_2ClO_4$	Cu:	Org.Comp.4-6

C_2ClF_6PS	$(CF_3)_2P(S)Cl$	F:	PFHOrg.SVol.1-140/5
$C_2ClF_6PS_2$	$(CF_3S)_2PCl$	F:	PFHOrg.SVol.2-243, 246
C_2ClF_6Sb	$(CF_3)_2SbCl$	F:	PFHOrg.SVol.1-178/9
		Sb:	Org.Comp.2-9
$C_2ClF_7N_2$	$NF_2CF_2CFClNF_2$	F:	PFHOrg.8-164, 189
C_2ClF_7Si	$ClCF_2CF_2SiF_3$	F:	PFHOrg.SVol.1-41/8
C_2ClF_8NS	$CF_3CClNSF_5$	F:	PFHOrg.9-8, 26
C_2ClF_9NP	$(CF_3)_2NPF_3Cl$	F:	PFHOrg.9-39, 41
C_2ClF_9S	$C_2F_5SF_4Cl$	F:	PFHOrg.SVol.3-187/8, 195
C_2ClF_9Se	$C_2F_5SeF_4Cl$	F:	PFHOrg.SVol.3-216/7, 225, 242
C_2ClF_9Te	$C_2F_5TeF_4Cl$	F:	PFHOrg.SVol.3-248
$C_2ClF_{10}NS$	$CF_3SF_4NClCF_3$	F:	PFHOrg.SVol.3-192/3, 203
C_2ClGaH_6	$Ga(CH_3)_2Cl$	Ga:	Org.Comp.1-128, 132/3
C_2ClGaH_6O	$Ga(CH_3)(Cl)OCH_3$	Ga:	Org.Comp.1-225
$C_2ClGaH_6O_4$	$Ga(CH_3)_2ClO_4$	Ga:	Org.Comp.1-213/4
C_2ClGaH_6S	$GaCH_3(Cl)SCH_3$	Ga:	Org.Comp.1-238
C_2ClGaH_9N	$Ga(CH_3)_2Cl \cdot NH_3$	Ga:	Org.Comp.1-129, 135
$C_2ClGaH_{12}N_2$	$Ga(CH_3)_2Cl \cdot 2 NH_3$	Ga:	Org.Comp.1-129, 135
$C_2ClGdH_2O_2{}^{2+}$	$[Gd(CH_2ClCOO)]^{2+}$	Sc:	MVol.D5-53/6
C_2ClGdO_4	$GdCl(C_2O_4)$	Sc:	MVol.D5-146/7
−	$GdCl(C_2O_4) \cdot 3 H_2O$	Sc:	MVol.D5-146/7
C_2ClHN_2OS	SN_2C_2OHCl	S:	S-N Comp.3-93/8, 153/4
C_2ClHN_2S	SN_2C_2HCl	S:	S-N Comp.3-93/8, 135/6
$C_2ClH_2HoO_2{}^{2+}$	$[Ho(CH_2ClCOO)]^{2+}$	Sc:	MVol.D5-53/6
$C_2ClH_2LaO_2{}^{2+}$	$[La(CH_2ClCOO)]^{2+}$	Sc:	MVol.D5-53/6
$C_2ClH_2LuO_2{}^{2+}$	$[Lu(CH_2ClCOO)]^{2+}$	Sc:	MVol.D5-53/6
$C_2ClH_2MnO_2{}^{+}$	$[Mn(CH_2ClCOO)]^{+}$	Mn:	MVol.D2-61/2
$C_2ClH_2NO_2S$	$[CH_2OS(O)NCCl]$	S:	S-N Comp.3-286/7
$C_2ClH_2N_3S$	$SN_2C_2NH_2Cl$	S:	S-N Comp.3-93/8, 145
$C_2ClH_2NdO_2{}^{2+}$	$[Nd(CH_2ClCOO)]^{2+}$	Sc:	MVol.D5-53/6
$C_2ClH_2O_2Pr^{2+}$	$[Pr(CH_2ClCOO)]^{2+}$	Sc:	MVol.D5-53/6
$C_2ClH_2O_2Sm^{2+}$	$[Sm(CH_2ClCOO)]^{2+}$	Sc:	MVol.D5-53/6
$C_2ClH_2O_2Tb^{2+}$	$[Tb(CH_2ClCOO)]^{2+}$	Sc:	MVol.D5-53/6
$C_2ClH_2O_2Tm^{2+}$	$[Tm(CH_2ClCOO)]^{2+}$	Sc:	MVol.D5-53/6
$C_2ClH_2O_2Y^{2+}$	$[Y(CH_2ClCOO)]^{2+}$	Sc:	MVol.D5-53/6
$C_2ClH_2O_2Yb^{2+}$	$[Yb(CH_2ClCOO)]^{2+}$	Sc:	MVol.D5-53/6
$C_2ClH_2O_4U^{+}$	$[UO_2(CH_2ClCOO)]^{+}$	U:	SVol.D1-198/201
$C_2ClH_4N_3S$	$SN_2C_2HNH_2 \cdot HCl$	S:	S-N Comp.3-93/8, 142
$C_2ClH_5NO_3Pr$	$Pr(NH_2CH_2COO)(OH)Cl \cdot 4 H_2O$	Sc:	MVol.D1-105
$C_2ClH_5O_6Se$	$H_2SeO_4 \cdot CH_2ClCOOH$	Se:	SVol.B1-292
C_2ClH_6ISn	$(CH_3)_2SnICl$	Sn:	Org.Comp.8-142
$C_2ClH_6N_5S_4$	$[S_4N_5(CH_3)_2]Cl$	S:	S-N Comp.2-97
$C_2ClH_6RhS_2$	$[Rh(SCH_3)_2Cl]_n$	Rh:	SVol.B3-2
C_2ClH_6Sb	$(CH_3)_2SbCl$	Sb:	Org.Comp.2-4/7
$C_2ClH_9S_2$	$2 H_2S \cdot C_2H_5Cl \cdot 17 H_2O$	S:	SVol.4a/b-326
$C_2ClH_9Se_2$	$C_2H_5Cl \cdot 2 H_2Se \cdot 7 H_2O$		
	$= 8 C_2H_5Cl \cdot 16 H_2Se \cdot 136 H_2O$	Se:	SVol.B1-62

$C_2ClH_{10}N_6ORhS_2$ . . .	$[Rh(NH_2NHC(S)NH)_2Cl(H_2O)] \cdot H_2O$	Rh:	SVol.B3-41
$C_2ClH_{12}N_4O_8Rh$	$[Rh(NH_3)_4(C_2O_4)](ClO_4) \cdot H_2O$	Rh:	SVol.B2-149
$C_2ClH_{12}P_2Rh$.	$Rh(PH_3)_2(C_2H_5)HCl$.	Rh:	SVol.B3-67
$-$	$Rh(PH_3)_2H_2Cl(C_2H_4)$	Rh:	SVol.B3-67
$C_2ClH_{18}O_{15}P_6Rh_3$ . . .	$Rh_3[P(OC_2H_5)(OH)O][P(OH)_3]_2[P(OH)_2]_3Cl$	Rh:	SVol.B3-79
C_2ClLaO_4	$LaCl(C_2O_4)$	Sc:	MVol.D5-146/7
$-$	$LaCl(C_2O_4) \cdot 3 H_2O$	Sc:	MVol.D5-146/7
$-$	$LaCl(C_2O_4) \cdot x H_2O$	Sc:	MVol.D5-146/7
C_2ClNdO_4	$NdCl(C_2O_4)$	Sc:	MVol.D5-146/7
$-$	$NdCl(C_2O_4) \cdot 3 H_2O$	Sc:	MVol.D5-146/7
$-$	$NdCl(C_2O_4) \cdot x H_2O$	Sc:	MVol.D5-146/7
C_2ClO_4Pr	$PrCl(C_2O_4)$	Sc:	MVol.D5-146/7
$-$	$PrCl(C_2O_4) \cdot 3 H_2O$	Sc:	MVol.D5-146/7
$-$	$PrCl(C_2O_4) \cdot x H_2O$	Sc:	MVol.D5-146/7
C_2ClO_4Sm	$SmCl(C_2O_4)$.	Sc:	MVol.D5-146/7
$-$	$SmCl(C_2O_4) \cdot 3 H_2O$	Sc:	MVol.D5-146/7
$-$	$SmCl(C_2O_4) \cdot x H_2O$	Sc:	MVol.D5-146/7
$C_2Cl_2CuH_2^-$	$[(C_2H_2)CuCl_2]^-$	Cu:	Org.Comp.4-44
$C_2Cl_2CuH_4^-$	$(CH_2CH_2CuCl_2)^-$	Cu:	Org.Comp.4-24
$C_2Cl_2CuH_4O_8$	$[CH_2CH_2Cu](ClO_4)_2$	Cu:	Org.Comp.4-16
$C_2Cl_2Cu_2H_2$.	$(C_2H_2)Cu_2Cl_2 = C_2H_2 \cdot 2 CuCl$	Cu:	Org.Comp.4-44/5
$C_2Cl_2Cu_2H_6O_{13}$	$[(CO)_2Cu_2(H_2O)_3](ClO_4)_2$	Cu:	Org.Comp.4-119
$C_2Cl_2Cu_4$.	$Cu_2C_2 \cdot 2 CuCl \cdot H_2O$	Cu:	Org.Comp.4-151
$C_2Cl_2Cu_4H_2O$	$C_2H_2 \cdot 2 CuCl \cdot Cu_2O$.	Cu:	Org.Comp.4-51
$C_2Cl_2EuHO_2^{2+}$	$[Eu(CHCl_2COO)]^{2+}$	Sc:	MVol.D5-54
$C_2Cl_2FH_{17}N_5O_{10}Rh$. .	$[Rh(NH_3)_5(OCOCH_2F)](ClO_4)_2$	Rh:	SVol.B2-133
C_2Cl_2FN.	Cl_2CFCN	F:	PFHOrg.9-50/1, 67, 82/4
C_2Cl_2FNO	$FC(O)NCCl_2$.	F:	PFHOrg.9-127/8, 138
$C_2Cl_2FNO_3$	$NO_2CFClCOCl$.	F:	PFHOrg.8-37/8
C_2Cl_2FNS	Cl_2CFNCS	F:	PFHOrg.9-107/8, 114, 121
$C_2Cl_2FNS_2$.	$CFCl_2SNCS$	F:	PFHOrg.SVol.2-102, 107
C_2Cl_2FNSe	$CFCl_2SeCN$	F:	PFHOrg.SVol.3-221, 235
$C_2Cl_2F_2H_{16}N_5O_{10}Rh$	$[Rh(NH_3)_5(OCOCHF_2)](ClO_4)_2$	Rh:	SVol.B2-133
$C_2Cl_2F_2N_2O_4$	$NO_2CFClCFClNO_2$.	F:	PFHOrg.8-7/8, 23, 47
$-$	$NO_2CF_2CCl_2NO_2$.	F:	PFHOrg.8-7/8, 23
$C_2Cl_2F_2OS_2$.	$CFCl_2SSC(O)F$.	F:	PFHOrg.SVol.2-175/6, 187, 206
$-$	$CF_2ClSSC(O)Cl$	F:	PFHOrg.SVol.2-175/6, 187, 206
$C_2Cl_2F_2S_2$	$ClFCS_2CFCl$	F:	PFHOrg.SVol.2-1, 5
$C_2Cl_2F_3H_{15}N_5O_{10}Rh$	$[Rh(NH_3)_5(OCOCF_3)](ClO_4)_2$.	Rh:	SVol.B2-133
$C_2Cl_2F_3N$.	CF_3NCCl_2	F:	PFHOrg.9-127/8, 138
$-$	$ClCF_2NCFCl$	F:	PFHOrg.9-127/8, 138
$-$	$Cl_2CFCFNF$	F:	PFHOrg.8-169/70, 195
$C_2Cl_2F_3NO$	$CF_3C(O)NCl_2$	F:	PFHOrg.8-159, 183
$C_2Cl_2F_3NOS$	$CF_3C(O)NSCl_2$.	F:	PFHOrg.9-5/6, 21, 33
$C_2Cl_2F_3NO_2$.	$CF_3CCl_2NO_2$	F:	PFHOrg.8-5, 22, 35
$-$	$ClCF_2CFClNO_2$	F:	PFHOrg.8-5, 22
$-$	$Cl_2CFCF_2NO_2$	F:	PFHOrg.8-5, 22
$C_2Cl_2F_3NO_2S$	$CF_3CClNSO_2Cl$	F:	PFHOrg.9-9, 30

$C_2Cl_2HO_4U^+$	$[UO_2(CHCl_2COO)]^+$	U:	SVol.D1-201
$C_2Cl_2H_2MnN_2O$	$Mn[ON(C_2H_2)N]Cl_2$	Mn:	MVol.D4-54
$C_2Cl_2H_3NOTh$	$ThOCl_2 \cdot CH_3CN$	Th:	SVol.E-24
$C_2Cl_2H_3Sb$	$CH_2CHSbCl_2$	Sb:	Org.Comp.2-94
$C_2Cl_2H_4MnN_4$	$Mn(NCNC(NH_2)_2)Cl_2$	Mn:	MVol.D5-158
$C_2Cl_2H_4O_6Se$	$H_2SeO_4 \cdot CHCl_2COOH$	Se:	SVol.B1-294
$C_2Cl_2H_5MnNO_2$	$Mn(NH_3CH_2COO)Cl_2$	Mn:	MVol.D4-252/5
–	$Mn(NH_3CH_2COO)Cl_2 \cdot H_2O$	Mn:	MVol.D4-253/4, 255
–	$Mn(NH_3CH_2COO)Cl_2 \cdot 2 H_2O$	Mn:	MVol.D4-253/5
–	$Mn(NH_3CH_2COO)Cl_2 \cdot 3 H_2O$	Mn:	MVol.D4-253/6
$C_2Cl_2H_5MnN_3O_2$	$MnCl_2(HN(CONH_2)_2)$	Mn:	MVol.D5-157
$C_2Cl_2H_5Sb$	$C_2H_5SbCl_2$	Sb:	Org.Comp.2-92
$C_2Cl_2H_6MnN_2O_2$	$Mn(OCHNH_2)_2Cl_2$	Mn:	MVol.D5-91
$C_2Cl_2H_6MnN_4O_2$	$Mn(H_2NNH(CO)_2NHNH_2)Cl_2 \cdot 2 H_2O$	Mn:	MVol.D5-199
$C_2Cl_2H_6N_2O_{14}U$	$UO_2(ClO_4)_2(CH_3NO_2)_2$	U:	SVol.A5-219
		U:	SVol.D2-60
$C_2Cl_2H_6O_2U$	$U(OCH_3)_2Cl_2$	U:	SVol.C13-61/2
$C_2Cl_2H_6O_3SSi$	$(CH_3)_2ClSiOSO_2Cl$	S:	SVol.3-320
$C_2Cl_2H_6O_6S_2Si$	$(CH_3)_2Si(OSO_2Cl)_2$	S:	SVol.3-320
$C_2Cl_2H_6O_6S_2Sn$	$(CH_3)_2Sn(OSO_2Cl)_2$	Sn:	Org.Comp.14-126, 129
$C_2Cl_2H_6O_8Sn$	$(CH_3)_2Sn(OClO_3)_2$	Sn:	Org.Comp.14-111
$C_2Cl_2H_6O_{10}Se_2$	$2 H_2SeO_4 \cdot CHCl_2COOH$	Se:	SVol.B1-294
$C_2Cl_2H_6Si$	$(CH_3)_2SiCl_2$		
	Pyrolysis	Si:	SVol.B3-129/32
$C_2Cl_2H_6Sn$	$(CH_3)_2SnCl_2$ systems		
	$(CH_3)_2SnCl_2-(CH_3)_2Sn(SCH_3)_2$	Sn:	Org.Comp.10-2/3
$C_2Cl_2H_7MnN$	$Mn[(CH_3)_2NH]Cl_2$	Mn:	MVol.D3-22
$C_2Cl_2H_7MnNO$	$Mn(NH_2C_2H_4OH)Cl_2$	Mn:	MVol.D4-223/4
$C_2Cl_2H_7N_5O_2Os$	$OsO_2(H_2NC(NH)NHC(NH)NH_2)Cl_2$	Os:	SVol.1-271
$C_2Cl_2H_8MnN_2$	$Mn(C_2H_8N_2)Cl_2$	Mn:	MVol.D3-40
–	$Mn(H_3CNHNHCH_3)Cl_2$	Mn:	MVol.D3-72
$C_2Cl_2H_8MnN_4O_2$	$Mn(OCHNHNH_2)_2Cl_2$	Mn:	MVol.D5-167/9
–	$Mn(OC(NH_2)_2)_2Cl_2$	Mn:	MVol.D5-137/8
–	$Mn(OC(NH_2)_2)_2Cl_2 \cdot H_2O$	Mn:	MVol.D5-138
$C_2Cl_2H_8S_2$	$2 H_2S \cdot CH_2ClCH_2Cl \cdot 17 H_2O$	S:	SVol.4a/b-326
$C_2Cl_2H_9MnNO_4$	$[Mn(NH_3CH_2COO)(H_2O)_2Cl_2]_n$	Mn:	MVol.D4-253
$C_2Cl_2H_9N_6RhS_2$	$Rh(NH_2NHC(S)NH)(NH_2NHC(S)NH_2)Cl_2 \cdot 2 H_2O$		
		Rh:	SVol.B3-41
$C_2Cl_2H_{10}MnN_2O_2$	$Mn(CH_3NHOH)_2Cl_2$	Mn:	MVol.D5-230
–	$Mn(CH_3ONH_2)_2Cl_2$	Mn:	MVol.D5-230
$C_2Cl_2H_{10}N_2Pt$	$(CH_3NH_2)_2PtCl_2$		
	Cytostatic activity	Pt:	SVol.A1-330/2
$C_2Cl_2H_{10}N_4O_5U$	$UO_2Cl_2(H_2O)((NH_2)_2CO)_2$	U:	SVol.A5-219
$C_2Cl_2H_{10}N_7O_3RhS_2$	$[Rh(NH_2NHC(S)NH_2)_2Cl_2]NO_3 \cdot 2 H_2O$	Rh:	SVol.B3-41
$C_2Cl_2H_{12}MnN_8O_2$	$Mn(OC(NHNH_2)_2)_2Cl_2$	Mn:	MVol.D5-210
$C_2Cl_2H_{15}N_5Na_2O_{12}Rh$			
	$[Rh(NH_3)_5(C_2O_4)] \cdot 2 NaClO_4 \cdot 3 H_2O$	Rh:	SVol.B2-134
$C_2Cl_2H_{16}N_5O_{12}Rh$	$[Rh(NH_3)_5(C_2O_4H)](ClO_4)_2$	Rh:	SVol.B2-134
$C_2Cl_2H_{18}N_5O_8RhS_2$	$[Rh(NH_3)_5(S_2CCH_3)](ClO_4)_2$	Rh:	SVol.B3-15

C$_2$Cl$_5$F$_2$PS$_2$	(CFCl$_2$S)$_2$PCl	F:	PFHOrg.SVol.2-243, 247
C$_2$Cl$_5$F$_3$NP	CF$_3$CCl$_2$NPCl$_3$	F:	PFHOrg.9-39, 42, 44
C$_2$Cl$_5$H$_{11}$N$_3$Rh	(NH$_4$)$_2$[RhCl$_5$(CH$_3$CN)]	Rh:	SVol.B2-120/1
C$_2$Cl$_5$H$_{15}$N$_5$O$_{10}$Rh . . .	[Rh(NH$_3$)$_5$(OCOCCl$_3$)](ClO$_4$)$_2$ . . .	Rh:	SVol.B2-133
C$_2$Cl$_6$Cu$_4$H$_2$K$_2$	C$_2$H$_2$ · 4 CuCl · 2 KCl	Cu:	Org.Comp.4-51
C$_2$Cl$_6$Cu$_6$H$_2$	C$_2$H$_2$ · 6 CuCl	Cu:	Org.Comp.4-46
C$_2$Cl$_6$H$_{12}$N$_2$Os	(CH$_3$NH$_3$)$_2$[OsCl$_6$]	Os:	SVol.1-134
C$_2$Cl$_6$Mn$_3$N$_2$S$_2$	Mn$_3$(SCN)$_2$Cl$_6$ · H$_2$O	Mn:	MVol.C7-237/8
C$_2$Cl$_6$N$_2$STi	SN$_2$C$_2$Cl$_2$ · TiCl$_4$	S:	S-N Comp.3-93/8, 140
C$_2$Cl$_6$N$_2$Th	ThCl$_4$ · 2 CNCl	Th:	SVol.E-24/5
C$_2$Cl$_7$Cu$_4$H$_{14}$N$_3$	C$_2$H$_2$ · 4 CuCl · 3 NH$_4$Cl	Cu:	Org.Comp.4-48
C$_2$Cl$_7$N$_2$SSb	SN$_2$C$_2$Cl$_2$ · SbCl$_5$	S:	S-N Comp.3-93/8, 140
C$_2$Cl$_8$Cu$_8$K$_2$	Cu$_2$C$_2$ · 6 CuCl · 2 KCl	Cu:	Org.Comp.4-151
C$_2$Cl$_9$Cu$_6$H$_{14}$N$_3$	6 CuCl · C$_2$H$_2$ · 3 NH$_4$Cl	Cu:	Org.Comp.4-48
C$_2$Cl$_{12}$H$_8$N$_4$OsS$_2$Sn$_4$$^{2-}$			
	[Os(SnCl$_3$)$_4$(CS(NH$_2$)$_2$)$_2$]$^{2-}$	Os:	SVol.1-344
C$_2$CoF$_6$O$_6$S$_2$	Co(OSO$_2$CF$_3$)$_2$	F:	PFHOrg.SVol.3-61, 79, 102/4
–	Co(OSO$_2$CF$_3$)$_2$ · 7 H$_2$O	F:	PFHOrg.SVol.3-61, 80
C$_2$CoH$_2$O$_4$	Co(HCOO)$_2$ solid solutions		
	Co$_{1-x}$Mn$_x$(HCOO)$_2$ · 2 H$_2$O	Mn:	MVol.D2-21
C$_2$CoH$_4$N$_4$O$_2$S$_4$	[Co((S$_2$N$_2$CHOH)$_2$)]	S:	S-N Comp.2-291/3
C$_2$CoH$_6$N$_4$S$_4$Si$_2$	[Co(S$_4$N$_4$Si$_2$(CH$_3$)$_2$)]	S:	S-N Comp.2-291
C$_2$CoU	UCoC$_2$	U:	SVol.C12-209/10
C$_2$CrF$_6$O$_8$S$_2$	CrO$_2$(OSO$_2$CF$_3$)$_2$	F:	PFHOrg.SVol.3-61, 79, 102/4
C$_2$CrH$_6$O$_4$Sn	(CH$_3$)$_2$SnOCrO$_3$	Sn:	Org.Comp.14-144/5
C$_2$CrU	UCrC$_2$	U:	SVol.C12-100, 190/2
C$_2$CsCuH$_6$	(CH$_3$)$_2$CuCs	Cu:	Org.Comp.2-1
C$_2$CsF$_6$N	(CF$_3$)$_2$NCs	F:	PFHOrg.9-40/1, 47/8
C$_2$CsF$_9$Se	C$_2$F$_5$SeF$_3$ · CsF	F:	PFHOrg.SVol.3-216/7, 226
C$_2$CsH$_6$NO$_{11}$SU	Cs(NH$_4$)[UO$_2$(C$_2$O$_4$)(SO$_4$)(H$_2$O)]	U:	SVol.C13-225/7
C$_2$CsH$_8$NO$_{12}$SU	Cs(NH$_4$)[UO$_2$(C$_2$O$_4$)(SO$_4$)(H$_2$O)$_2$]	U:	SVol.C13-225/7
C$_2$Cs$_2$H$_2$O$_{11}$SU	Cs$_2$[UO$_2$(C$_2$O$_4$)(SO$_4$)(H$_2$O)]	U:	SVol.C13-225/7
C$_2$Cs$_2$H$_4$O$_{12}$SU	Cs$_2$[UO$_2$(C$_2$O$_4$)(SO$_4$)(H$_2$O)$_2$]	U:	SVol.C13-225/7
C$_2$Cs$_2$O$_8$U	Cs$_2$[UO$_2$(CO$_3$)$_2$]	U:	SVol.C13-10/2
C$_2$Cs$_2$O$_9$SU	Cs$_2$[UO$_2$(C$_2$O$_4$)(SO$_3$)] · 3 H$_2$O	U:	SVol.C13-223/5
C$_2$Cs$_2$O$_{14}$S$_2$U$_2$	Cs$_2$[(UO$_2$)$_2$(C$_2$O$_4$)(SO$_3$)$_2$] · 3 H$_2$O	U:	SVol.C13-223/5
C$_2$Cs$_3$F$_3$H$_2$O$_7$U	Cs$_3$[UO$_2$(C$_2$O$_4$)F$_3$(H$_2$O)] · H$_2$O	U:	SVol.C13-219/21
C$_2$Cs$_3$F$_3$O$_6$U	Cs$_3$[UO$_2$(C$_2$O$_4$)F$_3$] · 2 H$_2$O	U:	SVol.C13-219/21
–	Cs$_3$[UO$_2$(C$_2$O$_4$)F$_3$] · x H$_2$O	U:	SVol.A5-139
C$_2$Cu	C$_2$Cu	Cu:	Org.Comp.4-1/2
C$_2$CuFH$_4$	CH$_2$CH$_2$CuF	Cu:	Org.Comp.4-5
C$_2$CuF$_3$I	F$_2$CCFCuI	Cu:	Org.Comp.1-466
C$_2$CuF$_3$O$_4$S	[(CO)Cu]O$_3$SCF$_3$	Cu:	Org.Comp.3-190
C$_2$CuF$_6$IZn	(CF$_3$)$_2$CuZnI	Cu:	Org.Comp.2-237
C$_2$CuF$_6$O$_6$S$_2$	Cu(OSO$_2$CF$_3$)$_2$	F:	PFHOrg.SVol.3-62, 80, 106
–	Cu(OSO$_2$CF$_3$)$_2$ · 5.5 H$_2$O	F:	PFHOrg.SVol.3-62

C_2CuH_5	C_2H_5Cu · x $LiMgBrCl_2$	Cu: Org.Comp.1-299, 304/5
–	C_2H_5Cu · x $LiMgBr_2Cl$	Cu: Org.Comp.1-299, 310
–	C_2H_5Cu · x $LiMgBr_2Hal$	Cu: Org.Comp.1-299, 307/8, 312/3
–	C_2H_5Cu · x $LiMgBr_3$	Cu: Org.Comp.1-299, 308/11
–	C_2H_5Cu · x $LiOCH(CH_3)C_6H_5$	Cu: Org.Comp.1-151
–	C_2H_5Cu · x $LiSC_6H_5$	Cu: Org.Comp.1-172/3
–	C_2H_5Cu · x $MgBrCl$	Cu: Org.Comp.1-210, 258, 268, 273, 279/80, 291
–	C_2H_5Cu · x $MgBrHal$	Cu: Org.Comp.1-210, 281/5
–	C_2H_5Cu · x $MgBrI$	Cu: Org.Comp.1-210, 290
–	C_2H_5Cu · x $MgBrOC_4H_9$-t	Cu: Org.Comp.1-318, 322
–	C_2H_5Cu · x $MgBr_2$	Cu: Org.Comp.1-210, 269/72, 281, 286, 293, 297
–	C_2H_5Cu · x $MgClI$	Cu: Org.Comp.1-211
–	C_2H_5Cu · x MgI_2	Cu: Org.Comp.1-211, 291
–	C_2H_5Cu · x $P(C_4H_9$-n$)_3$ · y LiI	Cu: Org.Comp.1-439
–	C_2H_5Cu · x $P(C_4H_9$-n$)_3$ · y $MgClI$	Cu: Org.Comp.1-439
–	C_2H_5Cu · x $P(OC_2H_5)_3$ · y $MgBr_2$	Cu: Org.Comp.1-440
–	C_2H_5Cu · x $S(CH_3)_2$ · y $MgBrI$	Cu: Org.Comp.1-359, 363/6
–	C_2H_5Cu · x $S(CH_3)_2$ · y $MgBr_2$	Cu: Org.Comp.1-363/6
$C_2CuH_5O_2S$	$CH_3SO_2CH_2Cu$ · x LiI	Cu: Org.Comp.1-56
C_2CuH_6	$(CH_3)_2Cu$	Cu: Org.Comp.2-2
–	C_2H_5CuH	Cu: Org.Comp.1-466
C_2CuH_6IMg	$(CH_3)_2CuMgI$	Cu: Org.Comp.2-212/3, 217/24
C_2CuH_6Li	$(CH_3)_2CuLi = (CH_3)_4Cu_2Li_2$	Cu: Org.Comp.2-6/7, 28, 33/173
		Cu: Org.Comp.4-103/4, 106
–	$CH_3(CD_3)CuLi$	Cu: Org.Comp.2-175, 205/8
–	$(CD_3)_2CuLi$	Cu: Org.Comp.2-6/7, 134
–	$(CH_3)_2CuLi$ · x $P(C_4H_9$-n$)_3$	Cu: Org.Comp.1-427
C_2CuH_6Na	$(CH_3)_2CuNa$	Cu: Org.Comp.2-1
$C_2CuH_{12}O_5^+$	$[C_2H_2Cu(H_2O)_5]^+$	Cu: Org.Comp.4-43
C_2CuNO	$(CO)CuCN$	Cu: Org.Comp.3-190
C_2CuO_2	$(CO)_2Cu$	Cu: Org.Comp.3-211
C_2Cu_2	Cu_2C_2	Cu: Org.Comp.3-2, 150
		Cu: Org.Comp.4-136/50
–	Cu_2C_2 · n H_2O (n = 0.5, 1 to 6)	Cu: Org.Comp.4-136, 142, 150
$C_2Cu_2H_2O$	$HCCCu$ · $CuOH$	Cu: Org.Comp.3-176
C_2Cu_2O	$OCCCu_2$	Cu: Org.Comp.4-127
–	$OCCCu_2$ · H_2O	Cu: Org.Comp.4-127
–	$OCCCu_2$ · x $CuCl$	Cu: Org.Comp.4-127
$C_2DyHN_2S_3^{2+}$	$Dy[(S)(HS)C_2N_2S]^{2+}$	Sc: MVol.D4-102
$C_2DyH_2IO_2^{2+}$	$[Dy(CH_2ICOO)]^{2+}$	Sc: MVol.D5-53/6
$C_2DyH_2N_3S_2^{2+}$	$Dy[(S)(H_2N)C_2N_2S]^{2+}$	Sc: MVol.D4-103
$C_2DyH_2O_4^+$	$[Dy(HCOO)_2]^+$	Sc: MVol.D5-4/5
$C_2DyH_3O_2^{2+}$	$[Dy(CH_3COO)]^{2+}$	Sc: MVol.D5-24, 26/9
$C_2DyH_3O_2S^{2+}$	$[Dy(OOCCH_2SH)]^{2+}$	Sc: MVol.D4-53/4

$C_2DyH_3O_3{}^{2+}$	$[Dy(HOCH_2COO)]^{2+}$	Sc:	MVol.D5-222/30
$C_2DyH_3O_4{}^{2+}$	$[Dy((HO)_2CHCOO)]^{2+}$	Sc:	MVol.D5-249/50
$C_2DyH_4NO_2{}^{2+}$	$Dy(NH_2CH_2COO)^{2+}$	Sc:	MVol.D1-103/4
$C_2DyH_5O_2{}^{2+}$	$[Dy(OCH_2CH_2OH)]^{2+}$	Sc:	MVol.D3-31/2
$C_2DyH_6O_6P_2{}^-$	$[Dy(CH_3PO_3)_2]^-$	Sc:	MVol.D4-143
$C_2DyH_8N_2{}^{3+}$	$[Dy(C_2H_8N_2)]^{3+}$	Sc:	MVol.D1-19/21
$C_2DyN_2S_2{}^+$	$[Dy(NCS)_2]^+$	Sc:	MVol.D4-340/3
$C_2DyO_4{}^+$	$[Dy(C_2O_4)]^+$	Sc:	MVol.D5-120/4
C_2Er	ErC_2 solid solutions		
	$(Er,U)C_2$	U:	SVol.C12-153
C_2ErFO_4	$ErF(C_2O_4)$	Sc:	MVol.D5-146/7
–	$ErF(C_2O_4) \cdot H_2O$	Sc:	MVol.D5-146/7
–	$ErF(C_2O_4) \cdot 2 H_2O$	Sc:	MVol.D5-146/7
–	$ErF(C_2O_4) \cdot 4 H_2O$	Sc:	MVol.D5-146/7
$C_2ErH_2IO_2{}^{2+}$	$[Er(CH_2ICOO)]^{2+}$	Sc:	MVol.D5-53/6
$C_2ErH_2O_4{}^+$	$[Er(HCOO)_2]^+$	Sc:	MVol.D5-4/5
$C_2ErH_3O_2{}^{2+}$	$[Er(CH_3COO)]^{2+}$	Sc:	MVol.D5-25/9
$C_2ErH_3O_2S^{2+}$	$[Er(OOCCH_2SH)]^{2+}$	Sc:	MVol.D4-53/4
$C_2ErH_3O_3{}^{2+}$	$[Er(HOCH_2COO)]^{2+}$	Sc:	MVol.D5-222/30
$C_2ErH_3O_4{}^{2+}$	$[Er((HO)_2CHCOO)]^{2+}$	Sc:	MVol.D5-249/50
$C_2ErH_4NO_2{}^{2+}$	$Er(NH_2CH_2COO)^{2+}$	Sc:	MVol.D1-103/4
$C_2ErH_5O_2{}^{2+}$	$[Er(OCH_2CH_2OH)]^{2+}$	Sc:	MVol.D3-31/2
$C_2ErH_6O_6P_2{}^-$	$[Er(CH_3PO_3)_2]^-$	Sc:	MVol.D4-143
$C_2ErH_8N_2{}^{3+}$	$[Er(C_2H_8N_2)]^{3+}$	Sc:	MVol.D1-19/21
$C_2ErN_2S_2{}^+$	$[Er(NCS)_2]^+$	Sc:	MVol.D4-340/3
$C_2ErN_3{}^{2+}$	$[ErN(CN)_2]^{2+}$	Sc:	MVol.D4-338
$C_2ErO_4{}^+$	$[Er(C_2O_4)]^+$	Sc:	MVol.D5-120/4
$C_2EuF_2H_3O_2$	$EuF_2(CH_3COO)$	Sc:	MVol.D5-50/1
$C_2EuH_2IO_2{}^{2+}$	$[Eu(CH_2ICOO)]^{2+}$	Sc:	MVol.D5-53/6
$C_2EuH_2O_4$	$Eu(HCOO)_2$	Sc:	MVol.D5-16
$C_2EuH_2O_4{}^+$	$[Eu(HCOO)_2]^+$	Sc:	MVol.D5-4/5
$C_2EuH_3O_2{}^{2+}$	$[Eu(CH_3COO)]^{2+}$	Sc:	MVol.D5-24/7
$C_2EuH_3O_2S^{2+}$	$[Eu(OOCCH_2SH)]^{2+}$	Sc:	MVol.D4-53/4
$C_2EuH_3O_3{}^{2+}$	$[Eu(HOCH_2COO)]^{2+}$	Sc:	MVol.D5-222/30
$C_2EuH_3O_4{}^{2+}$	$[Eu((HO)_2CHCOO)]^{2+}$	Sc:	MVol.D5-249/50
$C_2EuH_4NO_2{}^{2+}$	$Eu(NH_2CH_2COO)^{2+}$	Sc:	MVol.D1-103/4
C_2EuH_5I	$EuIC_2H_5$	Sc:	MVol.D6-147/8
$C_2EuH_5O_2{}^{2+}$	$[Eu(OCH_2CH_2OH)]^{2+}$	Sc:	MVol.D3-31/2
$C_2EuH_6O_6P_2{}^-$	$[Eu(CH_3PO_3)_2]^-$	Sc:	MVol.D4-143
$C_2EuH_8N_2{}^{3+}$	$[Eu(C_2H_8N_2)]^{3+}$	Sc:	MVol.D1-19/21
$C_2EuN_2S_2$	$Eu(NCS)_2$	Sc:	MVol.D4-342/3
$C_2EuN_2S_2{}^+$	$[Eu(NCS)_2]^+$	Sc:	MVol.D4-340/3
C_2EuO_2	$Eu(CO)_2$	Sc:	MVol.D6-166
C_2EuO_4	EuC_2O_4	Sc:	MVol.D5-120
–	$EuC_2O_4 \cdot H_2O$	Sc:	MVol.D5-120
–	$EuC_2O_4 \cdot 1.7 H_2O$	Sc:	MVol.D5-120
$C_2EuO_4{}^+$	$[Eu(C_2O_4)]^+$	Sc:	MVol.D5-120/4
C_2FGaH_6	$Ga(CH_3)_2F$	Ga:	Org.Comp.1-125
$C_2FH_2MnO_2{}^+$	$[Mn(CH_2FCOO)]^+$	Mn:	MVol.D2-61/2
$C_2FH_2N_3$	$NH_2C(NF)CN$	F:	PFHOrg.8-222

$C_2FH_3N_2O_5$	$(NO_2)_2CFCH_2OH$	F:	PFHOrg.8-42
$C_2FH_6O_3PSn$	$(CH_3)_2SnOP(O)(F)O$	Sn:	Org.Comp.14-136
C_2FH_7O	$HF \cdot C_2H_5OH$	F:	SVol.3-229
$C_2FH_8S_2$	$2 H_2S \cdot CH_3CHF \cdot 17 H_2O$	S:	SVol.4a/b-326
C_2FMnO_4	$[Mn(C_2O_4)F]$	Mn:	MVol.D2-121
C_2FNO	$FC(O)CN$	F:	PFHOrg.9-53, 69
C_2FNOS	$CF(O)NCS$	F:	PFHOrg.9-107/8, 114
C_2FNO_2	$CF(O)NCO$	F:	PFHOrg.9-105, 111/2,
			115, 119
C_2FNS_2	$FC(S)NCS$	F:	PFHOrg.SVol.1-192
$C_2FN_3O_4$	$CF(NO_2)_2CN$	F:	PFHOrg.8-5, 7, 22, 38
C_2FN_5	$NFC(CN)N_3$	F:	PFHOrg.8-222
C_2FNdO_4	$NdF(C_2O_4) \cdot 4 H_2O$	Sc:	MVol.D5-146/7
C_2FO_4Sc	$ScF(C_2O_4)$	Sc:	MVol.D5-117
—	$ScF(C_2O_4) \cdot H_2O$	Sc:	MVol.D5-117
$C_2F_2GaH_6N$	$Ga(CH_3)_2NF_2$	Ga:	Org.Comp.1-244
$C_2F_2GaH_6O_2P$	$Ga(CH_3)_2OP(O)F_2$	Ga:	Org.Comp.1-203/4,
			209/10
$C_2F_2HNO_2$	$CF_2NC(O)OH$	F:	PFHOrg.9-115/6
—	$FC(O)NHC(O)F$	F:	PFHOrg.9-115
$C_2F_2HNO_3S$	$CF_2(NO_2)C(O)SH$	F:	PFHOrg.8-5, 7, 23, 35, 38
$C_2F_2HNO_4$	$CF_2(NO_2)COOH$	F:	PFHOrg.8-5, 7, 22, 35
$C_2F_2HNO_7S$	$CF_2(NO_2)C(O)OSO_3H$	F:	PFHOrg.8-5, 7, 38
$C_2F_2H_3LaO_2$	$LaF_2(CH_3COO)$	Sc:	MVol.D5-50/1
$C_2F_2H_3NdO_2$	$NdF_2(CH_3COO)$	Sc:	MVol.D5-50/1
$C_2F_2H_3O_2Pr$	$PrF_2(CH_3COO)$	Sc:	MVol.D5-50/1
$C_2F_2H_3O_2Sm$	$SmF_2(CH_3COO)$	Sc:	MVol.D5-50/1
$C_2F_2H_4N_2$	$NHF_2 \cdot CH_3CN$	F:	SVol.5-55
$C_2F_2H_5NO$	$NHF_2 \cdot CH_2OCH_2$	F:	SVol.5-50
$C_2F_2H_6O_3SU$	$UO_2F_2 \cdot (CH_3)_2SO$	U:	SVol.A6-35
$C_2F_2H_6O_6S_2Sn$	$(CH_3)_2Sn(OSO_2F)_2$	Sn:	Org.Comp.14-125, 128/9
$C_2F_2H_7NO$	$NHF_2 \cdot (CH_3)_2O$	F:	SVol.5-49/50
$C_2F_2H_7NOS$	$NHF_2 \cdot (CH_3)_2SO$	F:	SVol.5-55
$C_2F_2HgN_4O_8$	$[CF(NO_2)_2]_2Hg$	F:	PFHOrg.8-4/5, 21, 42/3
C_2F_2IN	ICF_2CN	F:	PFHOrg.9-50/1, 67
$C_2F_2K_2O_6U$	$K_2[UO_2(C_2O_4)F_2]$	U:	SVol.C13-217
$C_2F_2Li_2$	F_2CCLi_2	F:	PFHOrg.SVol.1-1
$C_2F_2N_2$	$FC(CN)NF$	F:	PFHOrg.8-168/9, 194,
			222/3
$C_2F_2N_2O_2$	$CF_2(NO_2)CN$	F:	PFHOrg.8-5, 7, 22, 41
$C_2F_2N_2O_2S$	$F_2SNC(O)NCO$	F:	PFHOrg.9-35
$C_2F_2N_2S$	$SN_2C_2F_2$	S:	S-N Comp.3-93/8, 137
$C_2F_2N_4O_8$	$(NO_2)_2CFCF(NO_2)_2$	F:	PFHOrg.8-8/9, 23, 34, 35,
			47
$C_2F_2OS_2$	$F_2CS_2C(O)$	F:	PFHOrg.SVol.2-1, 5
$C_2F_2O_2S_2$	$FC(O)SSC(O)F$	F:	PFHOrg.SVol.2-175/6,
			204, 206
$C_2F_2O_4U$	$UF_2(C_2O_4) \cdot 1.5 H_2O$	U:	SVol.C8-36
		U:	SVol.C13-217/20
$C_2F_2O_6Rb_2U$	$Rb_2[UO_2(C_2O_4)F_2]$	U:	SVol.C13-219

C₂F₃HNP	CF₃PHCN.	F:	PFHOrg.SVol.1-82
C₂F₃HN₄	CF₃CHN₄	F:	PFHOrg.9-80
C₂F₃HOS	CF₃C(O)SH	F:	PFHOrg.SVol.2-62, 71/2
C₂F₃HO₂	CF₃C(O)OH	F:	PFHOrg.9-150
C₂F₃HO₂S	CF₂CFS(O)OH	F:	PFHOrg.SVol.3-3, 21
C₂F₃HO₃S	CF₂CFSO₃H.	F:	PFHOrg.SVol.3-48, 88
−	[CF₂CF(SO₃H)]ₙ	F:	PFHOrg.SVol.3-50
C₂F₃H₂K₃O₇U	K₃[UO₂(C₂O₄)F₃(H₂O)] · H₂O	U:	SVol.C13-218/21
C₂F₃H₂NO	CF₃C(O)NH₂	F:	PFHOrg.8-35
C₂F₃H₂NOS	CF₃SC(O)NH₂	F:	PFHOrg.SVol.2-211, 225, 234
C₂F₃H₂NO₂	CF₃NHC(O)OH.	F:	PFHOrg.9-115/6
C₂F₃H₂NS	CF₃C(S)NH₂	F:	PFHOrg.SVol.1-193/4
C₂F₃H₂NS₂	CF₃SC(S)NH₂	F:	PFHOrg.SVol.1-198/9, 201
C₂F₃H₂Na₃O₇U	Na₃[UO₂(C₂O₄)F₃(H₂O)] · 5 H₂O	U:	SVol.C13-217/21
C₂F₃H₂O₇Rb₃U	Rb₃[UO₂(C₂O₄)F₃(H₂O)] · H₂O	U:	SVol.C13-219/21
C₂F₃H₃N₂O	CF₃C(O)NHNH₂	F:	PFHOrg.8-65/6, 90
−	CH₃OC(NF)NF₂	F:	PFHOrg.8-222
C₂F₃H₃N₂S	CF₃C(S)NHNH₂	F:	PFHOrg.8-66
−	CF₃C(SH)NNH₂	F:	PFHOrg.8-66
C₂F₃H₃S	CF₃SCH₃	F:	PFHOrg.SVol.2-256
C₂F₃H₃S₂	CH₃SSCF₃	F:	PFHOrg.SVol.2-117, 126
C₂F₃H₄N	CF₃CH₂NH₂	F:	PFHOrg.9-79
C₂F₃H₄NS	CF₃SNHCH₃	F:	PFHOrg.SVol.2-113
C₂F₃H₅S₂	2 H₂S · CHFCF₂ · 17 H₂O.	S:	SVol.4a/b-326
C₂F₃H₆N₃S	CF₃C(SNH₄)NNH₂	F:	PFHOrg.8-66
C₂F₃H₆O₂U	U(OCH₃)₂F₃	U:	SVol.C13-72/3
C₂F₃H₁₂N₃O₆U	(NH₄)₃[UO₂(C₂O₄)F₃]	U:	SVol.C13-221
−	(NH₄)₃[UO₂(C₂O₄)F₃] · 6 H₂O	U:	SVol.C13-218
C₂F₃H₁₅N₅O₂Rh²⁺ . .	[Rh(NH₃)₅(OCOCF₃)]²⁺	Rh:	SVol.B2-134
C₂F₃H₁₅N₆Ru²⁺ . . .	[Ru(NH₃)₅(CF₃CN)]²⁺	F:	PFHOrg.9-87
C₂F₃INP.	CF₃PICN	F:	PFHOrg.SVol.1-107/8, 119
C₂F₃IO₃S	CF₂CFIOSO₂	F:	PFHOrg.SVol.3-125, 132/3
−	IC(O)CF₂SO₂F	F:	PFHOrg.SVol.3-152/3, 162
C₂F₃I₃	CF₃CI₃	F:	PFHOrg.8-106
C₂F₃K₃O₆U	K₃[UO₂(C₂O₄)F₃].	U:	SVol.C13-217/21
−	K₃[UO₂(C₂O₄)F₃] · 2 H₂O	U:	SVol.C13-218
−	K₃[UO₂(C₂O₄)F₃] · x H₂O	U:	SVol.A5-139
C₂F₃Li	F₂CCFLi.	F:	PFHOrg.SVol.1-1, 9, 11
C₂F₃LiO₂S	LiOS(O)CFCF₂	F:	PFHOrg.SVol.3-3
C₂F₃MnO₅	[MnO₃ · O₂CCF₃]	Mn:	MVol.D2-63
C₂F₃N	CF₃CN	F:	PFHOrg.9-50/1, 62/4, 67, 78/88, 95
−	CF₃NC	F:	PFHOrg.9-107, 113, 120/1
C₂F₃NO	CF₃CNO.	F:	PFHOrg.8-1
−	CF₃NCO.	F:	PFHOrg.8-203

C_2F_3NO	CF_3NCO	F:	PFHOrg.9-102/3, 109, 115/9
—	$FC(O)NCF_2$	F:	PFHOrg.9-127/8
C_2F_3NOS	CF_3SNCO	F:	PFHOrg.SVol.2-102
—	$CF_3S(O)CN$	F:	PFHOrg.SVol.3-11, 35
$C_2F_3NOS_2$	CF_3SSNCO	F:	PFHOrg.SVol.2-174, 240
C_2F_3NOSe	CF_3SeNCO	F:	PFHOrg.SVol.3-215, 224
$C_2F_3NO_2S$	$CF_3C(O)NSO$	F:	PFHOrg.9-6/7, 23, 33
—	$CF_3S(O)NCO$	F:	PFHOrg.SVol.3-6, 25, 41
$C_2F_3NO_2S_2$	CF_3SO_2NCS	F:	PFHOrg.SVol.3-140/1, 145/6
$C_2F_3NO_3$	$CF_2(NO_2)C(O)F$	F:	PFHOrg.8-5, 7, 22, 34, 37, 41
$C_2F_3NO_3S$	CF_3SO_2NCO	F:	PFHOrg.SVol.3-140/1, 145, 147
C_2F_3NS	CF_3NCS	F:	PFHOrg.9-107/8, 113, 121
$C_2F_3NS_2$	CF_3SSCN	F:	PFHOrg.SVol.2-174
C_2F_3NSe	CF_3SeCN	F:	PFHOrg.SVol.3-215, 234/5
$C_2F_3N_3O$	$CF_3C(O)N_3$	F:	PFHOrg.8-61, 87
$C_2F_3N_3OS_3$	$S_3N_2NCOCF_3$	F:	PFHOrg.9-7/8, 26
		S:	S-N Comp.2-48/9
$C_2F_3N_3O_2$	$CF_3C(NO_2)N_2$	F:	PFHOrg.8-58, 85, 102
$C_2F_3N_3O_6$	$NO_2CF_2CF(NO_2)_2$	F:	PFHOrg.8-8/9, 23
$C_2F_3N_3S_2$	$S_2N_3C(CF_3)$	S:	S-N Comp.4-1
$C_2F_3N_4S_3^+$	$[(SN)_3C(CF_3)N]^+$	S:	S-N Comp.4-126
$C_2F_3N_7O_4S_6$	$[CF_3CN_4S_3][N(SO_2N)_2S]$	S:	S-N Comp.4-126
$C_2F_3NaO_2$	$CF_3C(O)ONa$	F:	PFHOrg.9-150
$C_2F_3Na_3O_6U$	$Na_3[UO_2(C_2O_4)F_3] \cdot 6 H_2O$	U:	SVol.C13-217/21
—	$Na_3[UO_2(C_2O_4)F_3] \cdot x H_2O$	U:	SVol.A5-139
$C_2F_3O_6Rb_3U$	$Rb_3[UO_2(C_2O_4)F_3] \cdot 2 H_2O$	U:	SVol.C13-219/21
—	$Rb_3[UO_2(C_2O_4)F_3] \cdot x H_2O$	U:	SVol.A5-139
$C_2F_3O_6U^{3-}$	$[UO_2(C_2O_4)F_3]^{3-}$	U:	SVol.D2-378
C_2F_4HNO	$CF_3C(OH)NF$	F:	PFHOrg.8-169/70, 195
—	$CF_3NHC(O)F$	F:	PFHOrg.9-115
$C_2F_4HNOS_2$	$FC(O)S(CF_3S)NH$	F:	PFHOrg.SVol.2-86, 88, 90
$C_2F_4HNO_2$	$NF_2CF_2C(O)OH$	F:	PFHOrg.8-157/8, 180
$C_2F_4HN_3O$	$NF_2CF(NHF)NCO$	F:	PFHOrg.8-152/3
$C_2F_4H_2N_2$	$CF_3C(NF)NH_2$	F:	PFHOrg.8-152, 175
$C_2F_4H_2Na_2O_6U$	$Na_2[UF_4(HCO_3)_2]$	U:	SVol.C13-8/9
$C_2F_4H_3NU$	$UF_4 \cdot CH_3CN$	U:	SVol.C8-155, 160
$C_2F_4H_4N_2O$	$CH_3OCF(NHF)NF_2$	F:	PFHOrg.8-222
$C_2F_4H_6O_2U$	$UF_4(OCH_3)_2$	U:	SVol.C13-76, 82
		U:	SVol.D1-128
$C_2F_4H_6O_4P_2Sn$	$(CH_3)_2Sn(OP(O)F_2)_2$	Sn:	Org.Comp.14-136
$C_2F_4H_{10}N_2O_6U$	$(NH_4)_2[UF_4(HCO_3)_2]$	U:	SVol.C13-8/9
$C_2F_4K_2O_8U_2$	$K_2[(UO_2)_2(C_2O_4)F_4]$	U:	SVol.C13-217
$C_2F_4La_2O_4$	$La_2F_4(C_2O_4)$	Sc:	MVol.D5-146/7
$C_2F_4NO_4S^-$	$CF_3CF(NO)SO_3^-$	F:	PFHOrg.SVol.3-172

$C_2F_6HNS_2$	$(CF_3S)_2NH$	F:	PFHOrg.SVol.2-85, 87/8, 95/6
$C_2F_6HNS_3$	$CF_3S(CF_3SS)NH$	F:	PFHOrg.SVol.2-86, 174, 186, 195
$C_2F_6HNS_4$	$(CF_3SS)_2NH$	F:	PFHOrg.SVol.2-174, 185, 195
$C_2F_6HNSe_2$	$(CF_3Se)_2NH$	F:	PFHOrg.SVol.3-216, 224
C_2F_6HOP	$(CF_3)_2POH$	F:	PFHOrg.SVol.1-96/7
C_2F_6HOPS	$(CF_3)_2P(S)OH$	F:	PFHOrg.SVol.1-140/5
$C_2F_6HO_2P$	$(CF_3)_2P(O)OH$	F:	PFHOrg.SVol.1-96
C_2F_6HP	$(CF_3)_2PH$	F:	PFHOrg.SVol.1-82/6
$C_2F_6HPS_2$	$(CF_3)_2P(S)SH$	F:	PFHOrg.SVol.1-139/40, 144
C_2F_6HSb	$(CF_3)_2SbH$	F:	PFHOrg.SVol.1-178/9
$C_2F_6H_2HgN_2O_4S_2$	$Hg(NHSO_2CF_3)_2$	F:	PFHOrg.SVol.3-139, 143
$C_2F_6H_2NP$	$(CF_3)_2PNH_2$	F:	PFHOrg.SVol.1-101
$C_2F_6H_2N_2$	$CF_3NHNHCF_3$	F:	PFHOrg.8-63, 122
$C_2F_6H_2N_6$	$(NF_2)_2CFNNC(NF)NH_2$	F:	PFHOrg.8-97
$C_2F_6H_2P_2$	$(CF_3)_2PPH_2$	F:	PFHOrg.SVol.1-82
$-$	$CF_3PHPHCF_3$	F:	PFHOrg.SVol.1-82/6
$C_2F_6H_2Se_2Si$	$(CF_3Se)_2SiH_2$	F:	PFHOrg.SVol.3-221, 236, 244
$C_2F_6H_3NSi$	$(CF_3)_2NSiH_3$	F:	PFHOrg.9-40, 43
$C_2F_6H_3P$	$CH_3PF_3CF_3$	F:	PFHOrg.SVol.1-123/4
$C_2F_6H_3PS$	$CF_3PF_3SCH_3$	F:	PFHOrg.SVol.1-127/8
$C_2F_6H_3PSi$	$(CF_3)_2PSiH_3$	F:	PFHOrg.SVol.1-159
$C_2F_6H_4NPSi$	$(CF_3)_2PN(H)SiH_3$	F:	PFHOrg.SVol.1-103/4
$C_2F_6H_{10}N_2Sn$	$[C_2H_4(NH_3)_2]SnF_6$	F:	SVol.3-213
$C_2F_6HgO_4Se_2$	$Hg[OSe(O)CF_3]_2$	F:	PFHOrg.SVol.3-215/6, 224
$C_2F_6HgO_6S_2$	$Hg(OSO_2CF_3)_2$	F:	PFHOrg.SVol.3-61, 102/4
C_2F_6HgS	$Hg(SCF_3)CF_3$	F:	PFHOrg.SVol.2-251, 253
$C_2F_6HgS_2$	$Hg(SCF_3)_2$	F:	PFHOrg.SVol.2-251, 253, 254/6, 258/62
$C_2F_6HgSe_2$	$Hg(SeCF_3)_2$	F:	PFHOrg.SVol.3-215, 223, 241
C_2F_6IN	$(CF_3)_2NI$	F:	PFHOrg.8-161, 184, 208, 218/9
C_2F_6IP	$(CF_3)_2PI$	F:	PFHOrg.SVol.1-106, 112/30
C_2F_6IPS	$(CF_3)_2P(S)I$	F:	PFHOrg.SVol.1-140/5
C_2F_6ISb	$(CF_3)_2SbI$	F:	PFHOrg.SVol.1-178/9, 181
		Sb:	Org.Comp.2-25/6
$C_2F_6I_2N_4O_{12}$	$[(CF_3)_2I][I(NO_3)_4]$	F:	PFHOrg.SVol.3-265
$C_2F_6I_2Sn$	$(CF_3)_2SnI_2$	Sn:	Org.Comp.8-115, 116
$C_2F_6IrN_2$	$[IrN(CF_3)NCF_3]$	F:	PFHOrg.8-143
C_2F_6KN	$(CF_3)_2NK$	F:	PFHOrg.9-40/1
C_2F_6LiNS	$LiNS(CF_3)_2$	F:	PFHOrg.SVol.3-8, 38, 43

$C_2H_2N_2S$	SN_2C_2HD	S:	S–N Comp.3-99/118
–	$SN_2C_2D_2$	S:	S–N Comp.3-99/118
$C_2H_2N_2S^+$	$[SN_2C_2H_2]^+$	S:	S–N Comp.3-118
$C_2H_2N_2S^-$	$[SN_2C_2H_2]^-$	S:	S–N Comp.3-118/9
$C_2H_2N_3NdS_2^{2+}$	$Nd[(S)(H_2N)C_2N_2S]^{2+}$	Sc:	MVol.D4-103
$C_2H_2N_3S$	$[SN_3C_2H_2]$ (radical)	S:	S–N Comp.4-18
$C_2H_2N_3S_2Sm^{2+}$	$Sm[(S)(H_2N)C_2N_2S]^{2+}$	Sc:	MVol.D4-103
$C_2H_2N_3S_2Tb^{2+}$	$Tb[(S)(H_2N)C_2N_2S]^{2+}$	Sc:	MVol.D4-103
$C_2H_2N_4S_2$	$[(SNCHN)_2]$	S:	S–N Comp.4-134/5
$C_2H_2N_6S_4$	$(S_2N_3CH)_2$	S:	S–N Comp.4-2/3
$C_2H_2Na_2O_9U$	$Na_2[UO_2(OO)(C_2O_4)(H_2O)] \cdot 3 H_2O$	U:	SVol.C13-213/5
$C_2H_2Na_2O_{16}P_2U_2$ . . .	$Na_2[(UO_2)_2(C_2O_4)(HPO_4)_2] \cdot 4 H_2O$	U:	SVol.C14-129
$C_2H_2NdO_4^+$	$[Nd(HCOO)_2]^+$	Sc:	MVol.D5-4/5
$C_2H_2O_4Pr^+$	$[Pr(HCOO)_2]^+$	Sc:	MVol.D5-4/5
$C_2H_2O_4Sm^+$	$[Sm(HCOO)_2]^+$	Sc:	MVol.D5-4/5
$C_2H_2O_4Tb^+$	$[Tb(HCOO)_2]^+$	Sc:	MVol.D5-4/5
$C_2H_2O_4Tm^+$	$[Tm(HCOO)_2]^+$	Sc:	MVol.D5-4/5
$C_2H_2O_4Y^+$	$[Y(HCOO)_2]^+$	Sc:	MVol.D5-4/5
$C_2H_2O_4Yb^+$	$[Yb(HCOO)_2]^+$	Sc:	MVol.D5-4/5
$C_2H_2O_4Zn$	$Zn(HCOO)_2$ solid solutions		
	$Zn_{1-x}Mn_x(HCOO)_2 \cdot 2 H_2O$	Mn:	MVol.D2-20/1
$C_2H_2O_5U$	$UO(HCOO)_2$	U:	SVol.C13-98/100
–	$UO(HCOO)_2 \cdot H_2O$	U:	SVol.C13-98/100
–	$UO(HCOO)_2 \cdot 1.5 H_2O$	U:	SVol.C13-98, 100, 104
–	$UO(HCOO)_2 \cdot 2 H_2O$	U:	SVol.C13-98
–	$UO(HCOO)_2 \cdot 3 H_2O$	U:	SVol.C13-98, 100
$C_2H_2O_6U$	$U(OH)_2(C_2O_4) \cdot 5 H_2O$	U:	SVol.C13-210
–	$UO_2(HCOO)_2$	U:	SVol.A6-181
		U:	SVol.C13-99/101
		U:	SVol.D1-200
–	$UO_2(HCOO)_2 \cdot H_2O$	U:	SVol.A6-37, 181
		U:	SVol.C13-99/101
–	$UO_2(HCOO)_2 \cdot 2 H_2O$	U:	SVol.C13-99, 101
–	$UO_2(HCOO)_2 \cdot 3 H_2O$	U:	SVol.A6-66
$C_2H_2O_7Tl_2U$	$Tl_2UO_2(HCOO)_2(O) \cdot H_2O$		
	$= Tl_2UO_2(HCOO)_2(OH)_2$	U:	SVol.C13-105
$C_2H_2O_8U^{2-}$	$[U(OH)_2(CO_3)_2]^{2-}$	U:	SVol.D1-74
$C_2H_2O_9U_2$	$U_2O_5(HCOO)_2$	U:	SVol.C13-99, 102
$C_2H_2O_{10}U^{4-}$	$[UO_2(OH)_2(CO_3)_2]^{4-}$	U:	SVol.D1-177
$C_2H_2O_{10}U_2$	$(UO_2)_2(C_2O_4)(OH)_2 \cdot 2 H_2O$	U:	SVol.C13-210
$C_2H_2O_{11}Rb_2SU$	$Rb_2[UO_2(C_2O_4)(SO_4)(H_2O)]$	U:	SVol.C13-225/7
$C_2H_2O_{12}U_2^{2-}$	$[(UO_2)_2(OH)_2(CO_3)_2]^{2-}$	U:	SVol.D1-177
$C_2H_2O_{16}P_2Rb_2U_2$ . . .	$Rb_2[(UO_2)_2(C_2O_4)(HPO_4)_2] \cdot 4 H_2O$	U:	SVol.C14-129
$C_2H_3HoO_2^{2+}$	$[Ho(CH_3COO)]^{2+}$	Sc:	MVol.D5-24, 27/9
$C_2H_3HoO_2S^{2+}$	$[Ho(OOCCH_2SH)]^{2+}$	Sc:	MVol.D4-53/4
$C_2H_3HoO_3^{2+}$	$[Ho(HOCH_2COO)]^{2+}$	Sc:	MVol.D5-222/30
$C_2H_3HoO_4^{2+}$	$[Ho((HO)_2CHCOO)]^{2+}$	Sc:	MVol.D5-249/50
$C_2H_3K_3O_{10}U$	$K_3[UO_2OH(CO_3)_2(H_2O)]$	U:	SVol.C13-22/4
$C_2H_3K_3O_{11}U$	$K_3[UO_2HO_2(CO_3)_2(H_2O)]$	U:	SVol.C13-24/5
$C_2H_3LaO_2^{2+}$	$[La(CH_3COO)]^{2+}$	Sc:	MVol.D5-23, 25/8

$C_2H_3LaO_2S^{2+}$	$[La(OOCCH_2SH)]^{2+}$	Sc:	MVol.D4-53/4
$C_2H_3LaO_3^{2+}$	$[La(HOCH_2COO)]^{2+}$	Sc:	MVol.D5-222/30
$C_2H_3LaO_4^{2+}$	$[La((HO)_2CHCOO)]^{2+}$	Sc:	MVol.D5-249/50
$C_2H_3LuO_2^{2+}$	$[Lu(CH_3COO)]^{2+}$	Sc:	MVol.D5-25
$C_2H_3LuO_2S^{2+}$	$[Lu(OOCCH_2SH)]^{2+}$	Sc:	MVol.D4-53/4
$C_2H_3LuO_3^{2+}$	$[Lu(HOCH_2COO)]^{2+}$	Sc:	MVol.D5-222/30
$C_2H_3LuO_4^{2+}$	$[Lu((HO)_2CHCOO)]^{2+}$	Sc:	MVol.D5-249/50
$C_2H_3MnO_2^+$	$[Mn(CH_3COO)]^+$	Mn:	MVol.D2-30/1
$C_2H_3MnO_3^+$	$[Mn(CH_2(OH)COO)]^+$	Mn:	MVol.D2-146
$C_2H_3N_3S$	$SN_2C_2HNH_2$	S:	S-N Comp.3-93/8, 141/2
$C_2H_3N_3S_2$	$S_2N_3CCH_3$	S:	S-N Comp.4-1/2
$C_2H_3N_5OS_3$	$S_3N_3C(CONHNH_2)$	S:	S-N Comp.4-110
$C_2H_3NdO_2^{2+}$	$[Nd(CH_3COO)]^{2+}$	Sc:	MVol.D5-23/9
$C_2H_3NdO_2S^{2+}$	$[Nd(OOCCH_2SH)]^{2+}$	Sc:	MVol.D4-53/4
$C_2H_3NdO_3^{2+}$	$[Nd(HOCH_2COO)]^{2+}$	Sc:	MVol.D5-222/30
$C_2H_3NdO_4^{2+}$	$[Nd((HO)_2CHCOO)]^{2+}$	Sc:	MVol.D5-249/50
$C_2H_3O_2Pr^{2+}$	$[Pr(CH_3COO)]^{2+}$	Sc:	MVol.D5-23, 25/8
$C_2H_3O_2PrS^{2+}$	$[Pr(OOCCH_2SH)]^{2+}$	Sc:	MVol.D4-53/4
$C_2H_3O_2SSm^{2+}$	$[Sm(OOCCH_2SH)]^{2+}$	Sc:	MVol.D4-53/4
$C_2H_3O_2STb^{2+}$	$[Tb(OOCCH_2SH)]^{2+}$	Sc:	MVol.D4-53/4
$C_2H_3O_2STm^{2+}$	$[Tm(OOCCH_2SH)]^{2+}$	Sc:	MVol.D4-53/4
$C_2H_3O_2SY^{2+}$	$[Y(OOCCH_2SH)]^{2+}$	Sc:	MVol.D4-53/4
$C_2H_3O_2SYb^{2+}$	$[Yb(OOCCH_2SH)]^{2+}$	Sc:	MVol.D4-53/4
$C_2H_3O_2Sc^{2+}$	$Sc(CH_3COO)^{2+}$	Sc:	MVol.D5-21
$C_2H_3O_2Sm^{2+}$	$[Sm(CH_3COO)]^{2+}$	Sc:	MVol.D5-24, 26/9
$C_2H_3O_2Tb^{2+}$	$[Tb(CH_3COO)]^{2+}$	Sc:	MVol.D5-24
$C_2H_3O_2Tm^{2+}$	$[Tm(CH_3COO)]^{2+}$	Sc:	MVol.D5-25
$C_2H_3O_2U^{2+}$	$[U(CH_3COO)]^{2+}$	U:	SVol.D1-200
$C_2H_3O_2U^{3+}$	$[U(CH_3COO)]^{3+}$	U:	SVol.D1-200
$C_2H_3O_2Y^{2+}$	$[Y(CH_3COO)]^{2+}$	Sc:	MVol.D5-23, 27/8
$C_2H_3O_2Yb^{2+}$	$[Yb(CH_3COO)]^{2+}$	Sc:	MVol.D5-25/9
$C_2H_3O_3Pr$	$PrO(CH_3COO)$	Sc:	MVol.D5-43
$C_2H_3O_3Pr^{2+}$	$[Pr(HOCH_2COO)]^{2+}$	Sc:	MVol.D5-222/7
$C_2H_3O_3Sc^{2+}$	$[Sc(HOCH_2COO)]^{2+}$	Sc:	MVol.D5-222
$C_2H_3O_3Sm^{2+}$	$[Sm(HOCH_2COO)]^{2+}$	Sc:	MVol.D5-222/30
$C_2H_3O_3Tb^{2+}$	$[Tb(HOCH_2COO)]^{2+}$	Sc:	MVol.D5-222/30
$C_2H_3O_3Tm^{2+}$	$[Tm(HOCH_2COO)]^{2+}$	Sc:	MVol.D5-222/30
$C_2H_3O_3U^{2+}$	$[U(HOCH_2COO)]^{2+}$	U:	SVol.D1-202
$C_2H_3O_3Y^{2+}$	$[Y(HOCH_2COO)]^{2+}$	Sc:	MVol.D5-222/9
$C_2H_3O_3Yb^{2+}$	$[Yb(HOCH_2COO)]^{2+}$	Sc:	MVol.D5-222/30
$C_2H_3O_4Pr^{2+}$	$[Pr((HO)_2CHCOO)]^{2+}$	Sc:	MVol.D5-249/50
$C_2H_3O_4SU$	$UO_2(HSCH_2COO)$	U:	SVol.C13-154
$C_2H_3O_4SU^+$	$[UO_2(HSCH_2COO)]^+$	U:	SVol.D1-198, 202
$C_2H_3O_4Sc$	$Sc(OH)(OCH_2COO) \cdot 2 H_2O$	Sc:	MVol.D5-233
$C_2H_3O_4Sm^{2+}$	$[Sm((HO)_2CHCOO)]^{2+}$	Sc:	MVol.D5-249/50
$C_2H_3O_4Tb^{2+}$	$[Tb((HO)_2CHCOO)]^{2+}$	Sc:	MVol.D5-249/50
$C_2H_3O_4Tm^{2+}$	$[Tm((HO)_2CHCOO)]^{2+}$	Sc:	MVol.D5-249/50
$C_2H_3O_4U^+$	$[UO_2(CH_3COO)]^+$	U:	SVol.D1-198/200
		U:	SVol.D4-127
$C_2H_3O_4Y^{2+}$	$[Y((HO)_2CHCOO)]^{2+}$	Sc:	MVol.D5-249/50

C$_2$H$_3$O$_4$Yb^{2+} . . . . [Yb((HO)$_2$CHCOO)]$^{2+}$ . . . . Sc: MVol.D5-249/50

C$_2$H$_3$O$_5$PPm$^+$ . . . . . . Pm[H(OOCCH$_2$PO$_3$)]$^+$ . . . . Sc: MVol.D4-144

C$_2$H$_3$O$_5$TlU . . . . TlUO$_2$(CH$_3$COO)(O) · H$_2$O

                   = TlUO$_2$(CH$_3$COO)(OH)$_2$ . . . . U: SVol.C13-141/2

C$_2$H$_3$O$_5$U$^+$ . . . . [UO$_2$(HOCH$_2$COO)]$^+$ . . . . U: SVol.D1-198, 203

C$_2$H$_3$O$_8$SU$^-$ . . . . [UO$_2$(SO$_4$)(CH$_3$COO)]$^-$ . . . . U: SVol.D1-154/8

C$_2$H$_3$O$_{12}$S$_2$U^{3-} . . . . . [UO$_2$(SO$_4$)$_2$(CH$_3$COO)]$^{3-}$ . . . . U: SVol.D1-154/8

C$_2$H$_4$HoNO$_2$$^{2+}$ . . . . . . Ho(NH$_2$CH$_2$COO)$^{2+}$ . . . . Sc: MVol.D1-103/4

C$_2$H$_4$KNO$_7$Os . . . . . . K[OsNC$_2$O$_4$(OH)$_2$(H$_2$O)] · 2 H$_2$O . . . . Os: SVol.1-205, 211

C$_2$H$_4$K$_2$O$_{12}$SU . . . . . K$_2$[UO$_2$(C$_2$O$_4$)(SO$_4$)(H$_2$O)$_2$] · H$_2$O . . . . U: SVol.C13-225/7

C$_2$H$_4$LaNO$_2$$^{2+}$ . . . . . . La(NH$_2$CH$_2$COO)$^{2+}$ . . . . Sc: MVol.D1-103

C$_2$H$_4$LaN$_2$O$_2$S$_2$$^+$ . . . . [La(ONCHSH)$_2$]$^+$ . . . . Sc: MVol.D4-69

C$_2$H$_4$LuNO$_2$$^{2+}$ . . . . . . Lu(NH$_2$CH$_2$COO)$^{2+}$ . . . . Sc: MVol.D1-103/4

C$_2$H$_4$MnNO$_2$$^+$ . . . . . . Mn(NH$_2$CH$_2$COO)$^+$ . . . . Mn: MVol.D4-248/51

   − . . . . Mn(ONHC(O)CH$_3$)$^+$ . . . . Mn: MVol.D5-212

C$_2$H$_4$MnNO$_2$$^{2+}$ . . . . . Mn(NH$_2$CH$_2$COO)$^{2+}$ . . . . Mn: MVol.D4-249/50

C$_2$H$_4$MnN$_2$O$_4$ . . . . . . MnC$_2$O$_4$ · N$_2$H$_4$ . . . . Mn: MVol.D2-100

   − . . . . Mn(N$_2$H$_4$)C$_2$O$_4$ . . . . Mn: MVol.D3-68/9

C$_2$H$_4$MnN$_4$O$_2$ . . . . . . Mn(N$_2$(CONH$_2$)$_2$) · 2 H$_2$O . . . . Mn: MVol.D5-158

C$_2$H$_4$MnO$_{10}$S$^-$ . . . . . [Mn(C$_2$O$_4$)(SO$_4$)(H$_2$O)$_2$]$^-$ . . . . Mn: MVol.D2-121

C$_2$H$_4$NNdO$_2$$^{2+}$ . . . . . . Nd(NH$_2$CH$_2$COO)$^{2+}$ . . . . Sc: MVol.D1-103

C$_2$H$_4$NO$_2$Pm^{2+} . . . . . Pm(NH$_2$CH$_2$COO)$^{2+}$ . . . . Sc: MVol.D1-103

C$_2$H$_4$NO$_2$Pr^{2+} . . . . . . Pr(NH$_2$CH$_2$COO)$^{2+}$ . . . . Sc: MVol.D1-103

C$_2$H$_4$NO$_2$Sc^{2+} . . . . . . Sc(NH$_2$CH$_2$COO)$^{2+}$ . . . . Sc: MVol.D1-103

C$_2$H$_4$NO$_2$Sm^{2+} . . . . . Sm(NH$_2$CH$_2$COO)$^{2+}$ . . . . Sc: MVol.D1-103

C$_2$H$_4$NO$_2$Tb^{2+} . . . . . . Tb(NH$_2$CH$_2$COO)$^{2+}$ . . . . Sc: MVol.D1-103/4

C$_2$H$_4$NO$_2$U^{3+} . . . . [U(OOCCH$_2$NH$_2$)]$^{3+}$ . . . . U: SVol.D1-215

C$_2$H$_4$NO$_2$Y^{2+} . . . . Y(NH$_2$CH$_2$COO)$^{2+}$ . . . . Sc: MVol.D1-103

C$_2$H$_4$NO$_2$Yb^{2+} . . . . . . Yb(NH$_2$CH$_2$COO)$^{2+}$ . . . . Sc: MVol.D1-103/4

C$_2$H$_4$NO$_4$U$^+$ . . . . [UO$_2$(OOCCH$_2$NH$_2$)]$^+$ . . . . U: SVol.D1-216

C$_2$H$_4$NO$_8$SSc . . . . NH$_4$[Sc(SO$_4$)(C$_2$O$_4$)] · H$_2$O . . . . Sc: MVol.D5-118

C$_2$H$_4$N$_2$NdO$_2$S$_2$$^+$ . . . . [Nd(ONCHSH)$_2$]$^+$ . . . . Sc: MVol.D4-69

C$_2$H$_4$N$_2$O$_2$PrS$_2$$^+$ . . . . . [Pr(ONCHSH)$_2$]$^+$ . . . . Sc: MVol.D4-69

C$_2$H$_4$N$_2$O$_2$S$_2$Sm$^+$ . . . [Sm(ONCHSH)$_2$]$^+$ . . . . Sc: MVol.D4-69

C$_2$H$_4$N$_2$O$_2$S$_2$Y$^+$ . . . . . [Y(ONCHSH)$_2$]$^+$ . . . . Sc: MVol.D4-69

C$_2$H$_4$N$_2$O$_4$Os^{3-} . . . . . [Os(OH)$_4$(CN)$_2$]$^{3-}$ . . . . Os: SVol.1-178

C$_2$H$_4$N$_2$S . . . . SN$_2$C$_2$H$_4$ . . . . S: S-N Comp.3-91/3

C$_2$H$_4$N$_2$S$_3$ . . . . S(NSCH$_2$)$_2$ . . . . S: S-N Comp.4-110/1

C$_2$H$_4$N$_4$NiOS$_4$ . . . . . . [Ni((S$_2$N$_2$CH$_2$)$_2$O)] . . . . S: S-N Comp.2-291, 293,

                                             296

C$_2$H$_4$N$_4$NiO$_2$S$_4$ . . . . . [Ni((S$_2$N$_2$CHOH)$_2$)] . . . . S: S-N Comp.2-291/2, 296

C$_2$H$_4$N$_4$OS . . . . (O)SN$_2$C$_2$(NH$_2$)$_2$ . . . . S: S-N Comp.3-250/7

C$_2$H$_4$N$_4$S . . . . SN$_2$C$_2$(NH$_2$)$_2$ . . . . S: S-N Comp.3-93/8, 146/8

C$_2$H$_4$N$_6$Rh$_2$ . . . . Rh$_2$(CN)$_2$ · 2 N$_2$H$_2$ · 3.5 H$_2$O . . . . Rh: SVol.B1-197

C$_2$H$_4$N$_6$S$_2$ . . . . [(SNC(NH$_2$)N)$_2$] . . . . S: S-N Comp.4-135

C$_2$H$_4$Na$_2$O$_{10}$U . . . . . . Na$_2$[UO$_2$(CO$_3$)$_2$(H$_2$O)$_2$] . . . . U: SVol.C13-10/2

C$_2$H$_4$O$_2$ . . . . CH$_3$COOH systems

                 CH$_3$COOH-Mn(CH$_3$COO)$_2$-H$_2$O . . . . Mn: MVol.D2-50/2

C$_2$H$_4$O$_2$S$_2$U . . . . UO$_2$(SC$_2$H$_4$S) . . . . U: SVol.C13-310/2

C$_2$H$_4$O$_3$Sc$^+$ . . . . Sc(OH)(CH$_3$COO)$^+$ . . . . Sc: MVol.D5-21

$C_2H_4O_5PPm^{2+}$	$Pm[H_2(OOCCH_2PO_3)]^{2+}$	Sc:	MVol.D4-144
$C_2H_4O_5S$	$SO_3 \cdot CH_3COOH$	S:	SVol.3-315
$C_2H_4O_5U$	$UO_2(CH_3COO)(OH)$	U:	SVol.C13-121/2
–	$UO_2(CH_3COO)(OH) \cdot 2 H_2O$	U:	SVol.C13-121/2
$C_2H_4O_6U$	$U(OH)_2(HCOO)_2$	U:	SVol.C13-98, 100
$C_2H_4O_7P_2Y^-$	$Y[H(OC(PO_3)_2CH_3)]^-$	Sc:	MVol.D4-147
$C_2H_4O_8Tl_2U$	$Tl_2UO_2(HCOO)_2(OH)_2$		
	$= Tl_2UO_2(HCOO)_2(O) \cdot H_2O$	U:	SVol.C13-105
$C_2H_4O_{10}U^{2-}$	$[UO_2(CO_3)_2(H_2O)_2]^{2-}$	U:	SVol.D1-67
$C_2H_4O_{11}S_3$	$3 SO_3 \cdot CH_3COOH$	S:	SVol.3-315
$C_2H_4O_{12}Rb_2SU$	$Rb_2[UO_2(C_2O_4)(SO_4)(H_2O)_2]$	U:	SVol.C13-225/7
$C_2H_4O_{13}P_4Tc^{5-}$	$[TcO(CH_2(PO_3)_2)_2]^{5-}$	Tc:	SVol.2-212
$C_2H_5HoO_2^{2+}$	$[Ho(OCH_2CH_2OH)]^{2+}$	Sc:	MVol.D3-31/2
C_2H_5IYb	$YbIC_2H_5$	Sc:	MVol.D6-147/8
$C_2H_5I_2Sb$	$C_2H_5SbI_2$	Sb:	Org.Comp.2-115
$C_2H_5I_3Sn$	$C_2H_5SnI_3$	Sn:	Org.Comp.8-138
$C_2H_5LaOS^{2+}$	$La(HOC_2H_4S)^{2+}$	Sc:	MVol.D4-49
$C_2H_5LaO_2^{2+}$	$[La(OCH_2CH_2OH)]^{2+}$	Sc:	MVol.D3-31/2
$C_2H_5LaO_7P_2$	$La[H_2(OC(PO_3)_2CH_3)] \cdot 5 H_2O$	Sc:	MVol.D4-147
$C_2H_5LuO_2^{2+}$	$[Lu(OCH_2CH_2OH)]^{2+}$	Sc:	MVol.D3-31/2
$C_2H_5MnNO_2^{2+}$	$Mn(NH_3CH_2COO)^{2+}$	Mn:	MVol.D4-248/50
$C_2H_5MnNO_6S$	$Mn(NH_3CH_2COO)SO_4$	Mn:	MVol.D4-261/2
$C_2H_5MnN_3O_2^{2+}$	$Mn(HN(CONH_2)_2)^{2+}$	Mn:	MVol.D5-155
$C_2H_5MnN_3O_6S$	$MnSO_4(HN(CONH_2)_2)$	Mn:	MVol.D5-157
$C_2H_5NO_4U^{2+}$	$[UO_2H(OOCCH_2NH_2)]^{2+}$	U:	SVol.D1-198, 216
$C_2H_5NO_5U$	$UO_2(NH_2CH_2COO)(OH) \cdot 2 H_2O$	U:	SVol.C13-148/9
$C_2H_5NO_6SU$	$UO_2SO_3 \cdot CH_3CONH_2 \cdot 1.5 H_2O$	U:	SVol.C10-139
$C_2H_5NO_7SU$	$UO_2SO_4 \cdot CH_3CONH_2 \cdot 2 H_2O$	U:	SVol.C10-163
$C_2H_5NO_8SU$	$UO_2SO_4 \cdot NH_2CH_2COOH \cdot 2 H_2O$	U:	SVol.C10-163
$C_2H_5NO_8U$	$NH_4[UO_2(OOH)(C_2O_4)] \cdot x H_2O$	U:	SVol.C13-214
$C_2H_5N_2O_3U^+$	$[UO_2(HNCH_2CONH_2)]^+$	U:	SVol.D1-216
$C_2H_5N_3OS$	$SN_2C_2HONH_4$	S:	S-N Comp.3-93/8, 150
$C_2H_5N_3O_2Rh^{3+}$	$[Rh((NH_2CO)_2NH)]^{3+}$	Rh:	SVol.B2-321
$C_2H_5N_4RhS_4$	$Rh(S_2CNHNH_2)(S_2CNHNH) \cdot 2 H_2O$	Rh:	SVol.B3-15
$C_2H_5N_5Rh$	$Rh(CN)_2 \cdot N_2H_4 \cdot NH \cdot 3 H_2O$	Rh:	SVol.B1-197
$C_2H_5N_6Rh$	$Rh(CN)_2 \cdot N_2H_3 \cdot N_2H_2 \cdot 1.5 H_2O$	Rh:	SVol.B1-198
$C_2H_5NdO_2^{2+}$	$[Nd(OCH_2CH_2OH)]^{2+}$	Sc:	MVol.D3-31/2
$C_2H_5NdO_7P_2$	$Nd[H_2(OC(PO_3)_2CH_3)] \cdot 6 H_2O$	Sc:	MVol.D4-147
$C_2H_5OPrS^{2+}$	$Pr(HOC_2H_4S)^{2+}$	Sc:	MVol.D4-49
$C_2H_5OSSm^{2+}$	$Sm(HOC_2H_4S)^{2+}$	Sc:	MVol.D4-49
C_2H_5OSb	C_2H_5SbO	Sb:	Org.Comp.2-119
$C_2H_5OTh^{3+}$	$[Th(OC_2H_5)]^{3+}$	Th:	SVol.E-28
$C_2H_5O_2Pr^{2+}$	$[Pr(OCH_2CH_2OH)]^{2+}$	Sc:	MVol.D3-31/2
$C_2H_5O_2Sm^{2+}$	$[Sm(OCH_2CH_2OH)]^{2+}$	Sc:	MVol.D3-31/2
$C_2H_5O_2Tb^{2+}$	$[Tb(OCH_2CH_2OH)]^{2+}$	Sc:	MVol.D3-31/2
$C_2H_5O_2Tm^{2+}$	$[Tm(OCH_2CH_2OH)]^{2+}$	Sc:	MVol.D3-31/2
$C_2H_5O_2Y^{2+}$	$[Y(OCH_2CH_2OH)]^{2+}$	Sc:	MVol.D3-31/2
$C_2H_5O_2Yb^{2+}$	$[Yb(OCH_2CH_2OH)]^{2+}$	Sc:	MVol.D3-31/2
$C_2H_5O_3SU^+$	$[UO_2(SC_2H_4OH)]^+$	U:	SVol.D1-251

$C_2H_5O_6TlU$	$TlUO_2(CH_3COO)(OH)_2$		
	$= TlUO_2(CH_3COO)(O) \cdot H_2O$	U:	SVol.C13-141/2
$C_2H_5O_7P_2Pr$	$Pr[H_2(OC(PO_3)_2CH_3)] \cdot 5 H_2O$	Sc:	MVol.D4-147
$C_2H_5O_7P_2Sm$.	$Sm[H_2(OC(PO_3)_2CH_3)] \cdot 5 H_2O$	Sc:	MVol.D4-147
$C_2H_5O_7P_2Y$	$Y[H_2(OC(PO_3)_2CH_3)] \cdot 5 H_2O$	Sc:	MVol.D4-147
$C_2H_6HgO_3S$	$CH_3HgOSO_2CH_3$	S:	SVol.3-322
$C_2H_6HgO_6S_2$	$CH_3HgS_2O_6CH_3$	S:	SVol.3-322
$C_2H_6HoO_6P_2^-$	$[Ho(CH_3PO_3)_2]^-$	Sc:	MVol.D4-143
C_2H_6ISb	$(CH_3)_2SbI$	Sb:	Org.Comp.2-25
$C_2H_6I_2O_6Sn$	$(CH_3)_2Sn(OIO_2)_2$	Sn:	Org.Comp.14-111
$C_2H_6I_2Sn$	$(CH_3)_2SnI_2$	Sn:	Org.Comp.8-87/96
—	$(CH_3)_2SnI_2$ systems		
	$(CH_3)_2SnI_2-(CH_3)_2Sn(SCH_3)_2$	Sn:	Org.Comp.10-3
$C_2H_6LaO_4P^{2+}$	$La[(CH_3O)_2PO_2]^{2+}$	Sc:	MVol.D4-177
$C_2H_6LaO_6P_2^-$	$[La(CH_3PO_3)_2]^-$	Sc:	MVol.D4-143
$C_2H_6LuO_6P_2^-$	$[Lu(CH_3PO_3)_2]^-$	Sc:	MVol.D4-143
$C_2H_6MnN_4$	$Mn(NH_3)_2(CN)_2$	Mn:	MVol.D3-18
$C_2H_6MnN_4O_2$	$Mn(ONC)_2 \cdot 2 NH_3$	Mn:	MVol.D3-18/9
$C_2H_6MnN_4O_4$	$Mn(N_2H_3COO)_2$	Mn:	MVol.D5-87
—	$Mn(N_2H_3COO)_2 \cdot 2 H_2O$	Mn:	MVol.D3-68
		Mn:	MVol.D5-87/8
—	$Mn(OCNH_2NHO)_2$	Mn:	MVol.D5-147/8
$C_2H_6MoN_2O_5$	$MoO_3 \cdot 2 HCONH_2$	Mo:	SVol.B3a-25
$C_2H_6MoO_4Sn$	$(CH_3)_2SnOMoO_3$	Sn:	Org.Comp.14-145/6
$C_2H_6NO_5SU^+$	$[UO_2(O_3SC_2H_4NH_2)]^+$	U:	SVol.D1-252
$C_2H_6N_2O_2S_2Sn$	$S_2N_2 \cdot Sn(OCH_3)_2$	S:	S-N Comp.2-17, 188
$C_2H_6N_2O_6Sn$	$(CH_3)_2Sn(ONO_2)_2$	Sn:	Org.Comp.14-113, 114/7
$C_2H_6N_3PS_2$	$(SN)_2P(CH_3)_2N$	S:	S-N Comp.3-14/9
$C_2H_6N_4NiS_4$	$[Ni(S_2N_2CH_3)_2]$	S:	S-N Comp.2-292, 296
—	$[Ni(S_2N_2H)(S_2N_2C_2H_5)]$	S:	S-N Comp.2-292, 296
$C_2H_6N_4NiS_4Si_2$	$[Ni(S_4N_4Si_2(CH_3)_2)]$	S:	S-N Comp.2-291
$C_2H_6N_4O_2S_4$	$S_3N_2NSO_2N(CH_3)_2$	S:	S-N Comp.2-46
$C_2H_6N_5PS_3$	$S_3N_5P(CH_3)_2$	S:	S-N Comp.3-39/41
$C_2H_6N_6S_3Si_2$	$CH_3Si(NSN)_3SiCH_3$	S:	S-N Comp.3-11/2
$C_2H_6N_6S_5$	$S_5N_6(CH_3)_2$	S:	S-N Comp.2-97
$C_2H_6N_6Sn$	$(CH_3)_2Sn(N_3)_2$	Sn:	Org.Comp.8-204
$C_2H_6N_{12}S_9$	$(S_4N_5N)_2S(CH_3)_2$	S:	S-N Comp.2-270
$C_2H_6Na_2S_2Sn$	$(CH_3)_2Sn(SNa)_2$	Sn:	Org.Comp.10-28
$C_2H_6NdO_4P^{2+}$	$Nd[(CH_3O)_2PO_2]^{2+}$	Sc:	MVol.D4-177
—	$Nd[(CH_3O)_2PO_2](H_2O)_n^{2+}$	Sc:	MVol.D4-177
$C_2H_6NdO_6P_2^-$	$[Nd(CH_3PO_3)_2]^-$	Sc:	MVol.D4-143
$C_2H_6O_3SSn$	$(CH_3)_2SnOSO_2$	Sn:	Org.Comp.14-127, 130
$C_2H_6O_3SeSn$	$(CH_3)_2SnOSeO_2$	Sn:	Org.Comp.14-143/4
$C_2H_6O_4PPr^{2+}$	$Pr[(CH_3O)_2PO_2]^{2+}$	Sc:	MVol.D4-177
$C_2H_6O_4PYb^{2+}$	$Yb[(CH_3O)_2PO_2]^{2+}$	Sc:	MVol.D4-177
$C_2H_6O_4SSn$	$(CH_3)_2SnOSO_3$	Sn:	Org.Comp.14-127, 131
$C_2H_6O_4SeSn$	$(CH_3)_2SnOSeO_3$	Sn:	Org.Comp.14-144
$C_2H_6O_4SnW$	$(CH_3)_2SnOWO_3$	Sn:	Org.Comp.14-146/7
$C_2H_6O_4U$	$UO_2(OCH_3)_2$	U:	SVol.C13-85/6
$C_2H_6O_6Os$	$OsO_2(OH)_2(O_2C_2H_4)$	Os:	SVol.1-190

$C_2H_8N_2O_8U$	$(NH_4)_2[UO_2(OO)(C_2O_4)]$	U:	SVol.C13-214/5
–	$(NH_4)_2[UO_2(OO)(C_2O_4)] \cdot 3 H_2O$	U:	SVol.C13-213/5
$C_2H_8N_2O_9SU$	$(NH_4)_2[UO_2(C_2O_4)(SO_3)]$	U:	SVol.C13-223/5
–	$(NH_4)_2[UO_2(C_2O_4)(SO_3)] \cdot x H_2O$	U:	SVol.C13-223/5
$C_2H_8N_2O_{12}U_2$	$(NH_4)_2[(UO_2)_2(OO)_2(C_2O_4)] \cdot 2 H_2O$	U:	SVol.C13-213/4
–	$(NH_4)_2[(UO_2)_2(OO)_2(C_2O_4)] \cdot 4 H_2O$	U:	SVol.C13-213/4
–	$(NH_4)_2[(UO_2)_2(OO)_2(C_2O_4)] \cdot x H_2O$ (x = 3 to 4)		
		U:	SVol.C13-213/5
$C_2H_8N_2O_{14}S_2U_2$	$(NH_4)_2[(UO_2)_2(C_2O_4)(SO_3)_2] \cdot 2 H_2O$	U:	SVol.C13-223/5
$C_2H_8N_2Pr^{3+}$	$[Pr(C_2H_8N_2)]^{3+}$	Sc:	MVol.D1-19/21
$C_2H_8N_2Sm^{3+}$	$[Sm(C_2H_8N_2)]^{3+}$	Sc:	MVol.D1-19/21
$C_2H_8N_2Tb^{3+}$	$[Tb(C_2H_8N_2)]^{3+}$	Sc:	MVol.D1-19/21
$C_2H_8N_2Y^{3+}$	$[Y(C_2H_8N_2)]^{3+}$	Sc:	MVol.D1-19/21
$C_2H_8N_2Yb^{3+}$	$[Yb(C_2H_8N_2)]^{3+}$	Sc:	MVol.D1-19/21
$C_2H_8N_4Na_6O_{23}S_7Th_2$	$Na_6Th_2(SO_3)_7 \cdot 2 (NH_2)_2CO \cdot 15 H_2O$	Th:	SVol.C5-61
$C_2H_8N_4Nd_2O_{14}S_3$	$Nd_2(SO_4)_3 \cdot 2 OC(NH_2)_2$	Sc:	MVol.D2-222/3
$C_2H_8N_4O_6S_3U$	$UO_2SO_4 \cdot 2 (NH_2)_2CS$	U:	SVol.C10-163
$C_2H_8N_4O_7SU$	$UO_2SO_3 \cdot 2 (NH_2)_2CO$	U:	SVol.C10-139
$C_2H_8N_4O_8SU$	$UO_2SO_4 \cdot 2 (NH_2)_2CO$	U:	SVol.C10-163
–	$UO_2SO_4 \cdot 2 (NH_2)_2CO \cdot H_2O$	U:	SVol.C10-163
$C_2H_8N_4O_8U$	$UO_2(NO_3)_2(NH_2(CH_2)_2NH_2)$	U:	SVol.A5-217/8
$C_2H_8N_4O_{11}U$	$[(C_2H_5)NH_3]UO_2(NO_3)_3$	U:	SVol.C7-190/1
$C_2H_8N_4O_{14}Pr_2S_3$	$Pr_2(SO_4)_3 \cdot 2 CO(NH_2)_2$	Sc:	MVol.C8-133
		Sc:	MVol.D2-222/3
$C_2H_8N_4O_{14}S_3Sc_2$	$Sc_2(SO_4)_3 \cdot 2 CO(NH_2)_2 \cdot 4 H_2O$	Sc:	MVol.C8-70
$C_2H_8N_6O_{10}U$	$UO_2(NO_3)_2(CO(NH_2)_2)_2$	U:	SVol.A5-217/8
		U:	SVol.A6-65
$C_2H_8N_6O_{15}Th$	$[(CH_3)_2NH_2]Th(NO_3)_5 \cdot 8 H_2O$	Th:	SVol.C3-107/8, 118, 120/1
–	$(C_2H_5NH_3)Th(NO_3)_5 \cdot 5 H_2O$	Th:	SVol.C3-107/8, 118, 120/1
$C_2H_8N_6S_4$	$[(CH_3)_2NH_2]S_4N_5$	S:	S-N Comp.2-259, 263
$C_2H_8N_7O_{11}Pr$	$Pr(NO_3)_3 \cdot 2 OC(NH_2)_2$	Sc:	MVol.D2-220
$C_2H_8N_8O_{14}Th$	$Th(NO_3)_4 \cdot 2 (NH_2)_2CO \cdot H_2O$	Th:	SVol.E-38, 41
–	$Th(NO_3)_4 \cdot 2 (NH_2)_2CO \cdot 2 H_2O$	Th:	SVol.E-38, 41
–	$Th(NO_3)_4 \cdot 2 (NH_2)_2CO \cdot 6 H_2O$	Th:	SVol.E-38, 41
$C_2H_8N_8Rh_2$	$Rh_2(CN)_2 \cdot N_2H_4 \cdot 2 N_2H_2 \cdot H_2O$	Rh:	SVol.B1-197
$C_2H_8NdO_2^{3+}$	$[Nd(H_2O)_4(CH_3OH)_2]^{3+}$	Sc:	MVol.D3-17
$C_2H_8Nd_2O_{20}P_4$	$Nd_2(H_2PO_4)_4(C_2O_4) \cdot H_2O$	Sc:	MVol.D5-146/7
$C_2H_8O_2Sn$	$(CH_3)_2Sn(OH)_2$	Sn:	Org.Comp.14-24/5
$C_2H_8O_8P_2Pm^+$	$Pm(HOCH_2P(O)_2OH)_2^+$	Sc:	MVol.D4-143
$C_2H_8O_{20}P_4Pr_2$	$Pr_2(H_2PO_4)_4(C_2O_4) \cdot H_2O$	Sc:	MVol.D5-146/7
$C_2H_8O_{20}P_4Sm_2$	$Sm_2(H_2PO_4)_4(C_2O_4) \cdot H_2O$	Sc:	MVol.D5-146/7
$C_2H_9NO_{11}S_2Th_2$	$Th_2(OH)_4(SO_3)_2 \cdot CH_3CONH_2 \cdot 5 H_2O$	Th:	SVol.C5-58/9
$C_2H_{10}MnN_4O_4$	$Mn(N_2H_4)_2(HCOO)_2$	Mn:	MVol.D3-68
$C_2H_{10}MnN_8O_8$	$Mn(H_2NC(O)NHNH_2)_2(NO_3)_2$	Mn:	MVol.D5-208
$C_2H_{10}N_2O_6U$	$UO_2(NH_2O)_2(HOCH_2CH_2OH)$	U:	SVol.A6-33
$C_2H_{10}N_2O_9U$	$(NH_4)_2[UO_2(OO)(C_2O_4)(H_2O)]$	U:	SVol.C13-214/5
–	$(NH_4)_2[UO_2(OO)(C_2O_4)(H_2O)] \cdot 2 H_2O$	U:	SVol.C13-214/5
$C_2H_{10}N_2O_{10}S_2U$	$[C_2H_4(NH_3)_2]UO_2(SO_4)_2$	U:	SVol.C10-191
–	$[C_2H_4(NH_3)_2]UO_2(SO_4)_2 \cdot 4 H_2O$	U:	SVol.C10-192
$C_2H_{10}N_2O_{16}P_2U_2$	$(NH_4)_2[(UO_2)_2(C_2O_4)(HPO_4)_2] \cdot 4 H_2O$	U:	SVol.C14-129

$C_2H_{10}N_4O_8U$	$UO_2(NO_3)_2(CH_3NH_2)_2$	U:	SVol.A5-217/8
$C_2H_{10}N_6O_6U$	$UO_2(N_2H_3CO_2)_2 \cdot N_2H_4 \cdot H_2O$	U:	SVol.C13-34/5
$C_2H_{10}O_4P_2Sn$	$(CH_3)_2Sn(OP(O)H_2)_2$	Sn:	Org.Comp.14-134, 137
$C_2H_{10}S_3$	$2 H_2S \cdot (CH_3)_2S \cdot 17 H_2O$	S:	SVol.4a/b-326
$C_2H_{11}MnN_3O_8S$	$[MnSO_4(1,2,4-C_2H_2N_3H)(H_2O)_4]$	Mn:	MVol.D4-65
$C_2H_{11}NO_{10}U$	$NH_4[UO_2(C_2O_4)(OH)(H_2O)_3]$	U:	SVol.C13-209
$C_2H_{12}MnN_6O_2$	$Mn(NH_3)_4(ONC)_2$	Mn:	MVol.D3-18/9
$C_2H_{12}MnN_6S_2$	$Mn(NH_3)_4(NCS)_2$	Mn:	MVol.D3-19
$C_2H_{12}MoN_2O_4$	$(CH_3NH_3)_2MoO_4$	Mo:	SVol.B4-132
$C_2H_{12}Mo_2N_2O_7$	$(CH_3NH_3)_2O \cdot 2 MoO_3$	Mo:	SVol.B4-132
–	$(CH_3NH_3)_2O \cdot 2 MoO_3 \cdot 0.5 H_2O$	Mo:	SVol.B4-131/2
–	$(CH_3NH_3)_2O \cdot 2 MoO_3 \cdot H_2O$	Mo:	SVol.B4-131/2
–	$(CH_3NH_3)_2O \cdot 2 MoO_3 \cdot 2 H_2O$	Mo:	SVol.B4-131
$C_2H_{12}Mo_4N_2O_{13}$	$(CH_3NH_3)_2O \cdot 4 MoO_3$	Mo:	SVol.B4-132
$C_2H_{12}Mo_{5.87}N_2O_{18.61}$			
	$(CH_3NH_3)_2O \cdot 5.87 MoO_3 \cdot x H_2O$	Mo:	SVol.B4-43, 131/2
$C_2H_{12}Mo_{9.36}N_2O_{29.08}$			
	$(CH_3NH_3)_2O \cdot 9.36 MoO_3 \cdot x H_2O$	Mo:	SVol.B4-43, 131/2
$C_2H_{12}N_2O_{10}U$	$(NH_4)_2[UO_2(CO_3)_2(H_2O)_2]$	U:	SVol.C13-10/2
$C_2H_{12}N_2O_{14}S_3U$	$[C_2H_4(NH_3)_2]H_2UO_2(SO_4)_3$	U:	SVol.C10-192
$C_2H_{12}N_6O_9S_3Th$	$[(NH_2)_3C]_2Th(SO_3)_3$	Th:	SVol.C5-57, 60
–	$[(NH_2)_3C]_2Th(SO_3)_3 \cdot 2 H_2O$	Th:	SVol.C5-57, 60
–	$[(NH_2)_3C]_2Th(SO_3)_3 \cdot 10 H_2O$	Th:	SVol.C5-57, 60
–	$[(NH_2)_3C]_2Th(SO_3)_3 \cdot 12 H_2O$	Th:	SVol.C5-57, 60/2
$C_2H_{12}N_6O_{10}S_2U$	$[C(NH_2)_3]_2UO_2(SO_4)_2$	U:	SVol.C10-192
–	$[C(NH_2)_3]_2UO_2(SO_4)_2 \cdot 3 H_2O$	U:	SVol.A6-36
		U:	SVol.C10-191/2, 197, 199/200, 203/11
–	$[C(NH_2)_3]_2UO_2(SO_4)_2 \cdot 4 H_2O$	U:	SVol.C10-192
$C_2H_{12}N_6O_{15}S_5Th_2$	$[(NH_2)_3C]_2Th_2(SO_3)_5$	Th:	SVol.C5-60
$C_2H_{12}N_8O_{18}Th$	$(CH_3NH_3)_2Th(NO_3)_6$	Th:	SVol.C3-108, 118, 120/1
$C_2H_{14}MnN_8O_4$	$Mn(N_2H_3COO)_2 \cdot 2 N_2H_4$	Mn:	MVol.D5-87/8
$C_2H_{14}Si_4$	$(SiH_3)_3SiC_2H_5$	Si:	SVol.B1-175
$C_2H_{16}MnNO_8^+$	$[Mn(H_2O)_6](NH_2CH_2COO)^+$	Mn:	MVol.D4-248/50
$C_2H_{16}N_4Na_6O_{19}S_3Th_2$			
	$Na_6Th_2(OH)_8(SO_3)_3 \cdot 2 (NH_2)_2CO \cdot 20 H_2O$		
		Th:	SVol.C5-62
$C_2H_{16}N_4O_{14}S_2U$	$(NH_4)_4[(UO_2)(C_2O_4)(SO_4)_2]$	U:	SVol.C13-225/7
$C_2H_{16}Si_5$	$Si_2H_5Si(SiH_3)_2C_2H_5$	Si:	SVol.B1-175
$C_2H_{18}I_2N_4Sn$	$(CH_3)_2SnI_2 \cdot 4 NH_3$	Sn:	Org.Comp.8-95
$C_2H_{18}I_2N_5ORhS$	$[Rh(OSCCH_3)(NH_3)_5]I_2$	Rh:	SVol.B2-63
$C_2H_{18}I_2N_5RhS_2$	$[Rh(NH_3)_5(S_2CCH_3)]I_2$	Rh:	SVol.B3-15
$C_2H_{18}MnN_8$	$Mn(CN)_2 \cdot 6 NH_3$	Mn:	MVol.C7-229, 235
–	$Mn(NH_3)_6(CN)_2$	Mn:	MVol.D3-18
$C_2H_{18}N_5O_2Rh^{2+}$	$[Rh(NH_3)_5(OCOCH_3)]^{2+}$	Rh:	SVol.B2-134
$C_2H_{18}N_6Rh^{3+}$	$[Rh(NH_3)_5(NCCH_3)]^{3+}$	Rh:	SVol.B2-138/9
$C_2H_{19}N_6ORh^{2+}$	$[Rh(NH_3)_5(NHCOCH_3)]^{2+}$	Rh:	SVol.B2-139
$C_2H_{20}MnN_8O_6S$	$Mn(OC(NH_2)_2)_2(NH_3)_4SO_4$	Mn:	MVol.D5-144
$C_2H_{20}N_5Na_2O_{16}RhS_4$	$Na_2(NH_4)_3[Rh(NH_2C_2H_4NH_2)(SO_4)_4] \cdot 3 H_2O$		
		Rh:	SVol.B2-184

$C_2H_{20}N_5ORh^{2+}$	$[Rh(NH_3)_5(OC_2H_5)]^{2+}$	Rh:	SVol.B2-132/3
$C_2H_{20}N_6ORh^{3+}$	$[Rh(NH_3)_5(NH_2COCH_3)]^{3+}$	Rh:	SVol.B2-139
$C_2H_{21}N_5ORh^{3+}$	$[Rh(NH_3)_5(C_2H_5OD)]^{3+}$	Rh:	SVol.B2-133
$C_2H_{2x+2}O_{x+6}Rh_2{}^{2+}$	$[Rh_2(HCO_3)_2(H_2O)_n]^{2+}$	Rh:	SVol.B1-185
$C_2HoN_2S_2{}^+$	$[Ho(NCS)_2]^+$	Sc:	MVol.D4-340/3
C_2HoO_2	$Ho(CO)_2$.	Sc:	MVol.D6-166
$C_2HoO_4{}^+$	$[Ho(C_2O_4)]^+$	Sc:	MVol.D5-120/4
$C_2IMn_2N_2S_2$	$Mn_2(SCN)_2I$	Mn:	MVol.C7-239
$C_2I_2MnN_2S_2$	$Mn(SCN)_2I_2$	Mn:	MVol.C7-238/9
$C_2I_4O_2Os^{2-}$	$[OsI_4(CO)_2]^{2-}$	Os:	SVol.1-162
C_2IrU	$IrUC_2$	Pt:	SVol.A1-226
C_2IrU_2	U_2IrC_2	U:	SVol.C12-104, 203/4, 207, 209, 212
$C_2K_2MnN_2O_8$	$K_2[Mn(C_2O_4)(NO_2)_2] \cdot H_2O$	Mn:	MVol.D2-107
$C_2K_2MnO_7S_2$	$K_2[Mn(C_2O_4)(S_2O_3)] \cdot 2 H_2O$	Mn:	MVol.D2-107
$C_2K_2N_4O_8S_2U$	$K_2UO_2(NCS)_2(NO_3)_2$	U:	SVol.C7-196
		U:	SVol.C13-45, 48
$C_2K_2N_4O_8U$	$K_2UO_2(CN)_2(NO_3)_2$	U:	SVol.C7-196
		U:	SVol.C13-28/9
$C_2K_2N_6Ni_2S_4$	$K_2[Ni_2(S_2N_2)_2(CN)_2]$	S:	S-N Comp.2-299
$C_2K_2O_8U$	$K_2UO_2(CO_3)_2$.	U:	SVol.A5-125, 134, 139
		U:	SVol.A6-49/50, 55, 65
		U:	SVol.C13-10/2
–	$K_2[UO_2(OO)(C_2O_4)]$.	U:	SVol.C13-213/5
$C_2K_2O_9SU$	$K_2[UO_2(C_2O_4)(SO_3)]$	U:	SVol.C13-223/5
–	$K_2[UO_2(C_2O_4)(SO_3)] \cdot x H_2O$.	U:	SVol.C13-223/5
$C_2K_2O_{10}SU$	$K_2[UO_2(C_2O_4)(SO_4)]$	U:	SVol.C13-225/7
$C_2K_2O_{14}S_2U_2$	$K_2[(UO_2)_2(C_2O_4)(SO_3)_2] \cdot H_2O$	U:	SVol.C13-223/5
$C_2K_3MnN_4O_2$	$K_3[Mn(CN)_2(NO)_2]$.	Mn:	MVol.D2-279
$C_2K_4O_{10}U$	$K_4[UO_2_2(CO_3)_2]$	U:	SVol.C13-24/5
–	$K_4[UO_2_2(CO_3)_2] \cdot 3 H_2O$	U:	SVol.C13-24/5
C_2La	LaC_2 solid solutions		
	$(La,U)C_2$	U:	SVol.C12-93, 150
$C_2LaN_2S_2{}^+$	$[La(NCS)_2]^+$	Sc:	MVol.D4-340/3
$C_2LaO_4{}^+$	$[La(C_2O_4)]^+$	Sc:	MVol.D5-120/4
C_2Li_2	Li_2C_2.	B:	SVol.2-144
$C_2Li_2O_8U$	$Li_2[UO_2(CO_3)_2] \cdot x H_2O$	U:	SVol.C13-10/2
$C_2LuN_2S_2{}^+$	$[Lu(NCS)_2]^+$	Sc:	MVol.D4-340/3
$C_2LuO_4{}^+$	$[Lu(C_2O_4)]^+$	Sc:	MVol.D5-120/4
C_2Mn	MnC_2.	Mn:	MVol.C7-102
C_2MnN_2	$Mn(CN)_2$.	Mn:	MVol.C7-229
$C_2MnN_2S_2$	$Mn(NCS)_2$	Mn:	MVol.D2-284
–	$Mn(SCN)_2$	Mn:	MVol.C7-234/5
–	$Mn(SCN)_2 \cdot 3 H_2O$	Mn:	MVol.C7-235
$C_2MnN_2S_2Se$	$Mn(SCN)_2 \cdot Se$	Mn:	MVol.C7-235/6
$C_2MnN_2S_2Te$	$Mn(SCN)_2 \cdot Te$	Mn:	MVol.C7-235/6
$C_2MnN_2S_3$	$Mn(SCN)_2 \cdot S$	Mn:	MVol.C7-235/6
$C_2MnN_2Se_2$	$Mn(NCSe)_2$	Mn:	MVol.D2-293
–	$Mn(SeCN)_2$	Mn:	MVol.C7-239/40
$C_2MnN_4O_2{}^{3-}$	$[Mn(CN)_2(NO)_2]^{3-}$	Mn:	MVol.D2-279

C_2MnO_4	MnC_2O_4	Mn:	MVol.D2-89/94
–	$Mn^{12}C^{13}CO_4$	Mn:	MVol.D2-91
–	$MnC_2O_4 \cdot 2 H_2O$	Mn:	MVol.D2-85/6, 95/103
–	$MnC_2O_4 \cdot 3 H_2O$	Mn:	MVol.D2-85/6, 94/5
–	MnC_2O_4 solutions		
	MnC_2O_4–H_2O	Mn:	MVol.D2-86/9, 103/4
$C_2MnO_4^+$	$[MnC_2O_4]^+$	Mn:	MVol.D2-121
C_2MnU	$UMnC_2$	Mn:	MVol.C7-171/2
		U:	SVol.C12-200/1
$C_2Mn_2N_2S_2^{4+}$	$[Mn_2(SCN)_2]^{4+}$	Mn:	MVol.D2-292
$C_2Mn_2Na_6O_{14}P_2$	$3 Na_2O \cdot 2 MnO \cdot 2 CO_2 \cdot P_2O_5$		
	$= Na_3MnCO_3PO_4$	Mn:	MVol.C9-237/8
C_2Mn_5	Mn_5C_2	Mn:	MVol.C7-100/2, 110/2,
			115, 125/8
–	Mn_5C_2 solid solutions		
	$Mn_5(C_{1-x}B_x)_2$	Mn:	MVol.C7-240/3
C_2MoU	$UMoC_2$	U:	SVol.C12-101, 191/5
C_2Mo_3	Mo_3C_2	Mo:	SVol.A3-79
–	Mo_3C_2 solid solutions		
	$Mo_3(C,N)_2$	Mo:	SVol.A3-59
$C_2N_2NdS_2^+$	$[Nd(NCS)_2]^+$	Sc:	MVol.D4-340/3
$C_2N_2O_2S_2U$	$UO_2(SCN)_2$	U:	SVol.C13-40, 41
		U:	SVol.D1-170, 179/80
–	$UO_2(SCN)_2 \cdot H_2O$	U:	SVol.C13-40, 41
		U:	SVol.D2-338
–	$UO_2(SCN)_2 \cdot 3 H_2O$	U:	SVol.C13-40, 41
$C_2N_2O_4U$	$UO_2(CNO)_2$	U:	SVol.D1-170
$C_2N_2O_7S_2$	$S_2O_5(NCO)_2$	S:	SVol.3-316
$C_2N_2O_{12}U^{2-}$	$[UO_2(C_2O_4)(NO_3)_2]^{2-}$	U:	SVol.D2-378
C_2N_2Os	$Os(CN)_2$	Os:	SVol.1-173
$C_2N_2OsS_2^{2+}$	$[Os(SCN)_2]^{2+}$	Os:	SVol.1-168/9
C_2N_2Pd	$Pd(CN)_2$		
	Catalytic properties	Pt:	SVol.A1-307
$C_2N_2PrS_2^+$	$[Pr(NCS)_2]^+$	Sc:	MVol.D4-340/3
$C_2N_2S_2Sm^+$	$[Sm(NCS)_2]^+$	Sc:	MVol.D4-340/3
$C_2N_2S_2Tb^+$	$[Tb(NCS)_2]^+$	Sc:	MVol.D4-340/3
$C_2N_2S_2Tm^+$	$[Tm(NCS)_2]^+$	Sc:	MVol.D4-340/3
$C_2N_2S_2U^{2+}$	$[U(SCN)_2]^{2+}$	U:	SVol.D1-170, 178/9
$C_2N_2S_2Y^+$	$[Y(NCS)_2]^+$	Sc:	MVol.D4-340/3
$C_2N_2S_2Yb^+$	$[Yb(NCS)_2]^+$	Sc:	MVol.D4-340/3
$C_2N_3Nd^{2+}$	$[NdN(CN)_2]^{2+}$	Sc:	MVol.D4-338
$C_2N_4O_4S_3$	$S(NSCNO_2)_2$	S:	S-N Comp.4-112/7
$C_2N_4O_{20}U_2^{2-}$	$[(UO_2)_2(C_2O_4)(NO_3)_4]^{2-}$	U:	SVol.D2-378
C_2N_4SSe	$SN_2C_2N_2Se$	S:	S-N Comp.3-240
$C_2N_4S_2$	$SN_2C_2N_2S$	S:	S-N Comp.3-189/90
$C_2N_4S_2^-$	$[SN_2C_2N_2S]^-$ (radical)	S:	S-N Comp.3-191
$C_2N_4S_4$	$(NSNSN)C_2(SNS)$	S:	S-N Comp.4-122/4
$C_2N_6O_4S_5$	$S_4N_4 \cdot C(O)NSO_2NCO$	S:	S-N Comp.2-224
C_2NaO_8SSc	$Na[Sc(SO_4)(C_2O_4)] \cdot 3 H_2O$	Sc:	MVol.D5-118

C₂U . . . . UC₂

Nuclear magnetic resonance	U:	SVol.A5-255/6
	U:	SVol.C12-140/1
Optical properties	U:	SVol.C12-141
Preparation	U:	SVol.C12-124/5
Production	U:	SVol.A3-120
Thermal properties	U:	SVol.C12-131/9
Thermodynamic properties	U:	SVol.A6-180
	U:	SVol.C12-131/4
Transport properties	U:	SVol.C12-135/7
Use as nuclear fuel	U:	SVol.A3-192/6

− . . . . UC₂ solid solutions

(U,Ce)C₂	U:	SVol.C12-94, 151/2
(U,Er)C₂	U:	SVol.C12-153
(U,Gd)C₂	U:	SVol.C12-95, 152/3
(U,La)C₂	U:	SVol.C12-93, 150
(U,Nd)C₂	U:	SVol.C12-94
(U,Th)C₂		
Formation	U:	SVol.C12-96, 154/9
Physical properties	U:	SVol.C12-159/66
Production of nuclear fuel pellets . . .	U:	SVol.A3-269
Use as nuclear fuel	U:	SVol.A3-181, 195
(U,Y)C₂	U:	SVol.C12-93, 150

− . . . . UC₂ systems

UC₂-Be₂C-UC	U:	SVol.C12-146/7
UC₂-Fe	U:	SVol.C12-203/4
UC₂-ThC₂	U:	SVol.C12-154/9
UC₂-UC	U:	SVol.C12-4/7, 49/55
UC₂-UC-US	U:	SVol.C12-264/6

C₂UV	UVC₂	U:	SVol.C12-99, 183
C₂UW	UWC₂	U:	SVol.C12-101, 191, 196/7
		W:	SVol.A7-300

C₂Y . . . . YC₂ solid solutions

(Y,U)C₂	U:	SVol.C12-93, 150

C₂.₁MnN₂.₁	Mn(CN)₂.₁	Mn:	MVol.C7-229/30
C₂.₂₄H₆.₂Mn₀.₈₆N₀.₂₈O₀.₃PS₃			
	Mn₀.₈₆PS₃[(C₂H₅)₄N]₀.₂₈[H₂O]₀.₃	Mn:	MVol.C9-227/8
C₃CaMgO₁₁U	MgCa[UO₂(CO₃)₃] · 12 H₂O	U:	SVol.C13-15, 18
C₃CaNa₂O₁₁U	Na₂Ca[UO₂(CO₃)₃]	U:	SVol.C13-14
−	Na₂Ca[UO₂(CO₃)₃] · 2 H₂O	U:	SVol.C13-14
−	Na₂Ca[UO₂(CO₃)₃] · 6 H₂O	U:	SVol.C13-12, 14/6, 18
C₃Ca₂O₁₁U	Ca₂[UO₂(CO₃)₃]	U:	SVol.C13-13, 17
−	Ca₂[UO₂(CO₃)₃] · 4 H₂O	U:	SVol.C13-13, 17
−	Ca₂[UO₂(CO₃)₃] · 8 H₂O	U:	SVol.C13-12/4
−	Ca₂[UO₂(CO₃)₃] · 9 H₂O	U:	SVol.C13-12/4, 16
−	Ca₂[UO₂(CO₃)₃] · 10 H₂O	U:	SVol.C13-12/8
−	Ca₂[UO₂(CO₃)₃] · x H₂O (x = 8 to 11) . . .	U:	SVol.C13-12/4
C₃Ca₃FNaO₁₅SU	NaCa₃[UO₂F(CO₃)₃(SO₄)] · 4 H₂O	U:	SVol.C13-26

$C_3ClCuH_6O_5$	$HOCH_2CHCH_2CuClO_4$	Cu:	Org.Comp.4-8
C_3ClCuH_7N	$NH_2CH_2CHCH_2CuCl$	Cu:	Org.Comp.4-7
$C_3ClCuH_8N_2O$	$[(CO)CuC_2H_8N_2]Cl$	Cu:	Org.Comp.3-191, 204
$C_3ClCuH_8O_6$	$HOCH_2CHCH_2Cu(H_2O)ClO_4$	Cu:	Org.Comp.4-8
$C_3ClCu_2H_4O$	$(OCHCHCH_2Cu_2Cl)_n$	Cu:	Org.Comp.4-255/6
$C_3ClCu_2H_9Mg$	$(CH_3)_3Cu_2MgCl$	Cu:	Org.Comp.4-106/7
$C_3ClCu_4H_9I_2Mg$	$3\ CH_3Cu \cdot CuCl \cdot MgI_2$	Cu:	Org.Comp.1-323, 325
$C_3ClEuH_4O_2{}^{2+}$	$[Eu(CH_3CHClCOO)]^{2+}$	Sc:	MVol.D5-79
$C_3ClFH_3NO_4$	$NO_2CFClC(O)OCH_3$	F:	PFHOrg.8-37/8
$C_3ClF_2H_2N_3S$	$CF_2Cl(NH_2)C_2N_2S$	F:	PFHOrg.SVol.2-43, 53
C_3ClF_2Li	$c\text{-}C_3F_2ClLi$	F:	PFHOrg.SVol.1-2
C_3ClF_2N	CF_2CClCN	F:	PFHOrg.9-51/2, 68, 81/2
—	$ClCFCFCN$	F:	PFHOrg.9-51/2, 68
C_3ClF_2NO	$NCCF_2C(O)Cl$	F:	PFHOrg.9-54/5
$C_3ClF_3H_3IO_2S$	$FSO_2CF_2(CH_2CHCl)I$	F:	PFHOrg.SVol.3-179
$C_3ClF_3H_3NO_2$	$NF_2CFClC(O)OCH_3$	F:	PFHOrg.8-207
$C_3ClF_3H_5NS$	$CF_3C(NH \cdot HCl)SCH_3$	F:	PFHOrg.9-82
$C_3ClF_3H_8N_2O_5$	$[CF_2(NFH)OC_2H_4NH_3]ClO_4$	F:	PFHOrg.8-223
$C_3ClF_3N_2S$	$ClC_2N_2SCF_3$	F:	PFHOrg.SVol.2-44, 50, 58
C_3ClF_3OS	$FClCSC(O)CF_2$	F:	PFHOrg.SVol.2-15
—	$F_2CSC(O)CFCl$	F:	PFHOrg.SVol.2-15
$C_3ClF_4H_3N_2S$	$ClCF_2CF_2NSNCH_3$	F:	PFHOrg.9-35
$C_3ClF_4H_8N_3O_5$	$[NF_2CF(NFH)OC_2H_4NH_3]ClO_4$ . . .	F:	PFHOrg.8-223
C_3ClF_4NO	$ClCF_2CF_2NCO$	F:	PFHOrg.9-103, 109, 117
$C_3ClF_4NO_2$	$CF_3C(O)NClC(O)F$	F:	PFHOrg.8-161, 185
—	$FC(O)N(CF_3)C(O)Cl$	F:	PFHOrg.9-165, 174
$C_3ClF_4NO_3$	$CF_3CF(NO_2)COCl$	F:	PFHOrg.8-10
—	$NO_2CF_2CF_2C(O)Cl$	F:	PFHOrg.8-9/10, 24
$C_3ClF_4NO_5S$	$CF_3C(O)NClC(O)OSO_2F$	F:	PFHOrg.8-161, 185
$C_3ClF_5HNO_3$	$CF_3CCl(OH)CF_2NO_2$	F:	PFHOrg.8-9/10, 24
$C_3ClF_5H_2OS$	$CF_3(CF_2Cl)C(OH)SH$	F:	PFHOrg.SVol.2-64/5, 68
$C_3ClF_5H_3NOS$	$ClCF_2CF_2NSFOCH_3$	F:	PFHOrg.9-34
$C_3ClF_5N_2$	$ClCF_2C(CF_3)N_2$	F:	PFHOrg.8-57/8, 84, 104
$C_3ClF_5OS_3$	$CF_3SCFClSSC(O)F$	F:	PFHOrg.SVol.2-176, 187, 206
$C_3ClF_5O_2S$	$CF_2ClSOC(O)CF_3$	F:	PFHOrg.SVol.2-84
$C_3ClF_5O_3S$	$CF_3C(O)CF_2SO_2Cl$	F:	PFHOrg.SVol.3-159, 168
C_3ClF_5S	$CF_3SC(F)CFCl$	F:	PFHOrg.SVol.2-213, 227
$C_3ClF_6HO_3S$	$C_3F_6ClSO_3H$	F:	PFHOrg.SVol.3-49
$C_3ClF_6H_4N$	$(CF_3)_2CHNH_2 \cdot HCl$	F:	PFHOrg.8-41
$C_3ClF_6KO_4S_2$	$KCCl(SO_2CF_3)_2$	F:	PFHOrg.SVol.3-121/2
C_3ClF_6N	$(CF_3)(CF_2Cl)CNF$	F:	PFHOrg.8-170/1, 196
—	$(CF_3)_2CNCl$	F:	PFHOrg.8-170/1, 196
—	$CF_3NCClCF_3$	F:	PFHOrg.9-128/9, 139
—	CF_3NCFCF_2Cl	F:	PFHOrg.9-128, 137
—	C_2F_5CFNCl	F:	PFHOrg.8-170, 195
—	$ClCF_2CF_2NCF_2$	F:	PFHOrg.9-128/9, 136/7
C_3ClF_6NO	$(CF_3)_2NC(O)Cl$	F:	PFHOrg.9-160, 168, 177
C_3ClF_6NOS	$(CF_3)_2CClNSO$	F:	PFHOrg.9-6/7, 23

$C_3ClF_6NOS_3$	$ClC(O)S(CF_3S)_2N$	F:	PFHOrg.SVol.2-87, 89/90, 97/8
$C_3ClF_6NO_2$	$CF_3CFClCF_2NO_2$	F:	PFHOrg.8-9, 24
$C_3ClF_6NO_2S$	$C_2F_5CClNSO_2F$	F:	PFHOrg.9-9, 30
C_3ClF_6NS	$(CF_3)_2CNSCl$	F:	PFHOrg.9-1/2, 12, 35/6
–	$CF_3SNCClCF_3$	F:	PFHOrg.SVol.2-101, 106, 111/2
$C_3ClF_6N_3O_3S$	$(CF_3)_2C(OSO_2Cl)N_3$	F:	PFHOrg.8-60, 86, 103
$C_3ClF_7H_4S$	$CF_3SF_4CH_2CH_2Cl$	F:	PFHOrg.SVol.3-207
C_3ClF_7IP	$CF_3(ClCF_2CF_2)PI$	F:	PFHOrg.SVol.1-106
$C_3ClF_7N_2$	$CF_3NNCF_2CF_2Cl$	F:	PFHOrg.8-74, 97
$C_3ClF_7N_2O$	$CF_3(N_2O)CF_2CF_2Cl$	F:	PFHOrg.8-55/6, 82
$C_3ClF_7N_2O_2$	$CF_3N(NO_2)CF_2CF_2Cl$	F:	PFHOrg.8-56/7, 84
$C_3ClF_7O_2S$	$(CF_3)_2CClSO_2F$	F:	PFHOrg.SVol.3-154, 162
–	$(CF_3)_2CFSO_2Cl$	F:	PFHOrg.SVol.3-158/9, 168
$C_3ClF_7O_3S$	$CF_2ClCF_2OSO_2CF_3$	F:	PFHOrg.SVol.3-57, 58, 72
C_3ClF_7S	C_3F_7SCl	F:	PFHOrg.SVol.2-118/9, 122, 127
$C_3ClF_7S_2$	$CF_3SSCF_2CF_2Cl$	F:	PFHOrg.SVol.2-177/8, 189
$C_3ClF_7S_3$	$(CF_3S)_2CFSCl$	F:	PFHOrg.SVol.2-118, 121, 127, 137, 176
C_3ClF_8N	$(CF_3)_2CFNFCl$	F:	PFHOrg.8-156/7, 180
–	$(C_2F_5)(CF_3)NCl$	F:	PFHOrg.8-162, 186, 203
C_3ClF_8NOS	$(CF_3)_2CClNS(O)F_2$	F:	PFHOrg.9-8/9, 28
–	$CF_3CFClCF_2NS(O)F_2$	F:	PFHOrg.9-8/9, 27
–	$CF_3CF[NS(O)F_2]CF_2Cl$	F:	PFHOrg.9-8/9, 27/8
$C_3ClF_8NO_4S_2$	$(CF_3)_2CClN(SO_2F)_2$	F:	PFHOrg.9-9, 29
–	$CF_3CFClCF_2N(SO_2F)_2$	F:	PFHOrg.9-9, 29/30
C_3ClF_8NS	$(CF_3)_2CClNSF_2$	F:	PFHOrg.9-3/5, 17
–	$CF_3CFClCF_2NSF_2$	F:	PFHOrg.9-3/4, 15
–	$CF_3CF(NSF_2)CF_2Cl$	F:	PFHOrg.9-3/5, 15
–	$CF_3S(Cl)NC_2F_5$	F:	PFHOrg.SVol.3-7, 27
$C_3ClF_8NS_4$	$(CF_3S)_2(CF_2ClSS)N$	F:	PFHOrg.SVol.2-174, 186, 195
C_3ClF_9Ge	$(CF_3)_3GeCl$	F:	PFHOrg.SVol.1-51/63
$C_3ClF_9H_2S$	$CF_3SF_4CH_2CF_2Cl$	F:	PFHOrg.SVol.3-207
$C_3ClF_{10}HS$	$CF_3SF_4CFHCF_2Cl$	F:	PFHOrg.SVol.3-207
$C_3ClF_{10}NS$	$CF_3SF_4NCClCF_3$	F:	PFHOrg.SVol.3-192/3, 204
$C_3ClF_{11}NP$	$CF_3PF_2ClN(CF_3)_2$	F:	PFHOrg.SVol.1-102/3
$C_3ClF_{11}S$	$CF_3SF_4CF_2CF_2Cl$	F:	PFHOrg.SVol.3-190, 199
–	$n\text{-}C_3F_7SF_4Cl$	F:	PFHOrg.SVol.3-187/8, 196
–	$i\text{-}C_3F_7SF_4Cl$	F:	PFHOrg.SVol.3-187/8, 196
$C_3ClF_{12}NS$	$CF_3SF_4NClC_2F_5$	F:	PFHOrg.SVol.3-192/3, 204
$C_3ClGaH_6O_2$	$Ga(CH_3)(Cl)OOCCH_3$	Ga:	Org.Comp.1-225

C_3ClGaH_8S	$GaCH_3(Cl)SC_2H_5$	Ga:	Org.Comp.1-239
C_3ClHN_2OS	SN_2C_2HCOCl	S:	S-N Comp.3-93/8, 172/3
$C_3ClHN_2O_2S$	$SN_2C_2(COCl)OH$	S:	S-N Comp.3-93/8, 180
$C_3ClH_3N_2OS$	$SN_2C_2ClOCH_3$	S:	S-N Comp.3-93/8, 157/8
$C_3ClH_3N_2S$	$SN_2C_2ClCH_3$	S:	S-N Comp.3-93/8, 136
–	$SN_2C_2HCH_2Cl$	S:	S-N Comp.3-93/8, 121
$C_3ClH_4MnO_2{}^+$	$[Mn(ClCH_2CH_2COO)]^+$	Mn:	MVol.D2-65
$C_3ClH_4O_4U^+$	$[UO_2(ClCH_2CH_2COO)]^+$	U:	SVol.D1-198, 201
$C_3ClH_6NO_8U$	$UO_2(CH_3CHNH_2COO)(ClO_4) \cdot 2 H_2O$	U:	SVol.C13-166/7
$C_3ClH_6N_3O_2S_2$	$[CH_2OS(O)NC(SC(NH)NH_2 \cdot HCl)]$	S:	S-N Comp.3-286/7
$C_3ClH_6N_5OS_3$	$[N_3S_3(NCH_3C(O)NCH_3)]Cl$	S:	S-N Comp.4-151/3
$C_3ClH_6N_5S_3$	$((CH_3)_2N)CN_4S_3Cl$	S:	S-N Comp.4-127/8
$C_3ClH_7NO_3Pr$	$Pr[CH_3CH(NH_2)COO](OH)Cl \cdot 4 H_2O$	Sc:	MVol.D1-108
$C_3ClH_9NRh^{2+}$	$[Rh(NH_2C_3H_7-n)Cl]^{2+}$	Rh:	SVol.B2-156/7
$C_3ClH_9O_3SSi$	$(CH_3)_3SiOSO_2Cl$	S:	SVol.3-320
$C_3ClH_9O_3SSn$	$(CH_3)_3SnOSO_2Cl$	Sn:	Org.Comp.11-147
$C_3ClH_9O_4Sn$	$(CH_3)_3SnClO_4$	Sn:	Org.Comp.11-126/7
C_3ClH_9Si	$(CH_3)_3SiCl$		
	Pyrolysis	Si:	SVol.B3-132/4
C_3ClH_9Sn	$(CH_3)_3SnCl$	Sn:	Org.Comp.9-26
$C_3ClH_{13}O_7SU$	$[UO_2(SCH_2CHOHCH_2OH)Cl(H_2O)_3] \cdot H_2O$	U:	SVol.C13-308/10
$C_3ClH_{15}N_2O_4Sn$	$[(CH_3)_3Sn(NH_3)_2]ClO_4$	Sn:	Org.Comp.11-127
C_3ClN_3S	SN_2C_2ClCN	S:	S-N Comp.3-93/8, 166
$C_3Cl_2CoF_3NO_3P$	$[Co(CO)_2(NO)(Cl_2PCF_3)]$	F:	PFHOrg.SVol.1-130
$C_3Cl_2CuH_5$	$ClCH_2CHCH_2CuCl$	Cu:	Org.Comp.4-9
$C_3Cl_2CuH_5O$	$HOCH_2CClCH_2CuCl$	Cu:	Org.Comp.4-9
$C_3Cl_2CuH_6{}^-$	$(CH_3CHCH_2CuCl_2)^-$	Cu:	Org.Comp.4-24
$C_3Cl_2CuH_6O^-$	$(HOCH_2CHCH_2CuCl_2)^-$	Cu:	Org.Comp.4-24
$C_3Cl_2Cu_2H_3N$	$[NCCHCH_2Cu_2Cl_2]_n$	Cu:	Org.Comp.4-255/6
$C_3Cl_2Cu_2H_4O_8$	$(CH_2CCH_2)Cu_2(ClO_4)_2 \cdot 3 H_2O$	Cu:	Org.Comp.4-175
C_3Cl_2FNS	Cl_2FC_3NS	F:	PFHOrg.SVol.2-41, 47, 52
$C_3Cl_2F_2OS$	$Cl_2CSC(O)CF_2$	F:	PFHOrg.SVol.2-15
–	$F_2CSC(O)CCl_2$	F:	PFHOrg.SVol.2-15
$C_3Cl_2F_3N$	$ClCF_2CFClCN$	F:	PFHOrg.9-51, 68
$C_3Cl_2F_3NO$	$ClCF_2CFClNCO$	F:	PFHOrg.9-103, 109
$C_3Cl_2F_3NO_3$	$CF_3CCl(NO_2)COCl$	F:	PFHOrg.8-9/10, 24, 35, 38
$C_3Cl_2F_4H_2OS$	$(CF_2Cl)_2C(OH)SH$	F:	PFHOrg.SVol.2-64/5, 68
$C_3Cl_2F_4O_2S$	$CFCl_2SOC(O)CF_3$	F:	PFHOrg.SVol.2-84
$C_3Cl_2F_5HO_3S$	$C_3F_5Cl_2SO_3H$	F:	PFHOrg.SVol.3-49
$C_3Cl_2F_5N$	$(CF_2Cl)_2CNF$	F:	PFHOrg.8-170/1, 196
–	$CF_3NCClCF_2Cl$	F:	PFHOrg.9-128/9, 138
$C_3Cl_2F_5NO_2$	$CF_3CF(NO_2)CFCl_2$	F:	PFHOrg.8-9/10, 24
$C_3Cl_2F_6H_3P$	$CH_3PCl_2(CF_3)_2$	F:	PFHOrg.SVol.1-123/4
$C_3Cl_2F_6NOP$	$(CF_3)_2CNP(O)Cl_2$	F:	PFHOrg.9-40, 42
$C_3Cl_2F_6N_2$	$Cl_2CNN(CF_3)_2$	F:	PFHOrg.8-70, 93
$C_3Cl_2F_6N_2O_2$	$CF_3N(NO_2)CF_2CFCl_2$	F:	PFHOrg.8-56/7
$C_3Cl_2F_6O_3S$	$CF_2ClCFClOSO_2CF_3$	F:	PFHOrg.SVol.3-57, 58, 73
$C_3Cl_2F_6O_4S_2$	$(CF_3SO_2)_2CCl_2$	F:	PFHOrg.SVol.3-121/2, 127

$C_3Cl_2H_{21}N_5O_{10}Rh...$	$[Rh(NH_3)_4(H_2O)((CH_3)_2NCHO)](ClO_4)_2.....$	Rh:	SVol.B2-149
$C_3Cl_3CoCuH_{16}N_5O_{14}$	$[((NH_3)_5CoOC(O)CCH)Cu](ClO_4)_3.........$	Cu:	Org.Comp.4-54
$C_3Cl_3CoCuH_{18}N_5O_{14}$	$(NH_3)_5CoO_2CCHCH_2Cu(ClO_4)_3..........$	Cu:	Org.Comp.4-9
$C_3Cl_3CuH_5^-$	$(ClCH_2CHCH_2CuCl_2)^-$	Cu:	Org.Comp.4-24
$C_3Cl_3F_2N.$.........	$ClCF_2CCl_2CN$	F:	PFHOrg.9-51, 68
$-$	$Cl_2CFCFClCN$	F:	PFHOrg.9-51
$C_3Cl_3F_2NO_3.$.......	$CF_2ClCCl(NO_2)COCl.$...............	F:	PFHOrg.8-9/10, 24, 35, 38
$C_3Cl_3F_3HNO$	$CCl_3C(O)NHCF_3$	F:	PFHOrg.9-151
$C_3Cl_3F_3H_2OS$	$CF_2Cl(CFCl_2)C(OH)SH$	F:	PFHOrg.SVol.2-64/5, 68
$C_3Cl_3F_4HN_2S_2.$.....	$CF_2ClSNHC(Cl)NSCF_2Cl.$.........	F:	PFHOrg.SVol.2-101, 107
$C_3Cl_3F_4HO_3S$	$C_3F_4Cl_3SO_3H$	F:	PFHOrg.SVol.3-49
$C_3Cl_3F_4N.$.........	$CF_3NCFCCl_3$	F:	PFHOrg.9-128/9, 138
$C_3Cl_3F_4NO_2.$......	$CF_3CCl(NO_2)CFCl_2.$...........	F:	PFHOrg.8-9/10, 24
$C_3Cl_3F_5O_3S.$......	$CF_2ClCCl_2OSO_2CF_3.$...........	F:	PFHOrg.SVol.3-57, 58, 73
$C_3Cl_3F_5S_2.$.......	$CF_3SSCFClCFCl_2.$.............	F:	PFHOrg.SVol.2-177/8, 190
$C_3Cl_3F_6NS.$........	$(CF_3)_2CClNSCl_2.$.............	F:	PFHOrg.9-5/6, 22
$C_3Cl_3F_6NS_4.$.......	$(CF_3S)_2(CCl_3SS)N.$.............	F:	PFHOrg.SVol.2-87, 89, 91, 96
$C_3Cl_3F_6PS_3.$.......	$P(CF_2ClS)_3.$...............	F:	PFHOrg.SVol.2-243, 247
$C_3Cl_3GdH_{15}N_3$	$GdCl_3 \cdot 3\ CH_3NH_2.$..........	Sc:	MVol.D1-16
$C_3Cl_3H_3NSSb$	$C_3H_2NSSbCl_2 \cdot HCl.$..........	Sb:	Org.Comp.2-112
$C_3Cl_3H_6O_3U^-$	$[UO_2Cl_3((CH_3)_2CO)]^-.$........	U:	SVol.D2-376
$C_3Cl_3H_6Sb.$........	$Sb(CH_2Cl)_3.$................	Sb:	Org.Comp.1-34
$C_3Cl_3H_7LaNO_2$	$La[CH_3CH(NH_2)COOH]Cl_3 \cdot 3\ H_2O$	Sc:	MVol.D1-108
$C_3Cl_3H_7NO_2Pr$	$Pr[CH_3CH(NH_2)COOH]Cl_3 \cdot 4\ H_2O$	Sc:	MVol.D1-108
$C_3Cl_3H_8OSc$	$ScCl_3 \cdot i-C_3H_7OH$	Sc:	MVol.D3-21
$C_3Cl_3H_9NORh$	$Rh(NH_2C_3H_5)(H_2O)Cl_3$	Rh:	SVol.B2-160
$C_3Cl_3H_{10}MnN$	$(CH_3)_3NHMnCl_3 \cdot 2\ H_2O$		
	Spectra	Mn:	MVol.C10-38
$C_3Cl_3H_{11}NORh$	$Rh(NH_2C_3H_7-n)(H_2O)Cl_3$	Rh:	SVol.B2-160
$C_3Cl_3H_{12}LaO_3$......	$LaCl_3 \cdot 3\ CH_3OH$	Sc:	MVol.D3-16
$C_3Cl_3H_{12}N_6RhS_3.$...	$Rh(SC(NH_2)_2)_3Cl_3$	Rh:	SVol.B3-9
$C_3Cl_3H_{12}NdO_3$	$NdCl_3 \cdot 3\ CH_3OH$	Sc:	MVol.D3-16/7
$C_3Cl_3H_{12}O_3Pr$	$PrCl_3 \cdot 3\ CH_3OH$	Sc:	MVol.D3-16
$C_3Cl_3H_{12}O_3Sc.$.....	$ScCl_3 \cdot 3\ CH_3OH$	Sc:	MVol.D3-15/6
$C_3Cl_3H_{12}O_3Sm$	$SmCl_3 \cdot 3\ CH_3OH$	Sc:	MVol.D3-16
$C_3Cl_3H_{12}O_3Y.$......	$YCl_3 \cdot 3\ CH_3OH$	Sc:	MVol.D3-16
$C_3Cl_3H_{15}LaN_3$	$LaCl_3 \cdot 3\ CH_3NH_2$	Sc:	MVol.D1-16
$C_3Cl_3H_{15}MnN_3$	$Mn(CH_3NH_2)_3Cl_3$	Mn:	MVol.D3-23
$C_3Cl_3H_{15}N_3Nd.$......	$NdCl_3 \cdot 3\ CH_3NH_2$	Sc:	MVol.D1-16
$C_3Cl_3H_{15}N_3Pr$	$PrCl_3 \cdot 3\ CH_3NH_2.$..........	Sc:	MVol.D1-16
$C_3Cl_3H_{15}N_3Sm$	$SmCl_3 \cdot 3\ CH_3NH_2.$..........	Sc:	MVol.D1-16
$C_3Cl_3H_{15}N_3Y.$......	$YCl_3 \cdot 3\ CH_3NH_2$	Sc:	MVol.D1-16
$C_3Cl_3H_{15}N_9RhS_3....$	$[Rh(NH_2NHC(S)NH_2)_3]Cl_3 \cdot 3\ H_2O.$........	Rh:	SVol.B3-40
$C_3Cl_3H_{18}N_6O_{12}Rh...$	$[Rh(NH_3)_5(NCCHCH_2)](ClO_4)_3$	Rh:	SVol.B2-138
$C_3Cl_3H_{20}N_6O_{12}Rh...$	$[Rh(NH_3)_5(NCC_2H_5)](ClO_4)_3$	Rh:	SVol.B2-138
$C_3Cl_4CoCuH_{22}N_6O_{16}$	$(NH_3)_5Co(NH_2CH_2CHCH_2)Cu(ClO_4)_4$	Cu:	Org.Comp.4-9
$C_3Cl_4CoF_6NO_2P_2$ ...	$[Co(CO)(NO)(Cl_2PCF_3)_2].$...............	F:	PFHOrg.SVol.1-130

$C_3Cl_4CuH_5{}^{2-}$	. . . . . .	$(ClCH_2CHCH_2CuCl_3)^{2-}$	Cu: Org.Comp.4-24
$C_3Cl_4F_3HO_3S$	. . . . . .	$C_3F_3Cl_4SO_3H$	F: PFHOrg.SVol.3-49
$C_3Cl_4F_3NO_2$	. . . .	$ClCF_2CCl(NO_2)CFCl_2$	F: PFHOrg.8-9/10, 24
$C_3Cl_4F_6I_2O_{16}$	. . . . . .	$(ClO_4)_2ICF_2CF_2CF_2I(ClO_4)_2$	F: PFHOrg.SVol.3-254, 264
$C_3Cl_4F_6NP$	. . . .	$(CF_3)_2CClNPCl_3$	F: PFHOrg.9-39, 42, 44, 45
$C_3Cl_4F_6N_2$	. . . .	$Cl_2NCF_2CF_2CF_2NCl_2$	F: PFHOrg.8-165, 190, 204
$C_3Cl_4H_7NO_{18}Th$	. . .	$Th(ClO_4)_4 \cdot CH_3CH(NH_2)COOH \cdot 2 H_2O$ . .	Th: SVol.E-5, 6
–	. . . .	$Th(ClO_4)_4 \cdot NH_2C_2H_4COOH \cdot 2 H_2O$	Th: SVol.E-5, 6
$C_3Cl_4H_7OU$	. . . .	$U(OC_3H_7\text{-}i)Cl_4 \cdot CH_3COOC_3H_7\text{-}i$	U: SVol.C13-72/3
$C_3Cl_4H_{10}N_2NaRh$	. .	$Na[Rh(NH_2CH_2CH(CH_3)NH_2)Cl_4] \cdot H_2O$ . . .	Rh: SVol.B2-188
$C_3Cl_4N_3S$	. . . .	$[SN_3C_2ClCCl_3]$ (radical)	S: S-N Comp.4-19
$C_3Cl_5F_2HO_3S$	. . . . . .	$C_3F_2Cl_5SO_3H$	F: PFHOrg.SVol.3-49
$C_3Cl_5H_6N_5OS_3Sn$	. .	$[N_3S_3(NCH_3C(O)NCH_3)]SnCl_5$	S: S-N Comp.4-151/3
$C_3Cl_5H_6N_5OS_3Ti$	. . . .	$[N_3S_3(NCH_3C(O)NCH_3)]TiCl_5$	S: S-N Comp.4-151/3
$C_3Cl_5H_9Sb_2$	. . . .	$(CH_3)_3SbCl_2 \cdot SbCl_3$	Sb: Org.Comp.4-13, 17/8
$C_3Cl_5N_3S$	. . . .	$ClSN_3C_2ClCCl_3$	S: S-N Comp.4-54/5
$C_3Cl_6FHO_3S$	. . . .	$C_3FCl_6SO_3H$	F: PFHOrg.SVol.3-49
$C_3Cl_6F_3NS_3$	. . . .	$(CCl_3S)_2(CF_3S)N$	F: PFHOrg.SVol.2-86, 89
$C_3Cl_6H_6N_5OS_3Sb$	. . .	$[N_3S_3(NCH_3C(O)NCH_3)]SbCl_6$	S: S-N Comp.4-151/3
$C_3Cl_7H_6N_4S_2Sb$	. . . .	$[ClS_2N_3C(N(CH_3)_2)]SbCl_6$	S: S-N Comp.4-6
$C_3Cl_9H_{12}N_6OsS_3Sn_3{}^-$			
		$[Os(SnCl_3)_3(CS(NH_2)_2)_3]^-$	Os: SVol.1-344
C_3Cl_9OU	. . . .	$UCl_5 \cdot Cl_2CCClCOCl$	U: SVol.A5-244
			U: SVol.A6-23
C_3Cl_9Sb	. . . .	$Sb(CCl_3)_3$	Sb: Org.Comp.1-34
$C_3Cl_{10}F_6P_3Rh$	. . . . .	$Rh[PF_2CCl_3]_3Cl$	Rh: SVol.B3-78
$C_3Cl_{10}O_2S_{18}Sb_2$	. . . .	$S_{12}O_2 \cdot 2 SbCl_5 \cdot 3 CS_2$	S: SVol.3-12
$C_3Cl_{29}OU_5$	. . . .	$5 UCl_5 \cdot CCl_2CClCClO$	U: SVol.A5-122
$C_3CoF_5NO_3P$	. . . .	$[Co(CO)_2(NO)(F_2PCF_3)]$	F: PFHOrg.SVol.1-130
$C_3CoF_{10}NO_2P_2$	. . . . .	$[Co(CO)(NO)(F_2PCF_3)_2]$	F: PFHOrg.SVol.1-130
$C_3CoF_{15}NOP_3$	. . . . . .	$[Co(NO)(F_2PCF_3)_3]$	F: PFHOrg.SVol.1-130
$C_3Co_2H_{36}N_{14}O_{17}U$	. .	$[Co(NH_3)_6]_2[UO_2(CO_3)_3](NO_3)_2 \cdot H_2O$	U: SVol.C13-14, 18
–	. . . .	$[Co(NH_3)_6]_2[UO_2(CO_3)_3](NO_3)_2 \cdot 3 H_2O$ . . .	U: SVol.C13-14, 18
$C_3Co_2O_{11}U$	. . . .	$Co_2[UO_2(CO_3)_3] \cdot 2 H_2O$	U: SVol.C13-14, 16
$C_3CrF_9O_9S_3$	. . . .	$Cr(OSO_2CF_3)_3$	F: PFHOrg.SVol.3-61, 79,
			102/4
$C_3CrH_9O_4Sb$	. . . .	$(CH_3)_3SbCrO_4$	Sb: Org.Comp.4-111
$C_3CrH_{24}Mo_6N_3O_{24}$	. .	$(CH_3NH_3)_3[Cr(OH)_6Mo_6O_{18}] \cdot 8 H_2O$	Mo: SVol.B4-355/6
$C_3CrH_{24}Mo_6N_9O_{24}$	. .	$[C(NH_2)_3]_3[Cr(OH)_6Mo_6O_{18}] \cdot 4 H_2O$	Mo: SVol.B4-355/6
C_3Cr_2U	. . . .	UCr_2C_3	U: SVol.C12-190/2
C_3Cr_7	. . . .	Cr_7C_3 solid solutions	
		$(Cr,Mn)_7C_3$	Mn: MVol.C7-167/9
$C_3CsH_3O_8U$	. . . .	$Cs[UO_2(HCOO)_3]$	U: SVol.C13-105
$C_3CsH_4NO_8SU$	. . . . .	$Cs[UO_2(C_2O_4)(NCS)(H_2O)_2]$	U: SVol.C13-231/2
$C_3Cs_4O_{11}U$	. . . .	$Cs_4[UO_2(CO_3)_3]$	U: SVol.C13-13, 16/7
C_3CuFH_6	. . . .	CH_3CHCH_2CuF	Cu: Org.Comp.4-7
C_3CuF_3	. . . .	CF_3CCCu	Cu: Org.Comp.3-8, 11, 17, 42
$C_3CuF_3H_3NO_3S$	. . . .	$CuOSO_2CF_3 \cdot CH_3CN$	F: PFHOrg.SVol.3-106
$C_3CuF_3O_3$	. . . .	$(CO)CuO_2CCF_3$	Cu: Org.Comp.3-191
–	. . . .	$(CO)_4Cu_4(OOCCF_3)_4$	Cu: Org.Comp.4-216

$C_3DyH_3O_6$	$Dy(HCOO)_3$	Sc:	MVol.D5-6/7, 11, 14/5
−	$Dy(HCOO)_3 \cdot 0.1\ H_2O$	Sc:	MVol.D5-5/6
−	$Dy(HCOO)_3 \cdot 2\ H_2O$	Sc:	MVol.D5-6, 8, 11
−	$Dy(HOCH_2COO)CO_3$	Sc:	MVol.D5-239
$C_3DyH_4NO_2{}^{2+}$	$[Dy(ONCHCOCH_3)]^{2+}$	Sc:	MVol.D2-102
$C_3DyH_5NO_2S^+$	$[Dy(SCH_2CH(NH_2)COO)]^+$	Sc:	MVol.D4-59/60
$C_3DyH_5N_2O_2{}^{2+}$	$[Dy(CH_3C(NO)CHNOH)]^{2+}$	Sc:	MVol.D2-116/7
$C_3DyH_5O_2{}^{2+}$	$[Dy(C_2H_5COO)]^{2+}$	Sc:	MVol.D5-71/4
$C_3DyH_5O_2S^{2+}$	$Dy(OOCCHSHCH_3)^{2+}$	Sc:	MVol.D4-55/6
−	$Dy(OOCC_2H_4SH)^{2+}$	Sc:	MVol.D4-55/6
$C_3DyH_5O_3{}^{2+}$	$[Dy(CH_3OCH_2COO)]^{2+}$	Sc:	MVol.D5-69/70
−	$[Dy(HOC_2H_4COO)]^{2+}$	Sc:	MVol.D5-260/1
−	$[Dy(H_3CCHOHCOO)]^{2+}$	Sc:	MVol.D5-250/5
$C_3DyH_6NO_2{}^{2+}$	$[Dy(CH_3CH(NH_2)COO)]^{2+}$	Sc:	MVol.D1-107/8
$C_3DyH_6NO_2S^{2+}$	$Dy[HSCH_2CH(NH_2)COO]^{2+}$	Sc:	MVol.D1-114
$C_3DyH_6NO_3{}^{2+}$	$Dy[HOCH_2CH(NH_2)COO]^{2+}$	Sc:	MVol.D1-113/4
$C_3DyH_6O_6P$	$Dy[H(O_3POCH_2CH(O)CH_2O)] \cdot H_2O$	Sc:	MVol.D4-285
$C_3DyH_7O_2{}^{2+}$	$[Dy(OCH_2CHOHCH_3)]^{2+}$	Sc:	MVol.D3-31/2
$C_3DyH_7O_3{}^{2+}$	$[Dy(OC_3H_5(OH)_2)]^{2+}$	Sc:	MVol.D3-33/5
$C_3DyH_8O_7P$	$Dy[O_4PCH_2CH(OH)CH_2OH](OH)$	Sc:	MVol.D4-285
$C_3DyH_9NO_9P_3$	$Dy[N(CH_2P(O)_2OH)_3]$	Sc:	MVol.D4-149
−	$Dy[N(CH_2P(O)_2OH)_3] \cdot 2.5\ H_2O$	Sc:	MVol.D4-149/50
$C_3DyN_3S_3$	$Dy(NCS)_3$	Sc:	MVol.D4-340/3
$C_3Dy_2H_9O_9P_3$	$Dy_2(CH_3PO_3)_3$	Sc:	MVol.D4-143
$C_3ErF_9O_9S_3$	$Er(OSO_2CF_3)_3$	F:	PFHOrg.SVol.3-61, 104
−	$Er(OSO_2CF_3)_3 \cdot 9\ H_2O$	F:	PFHOrg.SVol.3-61, 87, 104
$C_3ErH_2O_4{}^+$	$[ErCH_2(COO)_2]^+$	Sc:	MVol.D5-151, 153
$C_3ErH_2O_6{}^+$	$[Er((HO)_2C(COO)_2)]^+$	Sc:	MVol.D5-316/7
C_3ErH_3	$ErCCCH_3$	Sc:	MVol.D6-161
$C_3ErH_3O_3{}^{2+}$	$[Er(H_3CCOCOO)]^{2+}$	Sc:	MVol.D5-269/70
$C_3ErH_3O_4{}^{2+}$	$[Er(CH_2(COO)_2H)]^{2+}$	Sc:	MVol.D5-152, 154
$C_3ErH_3O_6$	$Er(HCOO)_3$	Sc:	MVol.D5-6/7, 11, 15/6
−	$Er(HCOO)_3 \cdot 0.1\ H_2O$	Sc:	MVol.D5-5/6
−	$Er(HCOO)_3 \cdot 2\ H_2O$	Sc:	MVol.D5-5/15
−	$Er(HCOO)_3 \cdot x\ H_2O$	Sc:	MVol.D5-6
−	$Er(HOCH_2COO)CO_3$	Sc:	MVol.D5-239
C_3ErH_4	$Er(C_3H_4)$	Sc:	MVol.D6-161
$C_3ErH_4NO_2{}^{2+}$	$[Er(ONCHCOCH_3)]^{2+}$	Sc:	MVol.D2-102
$C_3ErH_5NO_2S^+$	$[Er(SCH_2CH(NH_2)COO)]^+$	Sc:	MVol.D4-59/60
$C_3ErH_5N_2O_2{}^{2+}$	$[Er(CH_3C(NO)CHNOH)]^{2+}$	Sc:	MVol.D2-116/7
$C_3ErH_5O_2{}^{2+}$	$[Er(C_2H_5COO)]^{2+}$	Sc:	MVol.D5-71/4
$C_3ErH_5O_2S^{2+}$	$Er(OOCCHSHCH_3)^{2+}$	Sc:	MVol.D4-55/6
−	$Er(OOCC_2H_4SH)^{2+}$	Sc:	MVol.D4-55/6
$C_3ErH_5O_3$	$Er[OCH(CH_2O)_2]$	Sc:	MVol.D3-33
$C_3ErH_5O_3{}^{2+}$	$[Er(CH_3OCH_2COO)]^{2+}$	Sc:	MVol.D5-69/70
−	$[Er(HOC_2H_4COO)]^{2+}$	Sc:	MVol.D5-260/1
−	$[Er(H_3CCHOHCOO)]^{2+}$	Sc:	MVol.D5-250/5
$C_3ErH_6NO_2{}^{2+}$	$[Er(CH_3CH(NH_2)COO)]^{2+}$	Sc:	MVol.D1-107/8
$C_3ErH_6NO_3{}^{2+}$	$Er[HOCH_2CH(NH_2)COO]^{2+}$	Sc:	MVol.D1-113/4

$C_3ErH_6O_6P$	$Er[H(O_3POCH_2CH(O)CH_2O)] \cdot H_2O$	Sc:	MVol.D4-285
$C_3ErH_7O_2{}^{2+}$	$[Er(OCH_2CHOHCH_3)]^{2+}$	Sc:	MVol.D3-31/2
$C_3ErH_7O_3{}^{2+}$	$[Er(OC_3H_5(OH)_2)]^{2+}$	Sc:	MVol.D3-33/5
$C_3ErH_8O_7P$	$Er[O_4PCH_2CH(OH)CH_2OH](OH)$	Sc:	MVol.D4-285
C_3ErH_9	$Er(CH_3)_3$	Sc:	MVol.D6-142
$C_3ErH_9NO_9P_3$	$Er[N(CH_2P(O)_2OH)_3]$	Sc:	MVol.D4-149
−	$Er[N(CH_2P(O)_2OH)_3] \cdot 3.5 H_2O$	Sc:	MVol.D4-149/50
$C_3ErH_9O_3$	$Er(OCH_3)_3$	Sc:	MVol.D3-17
$C_3ErH_{12}N_9O_{12}$	$Er(NO_3)_3 \cdot 3 OC(NH_2)_2 \cdot 2 H_2O$	Sc:	MVol.D2-220
$C_3ErN_3S_3$	$Er(NCS)_3$	Sc:	MVol.D4-340/3
$C_3ErO_9{}^{3-}$	$[Er(CO_3)_3]^{3-}$	Sc:	MVol.D4-338/40
$C_3Er_2H_2$	Er_2CCCH_2	Sc:	MVol.D6-161
$C_3Er_2H_4$	$(HEr)_2CCCH_2$	Sc:	MVol.D6-161
$C_3Er_2H_9O_9P_3$	$Er_2(CH_3PO_3)_3$	Sc:	MVol.D4-143
$C_3EuF_9O_9S_3$	$Eu(OSO_2CF_3)_3$	F:	PFHOrg.SVol.3-61, 104
−	$Eu(OSO_2CF_3)_3 \cdot 9 H_2O$	F:	PFHOrg.SVol.3-61, 87, 104
$C_3EuH_2O_4{}^{+}$	$[EuCH_2(COO)_2]^{+}$	Sc:	MVol.D5-151/3
$C_3EuH_2O_5{}^{+}$	$[Eu(HOCH(COO)_2)]^{+}$	Sc:	MVol.D5-316
$C_3EuH_2O_6{}^{+}$	$[Eu((HO)_2C(COO)_2)]^{+}$	Sc:	MVol.D5-316/7
$C_3EuH_3O_2{}^{2+}$	$[Eu(CH_2CHCOO)]^{2+}$	Sc:	MVol.D5-91/2
$C_3EuH_3O_3{}^{2+}$	$[Eu(H_3CCOCOO)]^{2+}$	Sc:	MVol.D5-269/70
$C_3EuH_3O_4{}^{2+}$	$[Eu(CH_2(COO)_2H)]^{2+}$	Sc:	MVol.D5-152, 154
$C_3EuH_3O_6$	$Eu(HCOO)_3$	Sc:	MVol.D5-6/7, 11/5
−	$Eu(HCOO)_3 \cdot 0.1 H_2O$	Sc:	MVol.D5-5/6
−	$Eu(HCOO)_3 \cdot x H_2O$	Sc:	MVol.D5-6
−	$Eu(HOCH_2COO)CO_3$	Sc:	MVol.D5-239
$C_3EuH_4NO_2{}^{2+}$	$[Eu(ONCHCOCH_3)]^{2+}$	Sc:	MVol.D2-102
$C_3EuH_5NO_2S^{+}$	$[Eu(SCH_2CH(NH_2)COO)]^{+}$	Sc:	MVol.D4-59/60
$C_3EuH_5N_2O_2{}^{2+}$	$[Eu(CH_3C(NO)CHNOH)]^{2+}$	Sc:	MVol.D2-116/7
$C_3EuH_5O_2{}^{2+}$	$[Eu(C_2H_5COO)]^{2+}$	Sc:	MVol.D5-71/4
$C_3EuH_5O_2S^{2+}$	$Eu(OOCCHSHCH_3)^{2+}$	Sc:	MVol.D4-55/6
−	$Eu(OOCC_2H_4SH)^{2+}$	Sc:	MVol.D4-55/6
$C_3EuH_5O_3{}^{2+}$	$[Eu(CH_3OCH_2COO)]^{2+}$	Sc:	MVol.D5-69/70
−	$[Eu(HOC_2H_4COO)]^{2+}$	Sc:	MVol.D5-260/1
−	$[Eu(H_3CCHOHCOO)]^{2+}$	Sc:	MVol.D5-250/5
$C_3EuH_5O_4{}^{2+}$	$[Eu(HOCH_2CHOHCOO)]^{2+}$	Sc:	MVol.D5-260/1
$C_3EuH_6NO_2{}^{2+}$	$[Eu(CH_3CH(NH_2)COO)]^{2+}$	Sc:	MVol.D1-107/8
$C_3EuH_6NO_3{}^{2+}$	$Eu[HOCH_2CH(NH_2)COO]^{2+}$	Sc:	MVol.D1-113
$C_3EuH_6O_6P$	$Eu[H(O_3POCH_2CH(O)CH_2O)] \cdot H_2O$	Sc:	MVol.D4-285
$C_3EuH_7O_2{}^{2+}$	$[Eu(OCH_2CHOHCH_3)]^{2+}$	Sc:	MVol.D3-31/2
$C_3EuH_7O_3{}^{2+}$	$[Eu(OC_3H_5(OH)_2)]^{2+}$	Sc:	MVol.D3-33/5
$C_3EuH_8O_7P$	$Eu[O_4PCH_2CH(OH)CH_2OH](OH)$	Sc:	MVol.D4-285
$C_3EuH_9NO_9P_3$	$Eu[N(CH_2P(O)_2OH)_3]$	Sc:	MVol.D4-149
−	$Eu[N(CH_2P(O)_2OH)_3] \cdot 3 H_2O$	Sc:	MVol.D4-149/50
$C_3EuN_3S_3$	$Eu(NCS)_3$	Sc:	MVol.D4-340/3
$C_3EuO_9{}^{3-}$	$[Eu(CO_3)_3]^{3-}$	Sc:	MVol.D4-338/40
$C_3Eu_2H_9O_9P_3$	$Eu_2(CH_3PO_3)_3$	Sc:	MVol.D4-143
C_3FGaH_9K	$K[Ga(CH_3)_3F]$	Ga:	Org.Comp.1-325
$C_3FH_4MnO_2{}^{+}$	$[Mn(FCH_2CH_2COO)]^{+}$	Mn:	MVol.D2-65

$C_3FH_4N_2O_4S_2$	$[CH_3NSNCHCO]FSO_3$	S:	S-N Comp.3-243
$C_3FH_4N_3O_5$	$(NO_2)_2CFC(NH)OCH_3$	F:	PFHOrg.8-38
$C_3FH_5OS_2$	$FC(O)SSC_2H_5$	F:	PFHOrg.SVol.2-128
$C_3FH_9O_3SSn$	$(CH_3)_3SnOSO_2F$	Sn:	Org.Comp.11-147/8
$C_3F_2GaH_9NOP$	$Ga(CH_3)_2OP(NCH_3)F_2$	Ga:	Org.Comp.1-205
$C_3F_2H_2N_2O_5$	$NO_2CF_2C(O)CH_2NO_2$.	F:	PFHOrg.8-42
$C_3F_2H_3NO_4$	$NO_2CF_2C(O)OCH_3$.	F:	PFHOrg.8-36/8
$C_3F_2H_5N_3O_2$	$NO_2CF_2C(NH)NHCH_3$	F:	PFHOrg.8-41
$C_3F_2H_7NO$	$NHF_2 \cdot CD_3COCD_3$.	F:	SVol.5-45
$C_3F_2H_8N_2O$	$NHF_2 \cdot HCON(CH_3)_2$	F:	SVol.5-55
$C_3F_2H_9N_3O_4S_3Sn$ . . .	$(CH_3)_3SnOSO(NS(O)F)_2N$	Sn:	Org.Comp.11-150
$C_3F_2H_9OPSSn$	$(CH_3)_3SnOP(S)F_2$	Sn:	Org.Comp.11-156
$C_3F_2H_9O_2PSn$	$(CH_3)_3SnOP(O)F_2$	Sn:	Org.Comp.11-155
$C_3F_2H_9PS_2Sn$	$(CH_3)_3SnSP(S)F_2$.	Sn:	Org.Comp.9-62
$C_3F_2H_9Sb$	$(CH_3)_3SbF_2$	Sb:	Org.Comp.4-1/2
$C_3F_2H_{16}N_6O_7U$ . . .`.	$(CN_3H_6)_2[UO_2F_2(CO_3)(H_2O)_2]$.	U:	SVol.C13-25
$C_3F_2N_2$	$CF_2(CN)_2$.	F:	PFHOrg.9-56/7, 65/6, 71
$C_3F_2N_6S_3$	$S_3N_2NC_3N_3F_2$	S:	S-N Comp.2-50
$C_3F_2O_4S$	$[OS(O)_2CFCFC(O)]$	F:	PFHOrg.SVol.3-126, 131
$C_3F_3GeH_3$	CF_3CCGeH_3	F:	PFHOrg.SVol.1-50, 56, 60
$C_3F_3HN_2S_2$	$HS(CF_3)C_2N_2S$.	F:	PFHOrg.SVol.2-44, 50, 59
$C_3F_3H_2N_3S$	$(NH_2)CF_3C_2N_2S$.	F:	PFHOrg.SVol.2-43, 49, 53/7
$C_3F_3H_2N_3Se$	$CF_3C[NNC(NH_2)Se]$.	F:	PFHOrg.SVol.3-210, 212
$C_3F_3H_3N_4S$	$HS(CF_3)NH_2C_2N_3$	F:	PFHOrg.SVol.2-64
$C_3F_3H_3O_2$	$CF_3C(O)OCH_3$	F:	PFHOrg.9-121
$C_3F_3H_3O_4S$	$CH_3OC(O)CF_2SO_2F$.	F:	PFHOrg.SVol.3-135
$C_3F_3H_3Si$	CF_3CCSiH_3	F:	PFHOrg.SVol.1-41, 44, 47
$C_3F_3H_4NO$	$CF_3C(NH)OCH_3$	F:	PFHOrg.9-81
$C_3F_3H_4NOS_2$	$CF_3SNHC(O)SCH_3$	F:	PFHOrg.SVol.2-113/4
$C_3F_3H_4NS$	$CF_3C(NH)SCH_3$	F:	PFHOrg.9-82
$C_3F_3H_5N_2$	$CF_3C(NH)NHCH_3$.	F:	PFHOrg.9-79
$-$	$CF_3C(ND)NDCH_3$.	F:	PFHOrg.9-80
$C_3F_3H_5N_2O$	$CF_3NHC(O)NHCH_3$	F:	PFHOrg.9-117
$C_3F_3H_6ISn$	$(CH_3)_2(CF_3)SnI$	Sn:	Org.Comp.8-51, 56
$C_3F_3H_6NS$	$CF_3SN(CH_3)_2$.	F:	PFHOrg.SVol.2-113
$C_3F_3H_6NS_2$	$(CH_3)_2NSSCF_3$.	F:	PFHOrg.SVol.2-199, 255
$C_3F_3H_6O_3PSe$	$CF_3SeP(O)(OCH_3)_2$	F:	PFHOrg.SVol.3-243
$C_3F_3H_6SSe$	$CF_3SSe(CH_3)_2$ (radical)	F:	PFHOrg.SVol.2-201
$C_3F_3H_6S_2$	$CF_3SS(CH_3)_2$ (radical)	F:	PFHOrg.SVol.2-201
$-$	$CF_3SS(CD_3)_2$ (radical)	F:	PFHOrg.SVol.2-201
$C_3F_3H_6Sb$	$(CH_3)_2SbCF_3$	Sb:	Org.Comp.1-109/10
$C_3F_3H_6SeTl$	$CF_3SeTl(CH_3)_2$	F:	PFHOrg.SVol.3-241
$C_3F_3H_9O_3U$	$UF_3(OCH_3)_3$	U:	SVol.C13-76, 82
		U:	SVol.D1-128
C_3F_3N	CF_2CFCN.	F:	PFHOrg.9-51/2, 68, 79, 81/2, 84
C_3F_3NO	CF_2CFNCO	F:	PFHOrg.9-103, 110, 117, 120
$-$	$[CF_3C(O)CN]_2$	F:	PFHOrg.9-53, 69, 79, 88

$C_3F_7HO_3S$	$(CF_3)_2CFSO_3H$.	F:	PFHOrg.SVol.3-48/9
–	$C_3F_7SO_3H$	F:	PFHOrg.SVol.3-48/9
C_3F_7HS	$(CF_3)_2CFSH$.	F:	PFHOrg.SVol.2-62
$C_3F_7H_2N$	$n-C_3F_7NH_2$	F:	PFHOrg.9-115/6
$C_3F_7H_2NO_2S$	$(CF_3)_2CFSO_2NH_2$	F:	PFHOrg.SVol.3-138, 143
$C_3F_7H_2NS$	$(CF_3)_2CFSNH_2$.	F:	PFHOrg.SVol.2-85, 88/90, 97
$C_3F_7H_2NS_3$	$(CF_3S)_2CFSNH_2$.	F:	PFHOrg.SVol.2-85, 88
$C_3F_7H_3N_2S$	$CF_3SF(NCH_3)NCF_3$	F:	PFHOrg.SVol.3-207
$C_3F_7H_4N_3O$	$CF_3NFC(OCH_3)(NF_2)(NHF)$	F:	PFHOrg.8-224
C_3F_7IO.	$CF_3CF_2CF_2IO$	F:	PFHOrg.SVol.3-253, 257, 263
$C_3F_7IO_2$.	$CF_3IF[OC(O)CF_3]$	F:	PFHOrg.SVol.3-254, 258, 265
$C_3F_7IO_3S$.	$ICF_2OCF_2CF_2SO_2F$	F:	PFHOrg.SVol.3-155/6, 163
$C_3F_7I_2P$	$i-C_3F_7PI_2$.	F:	PFHOrg.SVol.1-106, 117/20
C_3F_7K	$(CF_3)_2CFK$.	F:	PFHOrg.SVol.1-16
$C_3F_7KO_3S$	$KOSO_2CF(CF_3)_2$	F:	PFHOrg.SVol.3-77
–	$KOSO_2C_3F_7$	F:	PFHOrg.SVol.3-60, 77
C_3F_7N	$[CF_2CF(CF_2NF_2)]_n$	F:	PFHOrg.8-168, 193
–	$(CF_3)_2CNF$.	F:	PFHOrg.8-170/1, 195/6, 205, 220/1
–	CF_3NCFCF_3.	F:	PFHOrg.9-128, 138
–	$C_2F_5NCF_2$	F:	PFHOrg.9-128, 138, 149/50
–	$C_3F_5NF_2$	F:	PFHOrg.8-156/7, 179
C_3F_7NO	$(CF_3)_2NC(O)F$	F:	PFHOrg.9-160, 168, 177
–	$CF_3CF(NF_2)C(O)F$	F:	PFHOrg.8-158, 181, 207
–	$CF_3C(O)CF_2NF_2$	F:	PFHOrg.8-158, 181
–	$C_2F_5C(O)NF_2$.	F:	PFHOrg.8-159, 182/3
C_3F_7NOS.	$(CF_3)_2CFNSO$	F:	PFHOrg.9-6/7, 23
–	$CF_3S(F)NC(O)CF_3$	F:	PFHOrg.SVol.3-7, 26
–	$CF_3SN[C(O)F]CF_3$	F:	PFHOrg.SVol.2-101, 107, 111
$C_3F_7NOS_2$	$(CF_3)_2CFSNSO$	F:	PFHOrg.SVol.2-99, 103, 110
$C_3F_7NOS_3$	$FC(O)S(CF_3S)_2N$	F:	PFHOrg.SVol.2-87, 89/90, 97
$C_3F_7NO_2$	$(CF_3)_2CFNO_2$.	F:	PFHOrg.8-9/10, 24, 41
–	$(n-C_3F_7)NO_2$.	F:	PFHOrg.8-9, 24, 34/5
$C_3F_7NO_3$	$CF_3OCF(NO_2)CF_3$	F:	PFHOrg.8-12, 26
–	$CF_3O(CF_2)_2NO_2$.	F:	PFHOrg.8-12, 26
C_3F_7NS	$(CF_3)_2CNSF$.	F:	PFHOrg.9-1, 11, 32/3, 35/6
–	$CF_3SNCFCF_3$.	F:	PFHOrg.SVol.2-101, 106, 111
$C_3F_7N_3$	$NF_2CF_2CF(CN)NF_2$	F:	PFHOrg.8-166/7

$C_3F_9Hg_2NO_2S$	$(CF_3Hg)_2NSO_2CF_3$	F:	PFHOrg.SVol.3-141, 146
$C_3F_9HoO_9S_3$	$Ho(OSO_2CF_3)_3$	F:	PFHOrg.SVol.3-61, 104
–	$Ho(OSO_2CF_3)_3 \cdot 9\ H_2O$	F:	PFHOrg.SVol.3-61, 87, 104
C_3F_9I	$C_3F_7IF_2$	F:	PFHOrg.SVol.3-255/6, 259, 264
–	$i\text{-}C_3F_7IF_2$	F:	PFHOrg.SVol.3-255/6, 265
$C_3F_9IO_6S_2$	$C_3F_7I(OSO_2F)_2$	F:	PFHOrg.SVol.3-255, 264
$C_3F_9IO_9S_3$	$I(OSO_2CF_3)_3$	F:	PFHOrg.SVol.3-56, 70, 83/4, 102
C_3F_9ISn	$(CF_3)_3SnI$	Sn:	Org.Comp.8-33
$C_3F_9LaO_9S_3$	$La(OSO_2CF_3)_3$	F:	PFHOrg.SVol.3-61, 104
–	$La(OSO_2CF_3)_3 \cdot 9\ H_2O$	F:	PFHOrg.SVol.3-61, 87, 104
C_3F_9N	$(CF_3)_2CFNF_2$	F:	PFHOrg.8-156/7, 179, 203, 205
–	$(CF_3)_3N$	F:	PFHOrg.9-159, 166/7, 176/7
–	$(C_2F_5)(CF_3)NF$	F:	PFHOrg.8-162, 185
–	$n\text{-}C_3F_7NF_2$	F:	PFHOrg.8-156/7, 179
C_3F_9NO	$(CF_3)_2NOCF_3$	F:	PFHOrg.8-102
$C_3F_9NO_2$	$CF_3N(OCF_3)_2$	F:	PFHOrg.8-203
$C_3F_9NO_2S$	$CF_3S(O)ON(CF_3)_2$	F:	PFHOrg.SVol.3-4, 23, 38
$C_3F_9NO_2S_2$	$(CF_3)_2SNSO_2CF_3$	F:	PFHOrg.SVol.3-8, 29, 38
$C_3F_9NO_2S_3$	$CF_3SO_2N(SCF_3)_2$	F:	PFHOrg.SVol.3-143/4
$C_3F_9NO_3S$	$CF_3CF(NF_2)CF_2OSO_2F$	F:	PFHOrg.8-158, 182
–	$CF_3CF(OSO_2F)CF_2NF_2$	F:	PFHOrg.8-158, 182
C_3F_9NS	$(CF_3)_2CFNSF_2$	F:	PFHOrg.9-3/5, 16/7, 33, 35
–	$(CF_3)_2CFSF_2N$	F:	PFHOrg.SVol.3-192, 202, 207
–	$CF_3S(F)NC_2F_5$	F:	PFHOrg.SVol.3-7, 27, 43
–	$CF_3SN(CF_3)_2$	F:	PFHOrg.SVol.2-86, 89, 91, 101
–	$C_3F_7NSF_2$	F:	PFHOrg.9-34
$C_3F_9NS_2$	$(CF_3)_2CFSNSF_2$	F:	PFHOrg.SVol.2-99, 103, 110
–	$(CF_3)_2SNSCF_3$	F:	PFHOrg.SVol.3-8
$C_3F_9NS_3$	$N(SCF_3)_3$	F:	PFHOrg.SVol.2-86, 92/7
$C_3F_9NS_4$	$(CF_3S)_2(CF_3SS)N$	F:	PFHOrg.SVol.2-174, 186, 195
$C_3F_9NSe_3$	$N(SeCF_3)_3$	F:	PFHOrg.SVol.3-216, 224, 241/2
$C_3F_9N_2$	$(CF_3)_2NNCF_3$	F:	PFHOrg.8-103/5
$C_3F_9N_3O$	$(CF_3)_2NN(CF_3)NO$	F:	PFHOrg.8-56/7, 83/4
$C_3F_9N_3O_2$	$(CF_3)_2NN(CF_3)NO_2$	F:	PFHOrg.8-56/7, 84
$C_3F_9N_5O$	$(NF_2)_2CFNFCF(NCO)NF_2$	F:	PFHOrg.8-166/7
$C_3F_9NdO_9S_3$	$Nd(OSO_2CF_3)_3$	F:	PFHOrg.SVol.3-61, 78, 104

C$_3$F$_{10}$S.	CF$_2$C(CF$_3$)SF$_5$	F:	PFHOrg.SVol.3–189, 198, 206
–	(CF$_3$)$_2$CFSF$_3$	F:	PFHOrg.SVol.3–10, 44
–	C$_2$F$_5$(CF$_3$)SF$_2$	F:	PFHOrg.SVol.3–10, 33
C$_3$F$_{10}$Se.	C$_2$F$_5$(CF$_3$)SeF$_2$	F:	PFHOrg.SVol.3–216/7, 226
C$_3$F$_{10}$Si	CF$_3$CF$_2$CF$_2$SiF$_3$	F:	PFHOrg.SVol.1–41/4
C$_3$F$_{11}$Ge$^-$	[(CF$_3$)$_3$GeF$_2$]$^-$	F:	PFHOrg.SVol.1–61
C$_3$F$_{11}$GeH$_4$N	[NH$_4$][(CF$_3$)$_3$GeF$_2$]	F:	PFHOrg.SVol.1–52/3
C$_3$F$_{11}$GeNa	Na[(CF$_3$)$_3$GeF$_2$].	F:	PFHOrg.SVol.1–52/3
C$_3$F$_{11}$I	(CF$_3$)$_2$CFIF$_4$	F:	PFHOrg.SVol.3–256, 260, 264
–	CF$_3$(CF$_2$)$_2$IF$_4$	F:	PFHOrg.SVol.3–256, 260
C$_3$F$_{11}$IS	CF$_3$CF$_2$CFISF$_5$.	F:	PFHOrg.SVol.3–188/9, 197
C$_3$F$_{11}$NS	CF$_3$SF$_4$NCFCF$_3$	F:	PFHOrg.SVol.3–192/3, 204
C$_3$F$_{11}$N$_3$.	CF$_3$NFCF$_2$NFCF$_2$NF$_2$.	F:	PFHOrg.8–165, 190
C$_3$F$_{11}$N$_5$O	(NF$_2$)$_2$CFNFC(O)NFCF$_2$NF$_2$	F:	PFHOrg.8–167/8, 192
C$_3$F$_{11}$P.	(CF$_3$)$_3$PF$_2$	F:	PFHOrg.SVol.1–104, 109, 116, 123/8
C$_3$F$_{12}$HNS	CF$_3$SF$_4$NHC$_2$F$_5$	F:	PFHOrg.SVol.3–192, 203
C$_3$F$_{12}$O$_3$S$_2$.	(CF$_3$)$_2$C(SF$_5$)OSO$_2$F	F:	PFHOrg.SVol.3–189, 198
C$_3$F$_{12}$P$^-$	[(CF$_3$)$_3$PF$_3$]$^-$	F:	PFHOrg.SVol.1–137
C$_3$F$_{12}$S.	CF$_3$SF$_4$C$_2$F$_5$	F:	PFHOrg.SVol.3–189, 190, 199
–	n–C$_3$F$_7$SF$_5$.	F:	PFHOrg.SVol.3–187, 194
–	i–C$_3$F$_7$SF$_5$.	F:	PFHOrg.SVol.3–187, 194
C$_3$F$_{13}$NOS$_2$	(CF$_3$)$_2$CFS(O)F(NSF$_5$)	F:	PFHOrg.SVol.3–124, 129
C$_3$F$_{13}$NS	(CF$_3$)(C$_2$F$_5$)NSF$_5$	F:	PFHOrg.9–8, 26, 33
–	CF$_3$SF$_4$NFC$_2$F$_5$	F:	PFHOrg.SVol.3–192/3, 203
C$_3$F$_{13}$NS$_2$	(CF$_3$)$_2$CFSF$_4$NSF$_2$	F:	PFHOrg.9–4/5, 19
		F:	PFHOrg.SVol.3–192, 202
C$_3$F$_{13}$N$_5$.	NF$_2$CF$_2$NFCF$_2$NFCF(NF$_2$)$_2$	F:	PFHOrg.8–166, 191
C$_3$F$_{14}$N$_6$.	NF$_2$CF$_2$NFCF(NF$_2$)NFCF(NF$_2$)$_2$	F:	PFHOrg.8–166, 191
C$_3$F$_{16}$S$_2$.	SF$_5$CF$_2$CF$_2$CF$_2$SF$_5$	F:	PFHOrg.SVol.3–190, 191, 200
C$_3$GaH$_6$N.	Ga(CH$_3$)$_2$CN	Ga:	Org.Comp.1–168/9
C$_3$GaH$_7$I$_2$.	Ga(C$_3$H$_7$)I$_2$	Ga:	Org.Comp.1–162
C$_3$GaH$_7$O$_2$	Ga(CH$_3$)$_2$OOCH.	Ga:	Org.Comp.1–191
C$_3$GaH$_8$IS	GaCH$_3$(I)SC$_2$H$_5$	Ga:	Org.Comp.1–240
C$_3$GaH$_9$	Ga(CH$_3$)$_3$.	Ga:	Org.Comp.1–2/32
C$_3$GaH$_9$I$_2$S	Ga(CH$_3$)I$_2$ · C$_2$H$_5$SH	Ga:	Org.Comp.1–163/4
C$_3$GaH$_9$KN$_3$.	K[Ga(CH$_3$)$_3$N$_3$]	Ga:	Org.Comp.1–337
C$_3$GaH$_9$N$_2$O$_2$	Ga(CH$_3$)$_2$ONCH$_3$NO	Ga:	Org.Comp.1–202, 208
C$_3$GaH$_9$O.	Ga(CH$_3$)$_2$OCH$_3$	Ga:	Org.Comp.1–176, 178/80
–	Ga(CH$_3$)$_2$OCD$_3$	Ga:	Org.Comp.1–178/80
–	Ga(CD$_3$)$_2$OCH$_3$	Ga:	Org.Comp.1–178/80
–	Ga(CD$_3$)$_2$OCD$_3$	Ga:	Org.Comp.1–178/80

$C_3H_3O_4Tm^{2+}$	$[Tm(CH_2(COO)_2H)]^{2+}$	Sc:	MVol.D5-152, 154
$C_3H_3O_4Yb^{2+}$	$[Yb(CH_2(COO)_2H)]^{2+}$	Sc:	MVol.D5-152, 154
$C_3H_3O_5Sc$	$Sc(OH)(CH_2(COO)_2) \cdot 2 H_2O$	Sc:	MVol.D5-160/1
$C_3H_3O_5U^+$	$[UO_2(CH_3COCOO)]^+$	U:	SVol.D1-204
$C_3H_3O_6Pr$	$Pr(HCOO)_3$	Sc:	MVol.D5-6/7, 11/5
–	$Pr(HCOO)_3 \cdot 0.1 H_2O$	Sc:	MVol.D5-5/6
–	$Pr(HCOO)_3 \cdot 0.2 H_2O$	Sc:	MVol.D5-6
–	$Pr(HCOO)_3 \cdot x H_2O$	Sc:	MVol.D5-6
–	$Pr(OH)(HOCH(COO)_2) \cdot 4 H_2O$		
	$= [Pr(OCH(COO)_2)(H_2O)_3] \cdot 2 H_2O$	Sc:	MVol.D5-317
$C_3H_3O_6Sc$	$Sc(HCOO)_3$	Sc:	MVol.D5-2/4
–	$Sc(HOCH_2COO)CO_3$	Sc:	MVol.D5-239
$C_3H_3O_6Sm$	$Sm(HCOO)_3$	Sc:	MVol.D5-6/7, 11/5
–	$Sm(HCOO)_3 \cdot 0.1 H_2O$	Sc:	MVol.D5-5/6
–	$Sm(HCOO)_3 \cdot 0.2 H_2O$	Sc:	MVol.D5-6
–	$Sm(HCOO)_3 \cdot x H_2O$	Sc:	MVol.D5-6
–	$Sm(HOCH_2COO)CO_3$	Sc:	MVol.D5-239
$C_3H_3O_6Tb$	$Tb(HCOO)_3$	Sc:	MVol.D5-6/7, 11, 14/5
–	$Tb(HCOO)_3 \cdot 0.1 H_2O$	Sc:	MVol.D5-5/6
–	$Tb(HCOO)_3 \cdot 2 H_2O$	Sc:	MVol.D5-6, 8, 11
–	$Tb(HOCH_2COO)CO_3$	Sc:	MVol.D5-239
$C_3H_3O_6Tm$	$Tm(HCOO)_3$	Sc:	MVol.D5-6/7, 11, 15
–	$Tm(HCOO)_3 \cdot 0.1 H_2O$	Sc:	MVol.D5-5/6
–	$Tm(HCOO)_3 \cdot 2 H_2O$	Sc:	MVol.D5-5/8, 11/5
–	$Tm(HCOO)_3 \cdot x H_2O$	Sc:	MVol.D5-6
–	$Tm(HOCH_2COO)CO_3$	Sc:	MVol.D5-239
$C_3H_3O_6U$	$U(HCOO)_3$	U:	SVol.A5-73
		U:	SVol.A6-15, 20
		U:	SVol.C13-97/8
$C_3H_3O_6U^+$	$[UO_2H(CH_2(COO)_2)]^+$	U:	SVol.D1-204
$C_3H_3O_6Y$	$Y(HCOO)_3$	Sc:	MVol.D5-6/7, 11/5
–	$Y(HCOO)_3 \cdot 0.1 H_2O$	Sc:	MVol.D5-6
–	$Y(HCOO)_3 \cdot 2 H_2O$	Sc:	MVol.D5-5/15
–	$Y(HCOO)_3 \cdot x H_2O$	Sc:	MVol.D5-6
–	$Y(HOCH_2COO)CO_3$	Sc:	MVol.D5-239
$C_3H_3O_6Yb$	$Yb(HCOO)_3$	Sc:	MVol.D5-6, 11, 13, 15
–	$Yb(HCOO)_3 \cdot 2 H_2O$	Sc:	MVol.D5-5/15
–	$Yb(HCOO)_3 \cdot x H_2O$	Sc:	MVol.D5-6
–	$Yb(HOCH_2COO)CO_3$	Sc:	MVol.D5-239
$C_3H_3O_8TlU$	$Tl[UO_2(HCOO)_3] \cdot H_2O$	U:	SVol.C13-105, 107
$C_3H_3O_8U^-$	$UO_2(HCOO)_3^-$	U:	SVol.C13-105/6
		U:	SVol.D1-200
$C_3H_3O_{12}U^{5-}$	$[U(CO_3)_3(OH)_3]^{5-}$	U:	SVol.D1-169/72
C_3H_3Sm	$SmCCCH_3$	Sc:	MVol.D6-161
C_3H_3Yb	$YbCCCH_3$	Sc:	MVol.D6-161
$C_3H_4HoNO_2^{2+}$	$[Ho(ONCHCOCH_3)]^{2+}$	Sc:	MVol.D2-102
$C_3H_4IMnO_2^+$	$[Mn(ICH_2CH_2COO)]^+$	Mn:	MVol.D2-65
$C_3H_4KNO_8SU$	$K[UO_2(C_2O_4)(NCS)(H_2O)_2]$	U:	SVol.C13-230/2
$C_3H_4KN_3O_4S_3U$	$K[UO_2(NCS)_3(H_2O)_2]$	U:	SVol.C13-40, 41, 43
$C_3H_4K_2O_9U$	$K_2[UO_2(CH_2(COO)_2)(OO)(H_2O)] \cdot 2 H_2O$	U:	SVol.C13-238/40

$C_3H_4K_2O_{11}U$	$K_2[UO_2(C_2O_4)(CO_3)(H_2O)_2]$	U:	SVol.C13-228/9
$C_3H_4LaNO_2{}^{2+}$	$[La(ONCHCOCH_3)]^{2+}$	Sc:	MVol.D2-102
$C_3H_4LuNO_2{}^{2+}$	$[Lu(ONCHCOCH_3)]^{2+}$	Sc:	MVol.D2-102
$C_3H_4MnN_2{}^{2+}$	$[Mn(NHCHNCHCH)]^{2+}$	Mn:	MVol.D3-291
—	$Mn(NHN(CH)_3)^{2+}$	Mn:	MVol.D3-271
$C_3H_4MnN_4$	$NH_4Mn(CN)_3$	Mn:	MVol.D2-215
$C_3H_4MnO_3{}^+$	$[Mn(OCH(CH_3)COO)]^+$	Mn:	MVol.D2-151
$C_3H_4MnO_5$	$[Mn(H_2O)(CH_2(COO)_2)]$	Mn:	MVol.D2-126
$C_3H_4NNdO_2{}^{2+}$	$[Nd(ONCHCOCH_3)]^{2+}$	Sc:	MVol.D2-102
$C_3H_4NO_2Pr^{2+}$	$[Pr(ONCHCOCH_3)]^{2+}$	Sc:	MVol.D2-102
$C_3H_4NO_2Sm^{2+}$	$[Sm(ONCHCOCH_3)]^{2+}$	Sc:	MVol.D2-102
$C_3H_4NO_2Tb^{2+}$	$[Tb(ONCHCOCH_3)]^{2+}$	Sc:	MVol.D2-102
$C_3H_4NO_2Tm^{2+}$	$[Tm(ONCHCOCH_3)]^{2+}$	Sc:	MVol.D2-102
$C_3H_4NO_2Y^{2+}$	$[Y(ONCHCOCH_3)]^{2+}$	Sc:	MVol.D2-102
$C_3H_4NO_2Yb^{2+}$	$[Yb(ONCHCOCH_3)]^{2+}$	Sc:	MVol.D2-102
$C_3H_4NO_4OsS_2{}^+$	$[OsO_2(S_2CNHCH_2COOH)]^+$	Os:	SVol.1-290
$C_3H_4NO_8SU^-$	$[UO_2(C_2O_4)(NCS)(H_2O)_2]^-$	U:	SVol.C13-231
$C_3H_4N_2OS$	$SN_2C_2HCH_2OH$	S:	S-N Comp.3-93/8, 120
—	$SN_2C_2OHCH_3$	S:	S-N Comp.3-93/8, 150/1
$C_3H_4N_2O_2S$	$SN_2C_2OHOCH_3$	S:	S-N Comp.3-93/8, 155
$C_3H_4N_2S$	$SN_2C_2HCH_3$	S:	S-N Comp.3-93/8, 119/20
$C_3H_4N_3O_4RbS_3U$	$Rb[UO_2(NCS)_3(H_2O)_2]$	U:	SVol.C13-41, 45
$C_3H_4N_4NiOS_4$	$[Ni(S_2N_2CHC(OCH_3)S_2N_2)]$	S:	S-N Comp.2-293, 296
$C_3H_4N_4OS$	$SN_2C_2(CONH_2)NH_2$	S:	S-N Comp.3-93/8, 178
—	$SN_2C_2HCONHNH_2$	S:	S-N Comp.3-93/8, 174
—	$SN_2C_2HNHCONH_2$	S:	S-N Comp.3-93/8, 143
$C_3H_4Na_2O_9U$	$Na_2[UO_2(CH_2(COO)_2)(OO)(H_2O)] \cdot 5\ H_2O$ . .	U:	SVol.C13-238/40
$C_3H_4Na_2O_{11}U$	$Na_2[UO_2(C_2O_4)(CO_3)(H_2O)_2] \cdot H_2O$	U:	SVol.C13-228/9
$C_3H_4O_5U$	$UO_2[CH_3CH(O)COO]$	U:	SVol.C13-164
$C_3H_4O_7U$	$[UO_2(CH_2(COO)_2)(H_2O)] \cdot 2\ H_2O$	U:	SVol.C13-236/7
C_3H_4Sm	$Sm(C_3H_4)$	Sc:	MVol.D6-161
$C_3H_4Sm_2$	$(HSm)_2CCCH_2$	Sc:	MVol.D6-161
C_3H_4Yb	$Yb(C_3H_4)$	Sc:	MVol.D6-161
$C_3H_4Yb_2$	$(HYb)_2CCCH_2$	Sc:	MVol.D6-161
$C_3H_5HoNO_2S^+$	$[Ho(SCH_2CH(NH_2)COO)]^+$	Sc:	MVol.D4-59/60
$C_3H_5HoN_2O_2{}^{2+}$	$[Ho(CH_3C(NO)CHNOH)]^{2+}$	Sc:	MVol.D2-116/7
$C_3H_5HoO_2{}^{2+}$	$[Ho(C_2H_5COO)]^{2+}$	Sc:	MVol.D5-71/4
$C_3H_5HoO_2S^{2+}$	$Ho(OOCCHSHCH_3)^{2+}$	Sc:	MVol.D4-55/6
—	$Ho(OOCC_2H_4SH)^{2+}$	Sc:	MVol.D4-55/6
$C_3H_5HoO_3{}^{2+}$	$[Ho(CH_3OCH_2COO)]^{2+}$	Sc:	MVol.D5-69/70
—	$[Ho(HOC_2H_4COO)]^{2+}$	Sc:	MVol.D5-260/1
—	$[Ho(H_3CCHOHCOO)]^{2+}$	Sc:	MVol.D5-250/5
$C_3H_5IN_2S$	$[SN_2(CH_3)C_2H_2]I$	S:	S-N Comp.3-242/3
$C_3H_5K_2NO_9SU$	$K_2[UO_2(C_2O_4)(NCS)(OH)(H_2O)_2]$	U:	SVol.C13-231/3
$C_3H_5LaNO_2S^+$	$[La(SCH_2CH(NH_2)COO)]^+$	Sc:	MVol.D4-59/60
$C_3H_5LaN_2O_2{}^{2+}$	$[La(CH_3C(NO)CHNOH)]^{2+}$	Sc:	MVol.D2-116/7
$C_3H_5LaO_2{}^{2+}$	$[La(C_2H_5COO)]^{2+}$	Sc:	MVol.D5-71/4
$C_3H_5LaO_2S^{2+}$	$La(OOCCHSHCH_3)^{2+}$	Sc:	MVol.D4-55/6
—	$La(OOCC_2H_4SH)^{2+}$	Sc:	MVol.D4-55/6
$C_3H_5LaO_3{}^{2+}$	$[La(CH_3OCH_2COO)]^{2+}$	Sc:	MVol.D5-69/70

$C_3H_5LaO_3^{2+}$	$[La(HOC_2H_4COO)]^{2+}$	Sc:	MVol.D5-260/1
–	$[La(H_3CCHOHCOO)]^{2+}$	Sc:	MVol.D5-250/5
$C_3H_5LuN_2O_2^{2+}$	$[Lu(CH_3C(NO)CHNOH)]^{2+}$	Sc:	MVol.D2-116/7
$C_3H_5LuO_2^{2+}$	$[Lu(C_2H_5COO)]^{2+}$	Sc:	MVol.D5-71/4
$C_3H_5LuO_2S^{2+}$	$Lu(OOCCHSHCH_3)^{2+}$	Sc:	MVol.D4-55/6
–	$Lu(OOCC_2H_4SH)^{2+}$	Sc:	MVol.D4-55/6
$C_3H_5LuO_3^{2+}$	$[Lu(CH_3OCH_2COO)]^{2+}$	Sc:	MVol.D5-69/70
–	$[Lu(HOC_2H_4COO)]^{2+}$	Sc:	MVol.D5-260/1
–	$[Lu(H_3CCHOHCOO)]^{2+}$	Sc:	MVol.D5-250/5
$C_3H_5MnNO_2S$	$Mn[SCH_2CH(NH_2)COO]$	Mn:	MVol.D4-290
$C_3H_5MnNO_5S$	$Mn[O_3SCH_2CH(NH_2)COO]$	Mn:	MVol.D4-291
$C_3H_5MnO_2^+$	$[Mn(C_2H_5COO)]^+$	Mn:	MVol.D2-65
$C_3H_5MnO_3^+$	$[Mn(CH_3CH(OH)COO)]^+$	Mn:	MVol.D2-151
$C_3H_5NNdO_2S^+$	$[Nd(SCH_2CH(NH_2)COO)]^+$	Sc:	MVol.D4-59/60
$C_3H_5NO_2PrS^+$	$[Pr(SCH_2CH(NH_2)COO)]^+$	Sc:	MVol.D4-59/60
$C_3H_5NO_2SSm^+$	$[Sm(SCH_2CH(NH_2)COO)]^+$ ...	Sc:	MVol.D4-59/60
$C_3H_5NO_2STb^+$	$[Tb(SCH_2CH(NH_2)COO)]^+$ ...	Sc:	MVol.D4-59/60
$C_3H_5NO_2STm^+$	$[Tm(SCH_2CH(NH_2)COO)]^+$ ...	Sc:	MVol.D4-59/60
$C_3H_5N_2NdO_2^{2+}$	$[Nd(CH_3C(NO)CHNOH)]^{2+}$ ...	Sc:	MVol.D2-116/7
$C_3H_5N_2O_2Pr^{2+}$	$[Pr(CH_3C(NO)CHNOH)]^{2+}$ ...	Sc:	MVol.D2-116/7
$C_3H_5N_2O_2Sm^{2+}$	$[Sm(CH_3C(NO)CHNOH)]^{2+}$ ...	Sc:	MVol.D2-116/7
$C_3H_5N_2O_2Tb^{2+}$	$[Tb(CH_3C(NO)CHNOH)]^{2+}$ ...	Sc:	MVol.D2-116/7
$C_3H_5N_2O_2Tm^{2+}$	$[Tm(CH_3C(NO)CHNOH)]^{2+}$ ...	Sc:	MVol.D2-116/7
$C_3H_5N_2O_2Y^{2+}$	$[Y(CH_3C(NO)CHNOH)]^{2+}$ ...	Sc:	MVol.D2-116/7
$C_3H_5N_2O_2Yb^{2+}$	$[Yb(CH_3C(NO)CHNOH)]^{2+}$ ...	Sc:	MVol.D2-116/7
$C_3H_5N_2S^+$	$[SN_2C_2H_2CH_3]^+$	S:	S-N Comp.3-242/3
$C_3H_5N_3O_2S$	$(O)SN_2C_2(OCH_3)(NH_2)$	S:	S-N Comp.3-250/7
$C_3H_5N_3S$	$SN_2C_2HNHCH_3$	S:	S-N Comp.3-93/8, 142
$C_3H_5N_5OS$	$SN_2C_2(CONHNH_2)NH_2$	S:	S-N Comp.3-93/8, 178/9
$C_3H_5NdO_2^{2+}$	$[Nd(C_2H_5COO)]^{2+}$	Sc:	MVol.D5-71/4
$C_3H_5NdO_2S^{2+}$	$Nd(OOCCHSHCH_3)^{2+}$	Sc:	MVol.D4-55/6
–	$Nd(OOCC_2H_4SH)^{2+}$	Sc:	MVol.D4-55/6
$C_3H_5NdO_3^{2+}$	$[Nd(CH_3OCH_2COO)]^{2+}$	Sc:	MVol.D5-69/70
–	$[Nd(HOC_2H_4COO)]^{2+}$	Sc:	MVol.D5-260/1
–	$[Nd(H_3CCHOHCOO)]^{2+}$	Sc:	MVol.D5-250/5
$C_3H_5O_2Pr^{2+}$	$[Pr(C_2H_5COO)]^{2+}$	Sc:	MVol.D5-71/4
$C_3H_5O_2PrS^{2+}$	$Pr(OOCCHSHCH_3)^{2+}$	Sc:	MVol.D4-55/6
–	$Pr(OOCC_2H_4SH)^{2+}$	Sc:	MVol.D4-55/6
$C_3H_5O_2SSm^{2+}$	$Sm(OOCCHSHCH_3)^{2+}$	Sc:	MVol.D4-55/6
–	$Sm(OOCC_2H_4SH)^{2+}$	Sc:	MVol.D4-55/6
$C_3H_5O_2STb^{2+}$	$Tb(OOCCHSHCH_3)^{2+}$	Sc:	MVol.D4-55/6
–	$Tb(OOCC_2H_4SH)^{2+}$	Sc:	MVol.D4-55/6
$C_3H_5O_2STm^{2+}$	$Tm(OOCCHSHCH_3)^{2+}$	Sc:	MVol.D4-55/6
–	$Tm(OOCC_2H_4SH)^{2+}$	Sc:	MVol.D4-55/6
$C_3H_5O_2SY^{2+}$	$Y(OOCCHSHCH_3)^{2+}$	Sc:	MVol.D4-55/6
–	$Y(OOCC_2H_4SH)^{2+}$	Sc:	MVol.D4-55/6
$C_3H_5O_2SYb^{2+}$	$Yb(OOCCHSHCH_3)^{2+}$	Sc:	MVol.D4-55/6
–	$Yb(OOCC_2H_4SH)^{2+}$	Sc:	MVol.D4-55/6
$C_3H_5O_2Sm^{2+}$	$[Sm(C_2H_5COO)]^{2+}$	Sc:	MVol.D5-71/4
$C_3H_5O_2Tb^{2+}$	$[Tb(C_2H_5COO)]^{2+}$	Sc:	MVol.D5-71/4

$C_3H_9NNdO_9P_3$	$Nd[N(CH_2P(O)_2OH)_3] \cdot 5\ H_2O$	Sc:	MVol.D4-149/50
C_3H_9NOSSn	$(CH_3)_3SnNSO$	Sn:	Org.Comp.8-205
$C_3H_9NOS_2$	$S_2O \cdot N(CH_3)_3$	S:	SVol.3-32
$C_3H_9NO_2S$	$SO_2 \cdot N(CH_3)_3$	S:	SVol.3-32
$C_3H_9NO_2Sn$	$(CH_3)_3SnONO$	Sn:	Org.Comp.11-128, 131
$C_3H_9NO_3Sn$	$(CH_3)_3SnONO_2$	Sn:	Org.Comp.11-128, 131/4
—	$(CH_3)_3SnONO_2 \cdot H_2O$	Sn:	Org.Comp.11-128, 131/4
$C_3H_9NO_9P_3Pr$	$Pr[N(CH_2P(O)_2OH)_3]$	Sc:	MVol.D4-149
—	$Pr[N(CH_2P(O)_2OH)_3] \cdot 5\ H_2O$	Sc:	MVol.D4-149/50
$C_3H_9NO_9P_3Sm$	$Sm[N(CH_2P(O)_2OH)_3]$	Sc:	MVol.D4-149
—	$Sm[N(CH_2P(O)_2OH)_3] \cdot 3\ H_2O$	Sc:	MVol.D4-149/50
—	$Sm[N(CH_2P(O)_2OH)_3] \cdot 5\ H_2O$	Sc:	MVol.D4-149/50
$C_3H_9NO_9P_3Y$	$Y[N(CH_2P(O)_2OH)_3] \cdot x\ H_2O$	Sc:	MVol.D4-149/50
$C_3H_9NO_9P_3Yb$	$Yb[N(CH_2P(O)_2OH)_3]$	Sc:	MVol.D4-149
—	$Yb[N(CH_2P(O)_2OH)_3] \cdot 3\ H_2O$	Sc:	MVol.D4-149/50
$C_3H_9N_2O_4Rh$	$Rh(NH_2C_2H_4NH_2)(OCO_2)OH$	Rh:	SVol.B2-170
$C_3H_9N_2O_6Sb$	$(CH_3)_3Sb(NO_3)_2$	Sb:	Org.Comp.4-107/8
$C_3H_9N_3O_3SSi$	$N_3SO_2OSi(CH_3)_3$	S:	SVol.3-320
$C_3H_9N_3O_3S_3$	$S_3N_3(OCH_3)_3$	S:	S-N Comp.2-88
$C_3H_9N_3Sn$	$(CH_3)_3SnN_3$	Sn:	Org.Comp.8-194/7
—	$(CH_3)_3Sn^{15}N_3$	Sn:	Org.Comp.8-194
$C_3H_9N_6O_6Y$	$Y(H_2NNHCOO)_3 \cdot 3\ H_2O$	Sc:	MVol.D1-242
$C_3H_9N_6Rh$	$Rh(CN)_3(NH_3)_3$	Rh:	SVol.B1-198
$C_3H_9N_6Sb$	$(CH_3)_3Sb(N_3)_2$	Sb:	Org.Comp.4-102
C_3H_9NaSSn	$(CH_3)_3SnSNa$	Sn:	Org.Comp.9-66
$C_3H_9NdO_3$	$Nd(OCH_3)_3$	Sc:	MVol.D3-17
$C_3H_9Nd_2O_9P_3$	$Nd_2(CH_3PO_3)_3$	Sc:	MVol.D4-143
C_3H_9OSb	$(CH_3)_3SbO$	Sb:	Org.Comp.4-113/4
—	$(CD_3)_3SbO$	Sb:	Org.Comp.4-115/6
$C_3H_9O_2SSn$	$(CH_3)_3SnSO_2$	Sn:	Org.Comp.9-65
$C_3H_9O_2Sb$	$(CH_3)_3SbO_2$	Sb:	Org.Comp.4-130
—	$CH_3Sb(OCH_3)_2$	Sb:	Org.Comp.2-126
$C_3H_9O_2Sn$	$(CH_3)_3SnOO$ (radical)	Sn:	Org.Comp.11-166/9
$C_3H_9O_3Pr$	$Pr(OCH_3)_3$	Sc:	MVol.D3-17
$C_3H_9O_3STl$	$CH_3SO_3Tl(CH_3)_2$	S:	SVol.3-322
$C_3H_9O_3Sm$	$Sm(OCH_3)_3$	Sc:	MVol.D3-17
$C_3H_9O_3Yb$	$Yb(OCH_3)_3$	Sc:	MVol.D3-17
$C_3H_9O_4PS$	$(CH_3)_3PO \cdot SO_3$	S:	SVol.3-319
$C_3H_9O_4ReSn$	$(CH_3)_3SnOReO_3$	Sn:	Org.Comp.11-157
$C_3H_9O_4SSb$	$(CH_3)_3SbSO_4$	Sb:	Org.Comp.4-110
$C_3H_9O_4SbSe$	$(CH_3)_3SbSeO_4$	Sb:	Org.Comp.4-111
$C_3H_9O_9P_3Pr_2$	$Pr_2(CH_3PO_3)_3$	Sc:	MVol.D4-143
$C_3H_9O_9P_3Sm_2$	$Sm_2(CH_3PO_3)_3$	Sc:	MVol.D4-143
$C_3H_9O_9P_3Tb_2$	$Tb_2(CH_3PO_3)_3$	Sc:	MVol.D4-143
$C_3H_9O_9P_3Tm_2$	$Tm_2(CH_3PO_3)_3$	Sc:	MVol.D4-143
$C_3H_9O_9P_3Yb_2$	$Yb_2(CH_3PO_3)_3$	Sc:	MVol.D4-143
C_3H_9Pb	$(CH_3)_3Pb$ (radical)	Pb:	Org.Comp.1-127, 135, 156/7

$C_3H_9Pb^+$	$[Pb(CH_3)_3]^+$	Pb:	Org.Comp.1-125/6, 149, 156/7, 163, 185/6
$C_3H_9Pb^-$	$[Pb(CH_3)_3]^-$	Pb:	Org.Comp.1-147
C_3H_9SSb	$(CH_3)_3SbS$	Sb:	Org.Comp.4-199/200
C_3H_9Sb	$Sb(CH_3)_3$	Sb:	Org.Comp.1-10/8
$-$	$Sb(CD_3)_3$	Sb:	Org.Comp.1-10/3
$C_3H_9SbSe_2$	$(CH_3)_3SbSe_2$	Sb:	Org.Comp.4-209
C_3H_9Sc	$Sc(CH_3)_3 \cdot n\ C_4H_8O$	Sc:	MVol.D6-142, 145
C_3H_9Y	$Y(CH_3)_3 \cdot n\ C_4H_8O$	Sc:	MVol.D6-142, 145
$C_3H_{10}MnO_8$	$[Mn(H_2O)_4(CH_2(COO)_2)]$	Mn:	MVol.D2-126
$C_3H_{10}MnO_8^+$	$[Mn(CH_2(COO)_2)(H_2O)_4]^+$	Mn:	MVol.D2-129
$C_3H_{10}N_2O_8U$	$(NH_4)_2[UO_2(CH_2(COO)_2)(OO)] \cdot H_2O$	U:	SVol.C13-238/40
$C_3H_{10}N_2O_9SU$	$NH_4[UO_2(C_2O_4)(NCS)(H_2O)_3] \cdot x\ H_2O$	U:	SVol.C13-230/2
$C_3H_{10}N_2O_{12}U_2$	$(NH_4)_2[(UO_2)_2(CH_2(COO)_2)(OO)_2] \cdot 3\ H_2O$	U:	SVol.C13-238/41
$C_3H_{10}N_4O_8U$	$UO_2(NO_3)_2(NH_2(CH_2)_3NH_2)$	U:	SVol.A5-217/8
$C_3H_{10}OSn$	$(CH_3)_3SnOH$	Sn:	Org.Comp.11-44/9
$-$	$(CH_3)_3SnOH \cdot n\ H_2O$	Sn:	Org.Comp.11-45
$C_3H_{10}O_2Sn$	$(CH_3)_3SnOOH$	Sn:	Org.Comp.11-124/5
$C_3H_{10}O_3SeSn$	$(CH_3)_3SnOSe(O)OH$	Sn:	Org.Comp.11-153/8
$C_3H_{10}O_4SSn$	$[(CH_3)_3Sn(H_2SO_4)_n]HSO_4$	Sn:	Org.Comp.11-146
$C_3H_{10}O_5U$	$UO_2(OCH_3)_2 \cdot CH_3OH$	U:	SVol.C13-85/6
$C_3H_{10}SSn$	$(CH_3)_3SnSH$	Sn:	Org.Comp.9-24
$C_3H_{11}O_2PSn$	$(CH_3)_3SnOP(O)H_2$	Sn:	Org.Comp.11-155/9
$C_3H_{11}O_2Sb$	$(CH_3)_3Sb(OH)_2$	Sb:	Org.Comp.4-114/5
$C_3H_{11}O_4Sb$	$(CH_3)_3Sb(OOH)_2$	Sb:	Org.Comp.4-131
$C_3H_{11}O_{19}Tl_3U_2$	$Tl_3[(UO_2)_2OH(CO_3)_3(H_2O)_5]$	U:	SVol.C13-22/4
$C_3H_{12}INSn$	$(CH_3)_3SnI \cdot NH_3$	Sn:	Org.Comp.8-19
$C_3H_{12}MnN_6O_7S$	$Mn(OCHNHNH_2)_3SO_4$	Mn:	MVol.D5-167/9
$C_3H_{12}MnN_8O_9$	$Mn(OCHNHNH_2)_3(NO_3)_2$	Mn:	MVol.D5-167/9
$-$	$Mn(OC(NH_2)_2)_3(NO_3)_2 \cdot 3\ H_2O$	Mn:	MVol.D5-135
$C_3H_{12}N_2O_9U$	$(NH_4)_2[UO_2(CH_2(COO)_2)(OO)(H_2O)]$	U:	SVol.C13-238/40
$C_3H_{12}N_2O_{11}U$	$(NH_4)_2[UO_2(C_2O_4)(CO_3)(H_2O)_2]$	U:	SVol.C13-228/9
$C_3H_{12}N_6O_9SU$	$UO_2SO_4 \cdot 3\ (NH_2)_2CO$	U:	SVol.C10-163
$C_3H_{12}N_9RhS_3$	$Rh(NH_2NHC(S)NH)_3 \cdot 5\ H_2O$	Rh:	SVol.B3-40/1
$C_3H_{12}N_{10}O_{15}Th$	$Th(NO_3)_4 \cdot 3\ (NH_2)_2CO \cdot H_2O$	Th:	SVol.E-38, 41
$C_3H_{12}S_2$	$2\ H_2S \cdot C_3H_8 \cdot 17\ H_2O$	S:	SVol.4a/b-326
$C_3H_{14}N_{10}O_{16}Th$	$[Th(NO_3)_4((NH_2)_2CO)_3(H_2O)]$	Th:	SVol.E-41
$C_3H_{14}Na_2O_{18}U_2$	$Na_2[(UO_2)_2(CH_2(COO)_2)(OO)_2(H_2O)_6] \cdot H_2O$		
		U:	SVol.C13-238/41
$C_3H_{15}IN_2Sn$	$(CH_3)_3SnI \cdot 2\ NH_3$	Sn:	Org.Comp.8-19
$C_3H_{15}N_8RhS_3$	$[Rh(NH_3)_5(NCS)](SCN)_2$	Rh:	SVol.B1-201
$-$	$[Rh(NH_3)_5(SCN)](SCN)_2$	Rh:	SVol.B1-201
$C_3H_{15}N_{12}O_9RhS_3$ ...	$[Rh(NH_2NHC(S)NH_2)_3](NO_3)_3 \cdot 3\ H_2O$	Rh:	SVol.B3-40
$C_3H_{15}O_9RhS_3$	$Rh[S(O)_2CH_3]_3(H_2O)_3$	Rh:	SVol.B3-36
$C_3H_{16}N_2O_{15}U_2$	$(NH_4)_2[(UO_2)_2(CH_2(COO)_2)(OO)_2(H_2O)_3]$ ...	U:	SVol.C13-238/41
$C_3H_{16}N_4O_{11}U$	$(NH_4)_4UO_2(CO_3)_3$	U:	SVol.A6-33
		U:	SVol.C13-13, 15/7
$C_3H_{18}MnNO_8^+$	$[Mn(H_2O)_6](OOCCH(CH_3)NH_2)^+$	Mn:	MVol.D4-266/7
$C_3H_{18}MnN_{14}O_9$	$Mn(OC(NHNH_2)_2)_3(NO_3)_2$	Mn:	MVol.D5-209

$C_3H_{18}N_6O_9P_3RhS_3$ ..	$Rh[P(OH)_2O]_3[CS(NH_2)_2]_3$	Rh:	SVol.B1-211
		Rh:	SVol.B3-79
$C_3H_{20}N_5O_2Rh^{2+}$	$[Rh(NH_3)_5(OCOC_2H_5)]^{2+}$	Rh:	SVol.B2-133/4
$C_3H_{20}N_6Rh^{3+}$	$[Rh(NH_3)_5(NCC_2H_5)]^{3+}$	Rh:	SVol.B2-138/9
$C_3H_{23}N_3O_{19}U_2$	$(NH_4)_3[(UO_2)_2OH(CO_3)_3(H_2O)_5]$	U:	SVol.C13-22/4
$C_3H_{27}N_{12}O_9RhS_3$ ...	$[C(NH_2)_3]_3[Rh(SO_3)_3(NH_3)_3]$	Rh:	SVol.B1-149
$C_3H_{10+2x}O_{4+4x}S_{1+x}Sn$			
	$[(CH_3)_3Sn(H_2SO_4)_n]HSO_4$	Sn:	Org.Comp.11-146
$C_3HoN_3S_3$	$Ho(NCS)_3$	Sc:	MVol.D4-340/3
C_3HoO_3	$Ho(CO)_3$	Sc:	MVol.D6-166
$C_3I_2MnN_3S_3$	$Mn(SCN)_3I_2$	Mn:	MVol.C7-239
C_3KMnN_3	$KMn(CN)_3$	Mn:	MVol.D2-207, 215/6
C_3KMnN_4O	$K[Mn(CN)_3NO]$	Mn:	MVol.D2-276, 281
C_3KN_3OS	$SN_2C_2(CN)OK$	S:	S-N Comp.3-93/8, 168
$C_3K_3NO_{10}S_2U$	$K_3[UO_2(C_2O_4)(NCS)(SO_4)]$ · x H_2O (x = 0 to 1)		
		U:	SVol.C13-233/4
$C_3K_3NaO_{11}U$	$NaK_3[UO_2(CO_3)_3]$	U:	SVol.C13-14/7
–	$NaK_3[UO_2(CO_3)_3]$ · H_2O	U:	SVol.C13-12, 14, 17
$C_3K_4O_{11}U$	$K_4UO_2(CO_3)_3$	U:	SVol.A5-135
		U:	SVol.A6-65
		U:	SVol.C13-12, 16/7
$C_3LaN_3S_3$	$La(NCS)_3$	Sc:	MVol.D4-340/3
C_3La_2	La_2C_3 solid solutions		
	$(La,U)_2C_3$	U:	SVol.C12-93, 150
C_3LiN_3OS	$SN_2C_2(CN)OLi$	S:	S-N Comp.3-93/8, 167
$C_3Li_4O_{11}U$	$Li_4[UO_2(CO_3)_3]$ · x H_2O	U:	SVol.C13-12, 16/7
$C_3LuN_3S_3$	$Lu(NCS)_3$	Sc:	MVol.D4-340/3
$C_3Mg_2O_{11}U$	$Mg_2[UO_2(CO_3)_3]$ · 18 H_2O	U:	SVol.C13-12/3, 16/7
–	$Mg_2[UO_2(CO_3)_3]$ · 20 H_2O	U:	SVol.C13-13, 17
–	$Mg_2[UO_2(CO_3)_3]$ · x H_2O (x = 16 to 18)..	U:	SVol.C13-13, 16
C_3MnN_3Na	$NaMn(CN)_3$	Mn:	MVol.D2-215
$C_3MnN_3S_3^-$	$[Mn(NCS)_3]^-$	Mn:	MVol.D2-285
$C_3MnN_3Se_3^-$	$[Mn(NCSe)_3]^-$	Mn:	MVol.D2-293
$C_3MnN_4O^-$	$[Mn(CN)_3NO]^-$	Mn:	MVol.D2-281
$C_3Mn_2O_{11}U$	$Mn_2UO_2(CO_3)_3$	Mn:	MVol.C7-224
		U:	SVol.C13-14, 16/7
C_3Mn_7	Mn_7C_3	Mn:	MVol.C7-100/2, 110/1, 128/33
–	Mn_7C_3 solid solutions		
	$(Mn,Cr)_7C_3$	Mn:	MVol.C7-167/9
	$Mn_7(C_{1-x}B_x)_3$	Mn:	MVol.C7-240/4
–	Mn_7C_3 systems		
	$Mn_7C_3-Mn-Mn_2B$	Mn:	MVol.C7-240/2
C_3Mn_8	Mn_8C_3	Mn:	MVol.C7-102, 110/1, 124/5
C_3Mn_{11}	$Mn_{11}C_3$	Mn:	MVol.C7-101/2, 116
C_3N_3NaOS	$SN_2C_2(CN)ONa$	S:	S-N Comp.3-93/8, 167
$C_3N_3NdS_3$	$Nd(NCS)_3$	Sc:	MVol.D4-340/3
$C_3N_3O_2S_3U^-$	$[UO_2(SCN)_3]^-$	U:	SVol.D1-170, 179/80

$C_4CeH_7O_3^{2+}$	$[Ce((CH_3)_2C(OH)COO)]^{2+}$	Sc:	MVol.D5-261/6
$C_4CeH_7O_3^{3+}$	$[Ce((CH_3)_2C(OH)COO)]^{3+}$	Sc:	MVol.D5-267
$C_4CeH_7O_4^{2+}$	$[Ce(HOCH_2C(OH)(CH_3)COO)]^{2+}$	Sc:	MVol.D5-268/9
$C_4CeH_7O_5^{2+}$	$[Ce(HOC(CH_2OH)_2COO)]^{2+}$	Sc:	MVol.D5-268/9
$C_4CeH_8NO_3^{2+}$	$Ce[CH_3CHOHCH(NH_2)COO]^{2+}$	Sc:	MVol.D1-114
$C_4CeH_8N_2O_4^+$	$Ce(NH_2CH_2COO)_2^+$	Sc:	MVol.D1-103
$C_4CeH_8N_3O_{11}$	$Ce(NO_3)_3 \cdot C_4H_8O_2 \cdot 2 H_2O$	Sc:	MVol.D3-273, 275
$C_4CeH_8O^{4+}$	$[Ce(CH_3COC_2H_5)]^{4+}$	Sc:	MVol.D3-49
$C_4CeH_8O_3^{4+}$	$[Ce((CH_3)_2C(OH)COOH)]^{4+}$	Sc:	MVol.D5-267
$C_4CeH_{10}O_4P^{2+}$	$Ce[(C_2H_5O)_2PO_2]^{2+}$	Sc:	MVol.D4-177
$C_4CeH_{11}N_2O_6P_2$	$Ce[H((O_3PCH_2NH)_2C_2H_4)]$	Sc:	MVol.D4-151
$C_4CeH_{12}N_2O_6P_2^+$	$Ce[H_2((O_3PCH_2NH)_2C_2H_4)]^+$	Sc:	MVol.D4-151
$C_4CeH_{12}O_4$	$Ce(OCH_3)_4.$	Sc:	MVol.D3-17
$C_4CeH_{13}N_2O_6P_2^{2+}$	$Ce[H_3((O_3PCH_2NH)_2C_2H_4)]^{2+}$	Sc:	MVol.D4-151
$C_4CeH_{16}N_8O_{12}S_2$	$Ce(SO_4)_2 \cdot 4 CO(NH_2)_2 \cdot 4 H_2O$	Sc:	MVol.C8-122
$C_4CeH_{20}N_4O_8S_2$	$Ce(SO_4)_2 \cdot 4 CH_3NH_2$	Sc:	MVol.D1-19
C_4CeKO_8	$KCe(C_2O_4)_2 \cdot x H_2O$	Sc:	MVol.D5-141/5
C_4CeNaO_8	$NaCe(C_2O_4)_2 \cdot 6 H_2O$	Sc:	MVol.D5-141/5
$C_4CeO_4^+$	$[Ce(C_4O_4)]^+$	Sc:	MVol.D3-248/50
$C_4CeO_8^-$	$[Ce(C_2O_4)_2]^-$	Sc:	MVol.D5-120/4
$C_4CeO_{12}^{4-}$	$[Ce(CO_3)_4]^{4-}$	Sc:	MVol.D4-338/40
$C_4CeO_{12}^{5-}$	$[Ce(CO_3)_4]^{5-}$	Sc:	MVol.D4-338/40
$C_4Ce_2Cl_4H_4O_5$	$Ce_2OCl_2(CH_2ClCOO)_2.$	Sc:	MVol.D5-62
$C_4Ce_2H_4O_6^{4+}$	$[Ce_2(OOC(CHOH)_2COO)]^{4+}$	Sc:	MVol.D5-333
$C_4Ce_2H_6O_{10}$	$Ce_2(O_2)_3(CH_3COO)_2$	Sc:	MVol.D5-53
$C_4ClCoF_6NO_3P$	$[Co(CO)_2(NO)(ClP(CF_3)_2)].$	F:	PFHOrg.SVol.1-130
C_4ClCuH_4	$CHCCHCH_2CuCl.$	Cu:	Org.Comp.4-93
C_4ClCuH_4O	$C_2H_2 \cdot CuCl \cdot CH_2CO$	Cu:	Org.Comp.4-52
–	$ClCH_2CH(OH)CCCu$	Cu:	Org.Comp.3-19, 110
–	$OCH_2CH_2CClCCu \cdot x LiI$	Cu:	Org.Comp.1-58, 106, 108
$C_4ClCuH_4O_4$	$HOOCCHCHCOOHCuCl$	Cu:	Org.Comp.4-10, 11
$C_4ClCuH_4O_8$	$(HOOCCHCHCOOH)CuClO_4$	Cu:	Org.Comp.4-11
C_4ClCuH_6	$CH_2C(C_2H_4Cl)Cu \cdot x BF_3 \cdot y LiCN$	Cu:	Org.Comp.1-335, 343
–	$CH_2C(C_2H_4Cl)Cu \cdot x BF_3 \cdot y LiSC_6H_5$	Cu:	Org.Comp.1-335, 343
–	$CH_2C(C_2H_4Cl)Cu \cdot x LiCN$	Cu:	Org.Comp.1-140
–	$CH_2C(C_2H_4Cl)Cu \cdot x LiSC_6H_5$	Cu:	Org.Comp.1-186/7
–	$[(CH_3CCCH_3)CuCl]_4$	Cu:	Org.Comp.4-54, 222
$C_4ClCuH_6O_2$	$CH_3O_2CCClCH_3Cu \cdot x P(C_4H_9-n)_3$	Cu:	Org.Comp.1-447
–	$HOOCCHCHCH_3CuCl$	Cu:	Org.Comp.4-10
–	$HOOCCH_2CHCH_2CuCl$	Cu:	Org.Comp.4-10
$C_4ClCuH_6O_6$	$HOOCCH_2CHCH_2CuClO_4$	Cu:	Org.Comp.4-10
C_4ClCuH_8	$(CH_3)_2CCH_2CuCl.$	Cu:	Org.Comp.4-10
–	$C_2H_5CHCH_2CuCl.$	Cu:	Org.Comp.4-10
C_4ClCuH_8Mg	$(CH_2)_4CuMgCl.$	Cu:	Org.Comp.4-63
C_4ClCuH_8MgO	$ClMgO(CH_2)_4Cu \cdot x LiMgBrCl_2$	Cu:	Org.Comp.1-300
–	$ClMgO(CH_2)_4Cu \cdot x MgBr_2$	Cu:	Org.Comp.1-213, 284
C_4ClCuH_8O	$HOCH(CH_3)CHCH_2CuCl$	Cu:	Org.Comp.4-10
–	$HOCH_2C(CH_3)CH_2CuCl.$	Cu:	Org.Comp.4-10
–	$HOCH_2CHCHCH_3CuCl$	Cu:	Org.Comp.4-10
$C_4ClCuH_8O_4$	$(CH_3)_2CCH_2CuClO_4$	Cu:	Org.Comp.4-10, 17

C₄ClCuH₈O₄	C₂H₅CHCH₂CuClO₄	Cu:	Org.Comp.4-10
C₄ClCuH₈O₄S	[(CH₃CHCHClCH₃)Cu]SO₄	Cu:	Org.Comp.4-16
C₄ClCuH₁₀Mg	(C₂H₅)₂CuMgCl	Cu:	Org.Comp.2-213, 217/25
–	(i-C₃H₇)(CH₃)CuMgCl	Cu:	Org.Comp.2-228, 233
C₄ClCuH₁₀NO₄S	(CH₃CClCHCH₃CuNH₃)(SO₄)	Cu:	Org.Comp.4-29
C₄ClCuH₁₀O₆	HOCH₂C(CH₃)CH₂Cu(H₂O)ClO₄	Cu:	Org.Comp.4-10
C₄ClDyH₄O₆	Dy(OOC(CHOH)₂COO)Cl · 2 H₂O	Sc:	MVol.D5-337
C₄ClDyH₅NO₄	Dy[NH(CH₂COO)₂]Cl	Sc:	MVol.D1-123/4
–	Dy[NH(CH₂COO)₂]Cl · 3 H₂O	Sc:	MVol.D1-123/5
C₄ClErH₄O₆	Er(OOC(CHOH)₂COO)Cl · 3 H₂O	Sc:	MVol.D5-337
C₄ClErH₅NO₄	Er[NH(CH₂COO)₂]Cl	Sc:	MVol.D1-123/4
–	Er[NH(CH₂COO)₂]Cl · 3 H₂O	Sc:	MVol.D1-123/5
C₄ClEuH₅NO₄	Eu[NH(CH₂COO)₂]Cl	Sc:	MVol.D1-123/4
–	Eu[NH(CH₂COO)₂]Cl · 2 H₂O	Sc:	MVol.D1-123/4
–	Eu[NH(CH₂COO)₂]Cl · 3 H₂O	Sc:	MVol.D1-124/5
C₄ClEuH₉NO₂	Eu(HN(CH₂CH₂O)₂)Cl	Sc:	MVol.D2-2
C₄ClFH₅NO₃S	NO₂CFClC(O)SC₂H₅	F:	PFHOrg.8-37/8
C₄ClFH₅NO₄	NO₂CFClC(O)OC₂H₅	F:	PFHOrg.8-37/8
C₄ClFN₂	NCCClCFCN	F:	PFHOrg.9-57, 71
C₄ClF₃H₃NO₄	CF₃CCl(NO₂)C(O)OCH₃	F:	PFHOrg.8-38
C₄ClF₃H₅NO	CF₃C(NH)OC₂H₄Cl	F:	PFHOrg.9-81
C₄ClF₃H₅NO₂	CF₃NHC(O)OC₂H₄Cl	F:	PFHOrg.9-116
C₄ClF₄H₆NO₂S	ClCF₂CF₂NS(OCH₃)₂	F:	PFHOrg.9-34
C₄ClF₄NO	NCCF₂CF₂C(O)Cl	F:	PFHOrg.9-54/5
C₄ClF₄NO₃	NO₂CF₂CFCFC(O)Cl . . . . :	F:	PFHOrg.8-11, 25, 38
C₄ClF₄N₂S	ClF(CF₃)C₃N₂S	F:	PFHOrg.SVol.2-42, 48
C₄ClF₅H₆N₂S	ClCF₂CF₂NSFN(CH₃)₂	F:	PFHOrg.9-34/5
C₄ClF₆H₂N	(CF₃)₂NCHCHCl	F:	PFHOrg.8-211
C₄ClF₆H₄N	(CF₃)₂NC₂H₄Cl	F:	PFHOrg.8-209
C₄ClF₆H₄OP	(CF₃)₂POCH₂CH₂Cl	F:	PFHOrg.SVol.1-122
C₄ClF₆H₅N₂O	NF₂CFClCF(NF₂)OC₂H₅	F:	PFHOrg.8-207
C₄ClF₆N	(CF₃)₂CClCN	F:	PFHOrg.9-52/3, 68, 79
C₄ClF₆NO	(CF₃)₂CClNCO	F:	PFHOrg.9-104, 110
–	ClCF₂CF(CF₃)NCO	F:	PFHOrg.9-104, 110, 115
C₄ClF₆NO₃	NO₂(CF₃)₃C(O)Cl	F:	PFHOrg.8-11, 25
C₄ClF₆NO₃S	(CF₃)₂C(CN)OSO₂Cl	F:	PFHOrg.9-55, 71
C₄ClF₆NS	F₆C₄NSCl	F:	PFHOrg.SVol.3-13, 37
C₄ClF₇H₂N₂	C₃F₇C(NCl)NH₂	F:	PFHOrg.8-152/3, 176
C₄ClF₇H₃N	(CF₃)₂NCHFCH₂Cl	F:	PFHOrg.8-208, 209
–	(CF₃)₂NCH₂CHFCl	F:	PFHOrg.8-208, 209
C₄ClF₇H₃P	HCFClCH₂P(CF₃)₂	F:	PFHOrg.SVol.1-85
–	H₂CClCFHP(CF₃)₂	F:	PFHOrg.SVol.1-85
C₄ClF₇H₆S	CF₃SF₄CH₂CHClCH₃	F:	PFHOrg.SVol.3-207
C₄ClF₇N₂	CF₂NCF₂CFClNCF₂	F:	PFHOrg.9-130, 140
–	CF₃NCFCFNCF₂Cl	F:	PFHOrg.9-130/1, 140/1
C₄ClF₇S	(CF₃)₂CCFSCl	F:	PFHOrg.SVol.2-119, 122
C₄ClF₈IO₃S	ICFClCF₂OCF₂CF₂SO₂F	F:	PFHOrg.SVol.3-155/6, 164
C₄ClF₈N	(CF₃)₂NCFCFCl	F:	PFHOrg.9-161/2, 170
–	CF₃NCFCFClCF₃	F:	PFHOrg.9-130, 140

C₄ClF₈N	C₃F₇CClNF	F:	PFHOrg.8-170/1
–	C₃F₇CFNCl	F:	PFHOrg.8-170/1, 197
C₄ClF₈NO	CF₃NCFCF₂OCF₂Cl	F:	PFHOrg.9-130/1, 141
C₄ClF₈NOS	(CF₂)₄S(O)NCl	F:	PFHOrg.SVol.3-122/3, 129
C₄ClF₈NO₂S₂	(CF₂)₄SNS(O)₂Cl	F:	PFHOrg.SVol.3-9, 31
C₄ClF₈NO₃S⁻	CF₃CF(SO₂F)N(O)CF₂CFCl⁻ (radical anion)		
		F:	PFHOrg.SVol.3-154
C₄ClF₈NS	(CF₂)₄SNCl	F:	PFHOrg.SVol.3-9, 30, 38
C₄ClF₈N₃O₄	NO₂CF₂CF₂N(NO₂)CF₂CF₂Cl	F:	PFHOrg.8-56/7, 84
–	NO₂CF₂CF₂N(ONO)CF₂CF₂Cl	F:	PFHOrg.8-57
C₄ClF₉HN	(CF₃)₂NCF₂CHFCl	F:	PFHOrg.8-208, 210
–	(CF₃)₂NCHFCF₂Cl	F:	PFHOrg.8-208, 210
C₄ClF₉H₃P	CH₃PCl(CF₃)₃	F:	PFHOrg.SVol.1-123/5
C₄ClF₉OS	C₄F₉S(O)Cl	F:	PFHOrg.SVol.3-11, 34, 42
C₄ClF₉O₂S	C₄F₉SO₂Cl	F:	PFHOrg.SVol.3-158/9, 168, 173
C₄ClF₉O₃S	CF₂ClCF₂OCF₂CF₂SO₂F	F:	PFHOrg.SVol.3-155/6, 163
–	CF₃CFClCF₂OSO₂CF₃	F:	PFHOrg.SVol.3-57, 58, 73
–	ClOSO₂C₄F₉	F:	PFHOrg.SVol.3-55, 70, 103, 109
C₄ClF₉S	(CF₃)₃CSCl	F:	PFHOrg.SVol.2-119, 122
–	CF₃SCF(CF₃)CF₂Cl	F:	PFHOrg.SVol.2-212, 227
–	CF₃SCF₂CFClCF₃	F:	PFHOrg.SVol.2-212, 226
–	C₄F₉SCl	F:	PFHOrg.SVol.2-119, 122, 127
C₄ClF₉S₂	CF₃SSCF(CF₃)CF₂Cl	F:	PFHOrg.SVol.2-178, 190
–	CF₃SSCF₂CFClCF₃	F:	PFHOrg.SVol.2-178, 190
C₄ClF₉S₄	CF₂ClSSCF(SCF₃)₂	F:	PFHOrg.SVol.2-176/7, 188
–	(CF₃S)₃CSCl	F:	PFHOrg.SVol.2-118, 122
C₄ClF₁₀N	((CF₃)₂CF)(CF₃)NCl	F:	PFHOrg.8-162, 186
C₄ClF₁₀NOS	(CF₃)₂CFNS(O)ClCF₃	F:	PFHOrg.SVol.3-124, 129/30
C₄ClF₁₀NS	CF₃S(Cl)NCF(CF₃)₂	F:	PFHOrg.SVol.3-7, 28
–	ClCF₂CF₂CF₂CF₂NSF₂	F:	PFHOrg.9-3/4, 16
C₄ClF₁₂NS	(CF₃)₂CFNSF₂Cl(CF₃)	F:	PFHOrg.SVol.3-193, 204
–	C₃F₇CClNSF₅	F:	PFHOrg.9-8, 27
C₄ClF₁₃NP	CF₃CF₂PF₂ClN(CF₃)₂	F:	PFHOrg.SVol.1-102/3
C₄ClF₁₃S	CF₃SF₄CF(CF₃)CF₂Cl	F:	PFHOrg.SVol.3-190, 200
–	CF₃SF₄CF₂CFClCF₃	F:	PFHOrg.SVol.3-190, 199
–	n-C₄F₉SF₄Cl	F:	PFHOrg.SVol.3-187/8, 196
C₄ClGaH₆	Ga(CHCH₂)₂Cl	Ga:	Org.Comp.1-142
C₄ClGaH₁₀	Ga(C₂H₅)₂Cl	Ga:	Org.Comp.1-128, 133/4
C₄ClGaH₁₀N₂O	Ga(CH₃)(Cl)N(CH₃)CONHCH₃	Ga:	Org.Comp.1-292/3
C₄ClGaH₁₀O₂S	Ga(C₂H₅)(Cl)OS(O)C₂H₅	Ga:	Org.Comp.1-225/6
C₄ClGaH₁₀S	GaCH₃(Cl)SC₃H₇	Ga:	Org.Comp.1-239
–	GaCH₃(Cl)SC₃H₇-i	Ga:	Org.Comp.1-239

$C_4ClGaH_{11}O_{0.5}$	$Ga(CH_3)_2Cl \cdot 0.5\ O(C_2H_5)_2$	Ga:	Org.Comp.1-129
$C_4ClGaH_{13}NSi$	$Ga(CH_3)(Cl)NHSi(CH_3)_3$	Ga:	Org.Comp.1-291/2
$C_4ClGaH_{14}N_2$	$Ga(CH_3)_2Cl \cdot NH_2CH_2CH_2NH_2$	Ga:	Org.Comp.1-130, 136
$C_4ClGdH_4O_6$	$Gd(OOC(CHOH)_2COO)Cl \cdot 4\ H_2O$	Sc:	MVol.D5-337
$C_4ClGdH_5NO_4$	$Gd[NH(CH_2COO)_2]Cl$	Sc:	MVol.D1-123/4
$-$	$Gd[NH(CH_2COO)_2]Cl \cdot 2\ H_2O$	Sc:	MVol.D1-123/4
$-$	$Gd[NH(CH_2COO)_2]Cl \cdot 3\ H_2O$	Sc:	MVol.D1-123/5
$C_4ClGdH_9NO_2$	$Gd(HN(CH_2CH_2O)_2)Cl$	Sc:	MVol.D2-2
C_4ClHN_4OS	$SN_2C_4Cl(O)N_2H$	S:	S-N Comp.3-227/8
$-$	$SN_2C_4Cl(OH)N_2$	S:	S-N Comp.3-227/8, 229, 231
$C_4ClH_2N_4RhS_6$	$Rh(NNC(S)SC(SH))_2Cl$	Rh:	SVol.B3-47
$C_4ClH_3N_2OS$	$SN_2C_2HCH_2COCl$	S:	S-N Comp.3-93/8, 124
$C_4ClH_4HoO_6$	$Ho(OOC(CHOH)_2COO)Cl \cdot 3\ H_2O$	Sc:	MVol.D5-337
$C_4ClH_4LaO_6$	$La(OOC(CHOH)_2COO)Cl \cdot 3\ H_2O$	Sc:	MVol.D5-337
$C_4ClH_4N_5OS$	$SN_2C_4(NH_2)(O)N_2H \cdot HCl$	S:	S-N Comp.3-237
$C_4ClH_4N_6RhS_4$	$Rh[(S)(NH_2)C_2N_2S]_2Cl$	Rh:	SVol.B3-48
$C_4ClH_4NdO_4S$	$Nd(S(CH_2COO)_2)Cl \cdot 4\ H_2O$	Sc:	MVol.D4-63/4
$C_4ClH_4NdO_6$	$Nd(OOC(CHOH)_2COO)Cl \cdot 3\ H_2O$	Sc:	MVol.D5-337
$C_4ClH_4O_5U^+$	$[UO_2(OOC(CH_2)_2COCl)]^+$	U:	SVol.D1-209
$C_4ClH_4O_6Pr$	$Pr(OOC(CHOH)_2COO)Cl \cdot 3\ H_2O$	Sc:	MVol.D5-337
$C_4ClH_4O_6Sm$	$Sm(OOC(CHOH)_2COO)Cl \cdot 3\ H_2O$	Sc:	MVol.D5-337
$C_4ClH_4O_6Tb$	$Tb(OOC(CHOH)_2COO)Cl \cdot 3\ H_2O$	Sc:	MVol.D5-337
$C_4ClH_4O_6Y$	$Y(OOC(CHOH)_2COO)Cl \cdot 2\ H_2O$	Sc:	MVol.D5-337
$C_4ClH_4O_{12}Rh_2^-$	$[Rh_2(HCO_3)_4Cl]^-$	Rh:	SVol.B1-184, 186
$C_4ClH_5HoNO_4$	$Ho[NH(CH_2COO)_2]Cl$	Sc:	MVol.D1-123/4
$-$	$Ho[NH(CH_2COO)_2]Cl \cdot 3\ H_2O$	Sc:	MVol.D1-123/5
$C_4ClH_5LaNO_4$	$La[NH(CH_2COO)_2]Cl$	Sc:	MVol.D1-123/4
$-$	$La[NH(CH_2COO)_2]Cl \cdot 2\ H_2O$	Sc:	MVol.D1-123/4
$C_4ClH_5LuNO_4$	$Lu[NH(CH_2COO)_2]Cl$	Sc:	MVol.D1-123/4
$-$	$Lu[NH(CH_2COO)_2]Cl \cdot 3\ H_2O$	Sc:	MVol.D1-123/5
$C_4ClH_5NNdO_4$	$Nd[NH(CH_2COO)_2]Cl$	Sc:	MVol.D1-123/4
$-$	$Nd[NH(CH_2COO)_2]Cl \cdot 2\ H_2O$	Sc:	MVol.D1-123/4
$-$	$Nd[NH(CH_2COO)_2]Cl \cdot 3\ H_2O$	Sc:	MVol.D1-123/5
$C_4ClH_5NO_4Pr$	$Pr[NH(CH_2COO)_2]Cl$	Sc:	MVol.D1-123/4
$-$	$Pr[NH(CH_2COO)_2]Cl \cdot 2\ H_2O$	Sc:	MVol.D1-123/4
$-$	$Pr[NH(CH_2COO)_2]Cl \cdot 3\ H_2O$	Sc:	MVol.D1-124/5
$C_4ClH_5NO_4Sc$	$Sc[NH(CH_2COO)_2]Cl$	Sc:	MVol.D1-123/4
$-$	$Sc[NH(CH_2COO)_2]Cl \cdot 2\ H_2O$	Sc:	MVol.D1-123/4
$C_4ClH_5NO_4Sm$	$Sm[NH(CH_2COO)_2]Cl$	Sc:	MVol.D1-123/4
$-$	$Sm[NH(CH_2COO)_2]Cl \cdot 2\ H_2O$	Sc:	MVol.D1-123/4
$-$	$Sm[NH(CH_2COO)_2]Cl \cdot 3\ H_2O$	Sc:	MVol.D1-124/5
$C_4ClH_5NO_4Tb$	$Tb[NH(CH_2COO)_2]Cl$	Sc:	MVol.D1-123/4
$-$	$Tb[NH(CH_2COO)_2]Cl \cdot 3\ H_2O$	Sc:	MVol.D1-123/5
$C_4ClH_5NO_4Tm$	$Tm[NH(CH_2COO)_2]Cl$	Sc:	MVol.D1-123/5
$-$	$Tm[NH(CH_2COO)_2]Cl \cdot 3\ H_2O$	Sc:	MVol.D1-123/5
$C_4ClH_5NO_4Y$	$Y[NH(CH_2COO)_2]Cl$	Sc:	MVol.D1-123/4
$-$	$Y[NH(CH_2COO)_2]Cl \cdot 3\ H_2O$	Sc:	MVol.D1-123/4
$C_4ClH_5NO_4Yb$	$Yb[NH(CH_2COO)_2]Cl$	Sc:	MVol.D1-123/4
$-$	$Yb[NH(CH_2COO)_2]Cl \cdot 3\ H_2O$	Sc:	MVol.D1-123/5

$C_4ClH_5N_2OS$	$SN_2C_2ClOC_2H_5$	S:	S–N Comp.3-93/8, 157/8
$C_4ClH_5N_2S$	$SN_2C_2HCHClCH_3$	S:	S–N Comp.3-93/8, 123
–	$SN_2C_2CH_2ClCH_3$	S:	S–N Comp.3-93/8, 125
$C_4ClH_6NO_2S$	$[C(CH_3)_2OS(O)NCCl]$	S:	S–N Comp.3-286/7
C_4ClH_6Sb	$(CH_2CH)_2SbCl$	Sb:	Org.Comp.2-10
$C_4ClH_7N_2OS$	$NS(O)NHC(CH_3)_2CCl$	S:	S–N Comp.3-260
$C_4ClH_8NO_2Rh$	$Rh(O_2C(CH_2)_3NH_2)Cl$	Rh:	SVol.B2-64
–	$Rh(O_2C(CH_2)_3NH_2)Cl \cdot 2 H_2O$	Rh:	SVol.B2-63
$C_4ClH_8N_3S$	$[SN_2(CH_3)C_2H(NHCH_3)]Cl$	S:	S–N Comp.3-243/4
$C_4ClH_8N_8ORhS_4$	$Rh(NH_2C_2HN_3S_2)_2Cl(H_2O)$	Rh:	SVol.B3-54
C_4ClH_8Sb	$(CH_2)_4SbCl$	Sb:	Org.Comp.2-68
$C_4ClH_9LaNO_2$	$La(HN(CH_2CH_2O)_2)Cl$	Sc:	MVol.D2-2
$C_4ClH_9MnN_2O_4$	$[MnCl(OOCCH_2NHC(O)CH_2NH_2)(H_2O)]$	Mn:	MVol.D4-340/1
$C_4ClH_9NNdO_2$	$Nd(HN(CH_2CH_2O)_2)Cl$	Sc:	MVol.D2-2
$C_4ClH_9NO_2Pr$	$Pr(HN(CH_2CH_2O)_2)Cl$	Sc:	MVol.D2-2
$C_4ClH_9NO_2Sm$	$Sm(HN(CH_2CH_2O)_2)Cl$	Sc:	MVol.D2-2
$C_4ClH_9N_2O_2S$	$[CH_2OS(O)NC(N(CH_3)_2 \cdot HCl)]$	S:	S–N Comp.3-286/7
C_4ClH_9OSSn	$(CH_3)_2SnCl(SC(O)CH_3)$	Sn:	Org.Comp.10-204
$C_4ClH_{10}ISn$	$(C_2H_5)_2SnICl$	Sn:	Org.Comp.8-142
$C_4ClH_{10}MnNO_2$	$Mn(OC_2H_4NHC_2H_4OH)Cl$	Mn:	MVol.D4-224
$C_4ClH_{10}O_2Th^+$	$[ThCl(OC_2H_5)_2]^+$	Th:	SVol.E-28
$C_4ClH_{10}Sb$	$(C_2H_5)_2SbCl$	Sb:	Org.Comp.2-7
$C_4ClH_{11}I_2SiSn$	$(CH_3)_3SiCH_2SnI_2Cl$	Sn:	Org.Comp.8-144
$C_4ClH_{11}NO_2SU^+$	$[UO_2(SC_2H_4N(CH_3)_2 \cdot HCl)]^+$	U:	SVol.D1-251
$C_4ClH_{12}NSSn$	$(CH_3)_2SnCl(SCH_2CH_2NH_2)$	Sn:	Org.Comp.10-202
$C_4ClH_{12}Sb$	$(CH_3)_4SbCl$	Sb:	Org.Comp.3-43
$C_4ClH_{14}N_{12}O_4Rh$	$[Rh((NH_2CNH)_2NH)_2(NO_2)_2]Cl$	Rh:	SVol.B2-221
$C_4ClH_{14}N_{16}Rh$	$[Rh((NH_2CNH)_2NH)_2(N_3)_2]Cl$	Rh:	SVol.B2-221
$C_4ClH_{16}HgN_4Rh^{2+}$	$[Rh(NH_2C_2H_4NH_2)_2HgCl]^{2+}$	Rh:	SVol.B2-162
		Rh:	SVol.B3-238
$C_4ClH_{16}IN_4Rh^+$	$[Rh(NH_2C_2H_4NH_2)_2ClI]^+$	Rh:	SVol.B2-180, 182/3
$C_4ClH_{16}I_2N_4O_4Rh$	$[Rh(NH_2C_2H_4NH_2)_2I_2]ClO_4$	Rh:	SVol.B2-178
$C_4ClH_{16}N_5O_2Rh^+$	$[Rh(NH_2C_2H_4NH_2)_2Cl(NO_2)]^+$	Rh:	SVol.B2-172
$C_4ClH_{16}N_5O_4Rh$	$[Rh(NH_2C_2H_4NH_2)_2(NO_2)(O_2)]Cl$	Rh:	SVol.B2-172
$C_4ClH_{16}N_6O_5Rh$	$[Rh(NH_2C_2H_4NH_2)_2(NO_2)Cl]NO_3$	Rh:	SVol.B2-176
$C_4ClH_{16}N_8O_6Rh$	$[Rh(NH_2C_2H_4NH_2)_2(NO_2)(N_3)]ClO_4$	Rh:	SVol.B2-176
$C_4ClH_{16}N_{10}ORh^{2+}$	$[Rh((NH_2CNH)_2NH)_2Cl(H_2O)]^{2+}$	Rh:	SVol.B2-221
$C_4ClH_{16}N_{10}Rh$	$[Rh(NH_2C_2H_4NH_2)_2(N_3)_2]Cl$	Rh:	SVol.B2-176
$C_4ClH_{17}N_4ORh^+$	$[Rh(NH_2C_2H_4NH_2)_2(OH)Cl]^+$	Rh:	SVol.B2-173
$C_4ClH_{17}N_4Rh^+$	$[Rh(NH_2C_2H_4NH_2)_2HCl]^+$	Rh:	SVol.B2-167
$C_4ClH_{17}N_6Os^{2+}$	$[Os(C_4H_4N_2H)(NH_3)_4Cl]^{2+}$	Os:	SVol.1-247
$C_4ClH_{18}N_4ORh^{2+}$	$[Rh(NH_2C_2H_4NH_2)_2(OH_2)Cl]^{2+}$	Rh:	SVol.B2-172
$C_4ClH_{19}N_7O_6Rh$	$[Rh(NH_2C_2H_4NH_2)_2(NH_3)Cl](NO_3)_2$	Rh:	SVol.B2-174
$C_4ClH_{20}N_6Rh^{2+}$	$[Rh(NH_2C_2H_4NH_2)_2(N_2H_4)Cl]^{2+}$	Rh:	SVol.B2-175
$C_4Cl_2CsH_6RhS_4$	$Cs[Rh(S_2CCH_3)_2Cl_2]$	Rh:	SVol.B3-15
$C_4Cl_2Cs_2H_4O_8Rh_2$	$Cs_2[Rh_2(O_2CH)_4Cl_2]$	Rh:	SVol.B2-12
$C_4Cl_2Cs_2H_4O_{12}Rh_2$	$Cs_2[Rh_2(HCO_3)_4Cl_2]$	Rh:	SVol.B1-186
$C_4Cl_2Cs_4Na_2O_{12}Rh_2$	$Cs_4Na_2[Rh_2(CO_3)_4Cl_2] \cdot 8 H_2O$	Rh:	SVol.B1-185
C_4Cl_2CuHS	$SC_4HCl_2Cu \cdot x LiCl$	Cu:	Org.Comp.1-57, 112, 120
$C_4Cl_2Cu_2H_4$	$(HCCCHCH_2)Cu_2Cl_2$	Cu:	Org.Comp.4-175

$C_4Cl_2Cu_2H_5N$	$(NCCH_2CHCH_2Cu_2Cl_2)_n$	Cu:	Org.Comp.4-257/8
$C_4Cl_2Cu_2H_6$	$(CH_2CHCHCH_2)Cu_2Cl_2$	Cu:	Org.Comp.4-175
$C_4Cl_2Cu_2H_6O_8$	$(CH_2CHCHCH_2)Cu_2(ClO_4)_2 \cdot 4 H_2O$	Cu:	Org.Comp.4-176
$C_4Cl_2Cu_2S$........	$3,4\text{-}Cl_2\text{-}2,5\text{-}Cu_2C_4S \cdot x\ LiCl$	Cu:	Org.Comp.4-128, 130
—	$3,4\text{-}Cl_2\text{-}2,5\text{-}Cu_2C_4S \cdot x\ LiI$	Cu:	Org.Comp.4-128, 130
$C_4Cl_2DyH_4O_4{}^+$	$[Dy(CH_2ClCOO)_2]^+$	Sc:	MVol.D5-53/6
$C_4Cl_2ErH_4O_4{}^+$	$[Er(CH_2ClCOO)_2]^+$	Sc:	MVol.D5-53/6
$C_4Cl_2EuH_4O_4{}^+$	$[Eu(CH_2ClCOO)_2]^+$	Sc:	MVol.D5-53/6
$C_4Cl_2FH_{18}N_4O_9Rh$ ..	$[Rh(NH_2C_2H_4NH_2)_2(OH_2)F](ClO_4)_2$	Rh:	SVol.B2-172
$C_4Cl_2F_2H_3NO_4$	$CF_2ClCCl(NO_2)C(O)OCH_3$	F:	PFHOrg.8-38
$C_4Cl_2F_2S$..........	$Cl_2F_2C_4S$	F:	PFHOrg.SVol.2-17
$C_4Cl_2F_3NS$.........	$Cl_2(CF_3)C_3NS$	F:	PFHOrg.SVol.2-42
—	$Cl_2C_2NSC(CF_3)$	F:	PFHOrg.SVol.2-42, 48
$C_4Cl_2F_4HNO$	$(CF_2Cl)_2C(CN)OH$	F:	PFHOrg.9-53/4
$C_4Cl_2F_4HNO_2$	$(CF_2Cl)_2C(OH)NCO$	F:	PFHOrg.9-106, 112, 115, 117
$C_4Cl_2F_4H_6N_2S$......	$ClCF_2CF_2NSClN(CH_3)_2$	F:	PFHOrg.9-35
$C_4Cl_2F_4KNO$	$(CF_2Cl)_2C(CN)OK$	F:	PFHOrg.9-53/4
$C_4Cl_2F_4NNaO$	$(CF_2Cl)_2C(ONa)CN$	F:	PFHOrg.9-88
$C_4Cl_2F_4O_2S$........	$ClOCCF_2SCF_2COCl$	F:	PFHOrg.SVol.2-214/5, 228
$C_4Cl_2F_4S$..........	$Cl_2F_4C_4S$	F:	PFHOrg.SVol.2-17, 22
$C_4Cl_2F_5N$..........	$ClCF_2CFClCF_2CN$	F:	PFHOrg.9-52/3, 69
$C_4Cl_2F_5NO$........	$ClCF_2CCl(CF_3)NCO$	F:	PFHOrg.9-104, 110, 117
$C_4Cl_2F_6H_3N$.......	$(CF_3)_2NCH_2CHCl_2$	F:	PFHOrg.8-210
$C_4Cl_2F_6H_{12}NP_2Rh$...	$[N(CH_3)_4][Rh(PF_3)_2Cl_2]$	Rh:	SVol.B3-73
$C_4Cl_2F_6N_2$.........	$ClCF_2CFClCF_2CFN_2$	F:	PFHOrg.8-57/8, 84, 104
$C_4Cl_2F_6S$..........	$F_6Cl_2C_4S$	F:	PFHOrg.SVol.2-16, 22
$C_4Cl_2F_7HO_2S$	$ClCF_2CFClCF_2CFHSO_2F$	F:	PFHOrg.SVol.3-135
$C_4Cl_2F_7N$..........	$CF_3NCFCFClCF_2Cl$	F:	PFHOrg.9-130, 140
—	$ClCF_2CF_2NCFCF_2Cl$	F:	PFHOrg.9-130, 139/40
$C_4Cl_2F_8HP$.........	$HCCl_2CF_2P(CF_3)_2$	F:	PFHOrg.SVol.1-85
$C_4Cl_2F_8N_2$.........	$ClC_2F_4NNC_2F_4Cl$	F:	PFHOrg.8-74/5, 98, 203
$C_4Cl_2F_8O_3S$........	$Cl_2CFCF_2OCF_2CF_2SO_2F$	F:	PFHOrg.SVol.3-155/6, 164
$C_4Cl_2F_8S_4$.........	$CFCl_2SSCF(SCF_3)_2$	F:	PFHOrg.SVol.2-176/7, 188
$C_4Cl_2F_9N$..........	$CF_3CFClCCl(NF_2)CF_3$	F:	PFHOrg.8-156/7, 179
$C_4Cl_2F_9NOS$	$(CF_3)_2CClNS(O)ClCF_3$	F:	PFHOrg.SVol.3-124, 130
$C_4Cl_2F_9N_2O_2P$......	$CF_3NCFCF_2N(CF_3)OP(O)Cl_2$	F:	PFHOrg.9-130/1, 141, 150, 154
$C_4Cl_2F_{12}NP$........	$ClCF_2CF_2PF_2ClN(CF_3)_2$	F:	PFHOrg.SVol.1-102/3
$C_4Cl_2GaH_3S$	$Ga(C_4H_3S)Cl_2$	Ga:	Org.Comp.1-149
$C_4Cl_2GaH_9$	$Ga(C_4H_9)Cl_2$	Ga:	Org.Comp.1-148, 152
—	$Ga(C_4H_9\text{-}i)Cl_2$	Ga:	Org.Comp.1-148
$C_4Cl_2GaH_{10}O_2P$	$Ga(C_2H_5)_2OP(O)Cl_2$	Ga:	Org.Comp.1-207
$C_4Cl_2GaH_{11}Si$......	$Ga(CH_2Si(CH_3)_3)Cl_2$	Ga:	Org.Comp.1-148
$C_4Cl_2GaH_{12}N$	$Ga(CH_3)Cl_2 \cdot N(CH_3)_3$	Ga:	Org.Comp.1-149
$C_4Cl_2GaH_{15}N_2$	$Ga(C_4H_9)Cl_2 \cdot 2\ NH_3$	Ga:	Org.Comp.1-150
$C_4Cl_2GaH_{18}N_3$	$Ga(C_4H_9)Cl_2 \cdot 3\ NH_3$	Ga:	Org.Comp.1-150

$C_4Cl_2GdH_4O_4^+$	$[Gd(CH_2ClCOO)_2]^+$	Sc:	MVol.D5-53/6
$C_4Cl_2H_4HoO_4^+$	$[Ho(CH_2ClCOO)_2]^+$	Sc:	MVol.D5-53/6
$C_4Cl_2H_4K_2O_{12}Rh_2$	$K_2[Rh_2(HCO_3)_4Cl_2] \cdot H_2O$	Rh:	SVol.B1-186
$C_4Cl_2H_4LaO_4^+$	$[La(CH_2ClCOO)_2]^+$	Sc:	MVol.D5-53/6
$C_4Cl_2H_4LuO_4^+$	$[Lu(CH_2ClCOO)_2]^+$	Sc:	MVol.D5-53/6
$C_4Cl_2H_4MnN_2$	$Mn(1,2\text{-}C_4H_4N_2)Cl_2$	Mn:	MVol.D4-3
−	$Mn(1,3\text{-}C_4H_4N_2)Cl_2$	Mn:	MVol.D4-5
−	$Mn(1,4\text{-}C_4H_4N_2)Cl_2$	Mn:	MVol.D4-51/2
$C_4Cl_2H_4MnN_2O_2$	$Mn[2,4\text{-}(O)_2\text{-}1,3\text{-}C_4H_2N_2H_2]Cl_2 \cdot H_2O$	Mn:	MVol.D4-9
$C_4Cl_2H_4MnO_4$	$Mn(CH_2ClCOO)_2 \cdot 1.5 H_2O$	Mn:	MVol.D2-62
$C_4Cl_2H_4N_2S$	$SN_2C_2(CH_2Cl)_2$	S:	S-N Comp.3-93/8, 125
−	$SN_2C_2CHCl_2CH_3$	S:	S-N Comp.3-93/8, 125
$C_4Cl_2H_4Na_2O_8Rh_2$	$Na_2[Rh_2(O_2CH)_4Cl_2]$	Rh:	SVol.B2-12
−	$Na_2[Rh_2(O_2CH)_4Cl_2] \cdot 4 H_2O$	Rh:	SVol.B2-12
−	$Na_2[Rh_2(O_2CH)_4Cl_2] \cdot 6 H_2O$	Rh:	SVol.B2-12
$C_4Cl_2H_4NdO_4^+$	$[Nd(CH_2ClCOO)_2]^+$	Sc:	MVol.D5-53/6
$C_4Cl_2H_4O_4Pr^+$	$[Pr(CH_2ClCOO)_2]^+$	Sc:	MVol.D5-53/6
$C_4Cl_2H_4O_4Sm^+$	$[Sm(CH_2ClCOO)_2]^+$	Sc:	MVol.D5-53/6
$C_4Cl_2H_4O_4Tb^+$	$[Tb(CH_2ClCOO)_2]^+$	Sc:	MVol.D5-53/6
$C_4Cl_2H_4O_4Tm^+$	$[Tm(CH_2ClCOO)_2]^+$	Sc:	MVol.D5-53/6
$C_4Cl_2H_4O_4Y^+$	$[Y(CH_2ClCOO)_2]^+$	Sc:	MVol.D5-53/6
$C_4Cl_2H_4O_4Yb^+$	$[Yb(CH_2ClCOO)_2]^+$	Sc:	MVol.D5-53/6
$C_4Cl_2H_4O_5U$	$UO(ClCH_2COO)_2$	U:	SVol.C13-150
−	$UO(ClCH_2COO)_2 \cdot H_2O$	U:	SVol.C13-150/2
$C_4Cl_2H_4O_6U$	$UO_2(CH_2ClCOO)_2$	U:	SVol.C13-150/2
		U:	SVol.D1-201
		U:	SVol.D2-20, 45
−	$UO_2(CH_2ClCOO)_2 \cdot H_2O$	U:	SVol.C13-150, 152/3
$C_4Cl_2H_4O_{12}Rb_2Rh_2$	$Rb_2[Rh_2(HCO_3)_4Cl_2]$	Rh:	SVol.B1-186
$C_4Cl_2H_4O_{12}Rh_2^{2-}$	$[Rh_2(HCO_3)_4Cl_2]^{2-}$	Rh:	SVol.B1-184, 186
$C_4Cl_2H_5MnN_3$	$Mn(2\text{-}NH_2\text{-}1,3\text{-}N_2C_4H_3)Cl_2$	Mn:	MVol.D4-6
$C_4Cl_2H_5MnN_3O$	$Mn(C_3H_3N_2CONH_2)Cl_2$	Mn:	MVol.D5-132
$C_4Cl_2H_6MnN_2$	$Mn(CH_3NCHNCHCH)Cl_2$	Mn:	MVol.D3-297
−	$Mn[HNC(CH_3)NCHCH]Cl_2$	Mn:	MVol.D3-299/300
$C_4Cl_2H_6Mn_2O_6$	$Mn_2(CH_2(OH)COO)_2Cl_2 \cdot 4 H_2O$	Mn:	MVol.D2-148
$C_4Cl_2H_6O_4U$	$UCl_2(CH_3COO)_2$	U:	SVol.C13-112, 114
$C_4Cl_2H_6O_5U$	$UOCl_2(CH_3COO)_2 = UO_2Cl_2 \cdot (CH_3CO)_2O$		
		U:	SVol.C13-115
$C_4Cl_2H_6O_6U$	$U(OH)_2(ClCH_2COO)_2 \cdot H_2O$	U:	SVol.C13-150/2
$C_4Cl_2H_7MnNO$	$Mn(HN(CH_2)_3C(O))Cl_2$	Mn:	MVol.D3-77/9
$C_4Cl_2H_7O_2Sb$	$C_2H_5OC(O)CH_2SbCl_2$	Sb:	Org.Comp.2-93
$C_4Cl_2H_8MnN_2O_2$	$Mn(N_2H_2(COCH_3)_2)Cl_2 \cdot H_2O$	Mn:	MVol.D5-196
$C_4Cl_2H_8N_2O_4U$	$U(NH_2CH_2COO)_2Cl_2 \cdot 4 H_2O$	U:	SVol.A5-196
$C_4Cl_2H_8OYb$	$YbCl_2 \cdot C_4H_8O$	Sc:	MVol.C4a-31
$C_4Cl_2H_8O_3Se$	$SeOCl_2 \cdot C_4H_8O_2$	Se:	SVol.B2-200
$C_4Cl_2H_8O_6U$	$UO_2(CH_3COO)_2 \cdot 2 HCl$	U:	SVol.C13-121/2
$C_4Cl_2H_8O_8Se$	$H_2SeO_4 \cdot 2 CH_2ClCOOH$	Se:	SVol.B1-292
$C_4Cl_2H_9MnNO$	$Mn(OCCH_3N(CH_3)_2)Cl_2 \cdot H_2O$	Mn:	MVol.D5-101
$C_4Cl_2H_9MnNO_3$	$[Mn(HN(CCH_3O)_2)(H_2O)Cl_2]$	Mn:	MVol.D5-104/5
$C_4Cl_2H_9Sb$	$C_4H_9SbCl_2$	Sb:	Org.Comp.2-93

$C_4Cl_2H_9Sb$	$t-C_4H_9SbCl_2$	Sb:	Org.Comp.2-93
$C_4Cl_2H_{10}MnN_2O_2$	$Mn(OCHNHCH_3)_2Cl_2$	Mn:	MVol.D5-93/4
–	$Mn(OCCH_3NH_2)_2Cl_2$	Mn:	MVol.D5-99
–	$Mn(OCCH_3ND_2)_2Cl_2$	Mn:	MVol.D5-99
–	$Mn(OCCH_3NH_2)_2Cl_2 \cdot H_2O$	Mn:	MVol.D5-99
$C_4Cl_2H_{10}MnN_2O_4$	$Mn(NH_3CH_2COO)_2Cl_2$	Mn:	MVol.D4-254/7
–	$Mn(NH_3CH_2COO)_2Cl_2 \cdot 2 H_2O$	Mn:	MVol.D4-254, 257/8
–	$Mn(ND_3CD_2COO)_2Cl_2 \cdot 2 D_2O$	Mn:	MVol.D4-258
$C_4Cl_2H_{10}MnN_4O_2$	$Mn(C_2H_4(C(O)NHNH_2)_2)Cl_2$	Mn:	MVol.D5-204/5
–	$Mn(C_2H_4(C(O)NHNH_2)_2)Cl_2 \cdot 3 H_2O$	Mn:	MVol.D5-204/5
$C_4Cl_2H_{10}MnN_6O_4$	$Mn(HN(CONH_2)_2)_2Cl_2$	Mn:	MVol.D5-156
–	$Mn(HN(CONH_2)_2)_2Cl_2 \cdot H_2O$	Mn:	MVol.D5-156/7
$C_4Cl_2H_{10}N_2Pt$	$(C_2H_4NH)_2PtCl_2$		
	Cytostatic activity	Pt:	SVol.A1-332
$C_4Cl_2H_{10}O_8Sn$	$(C_2H_5)_2Sn(OClO_3)_2$	Sn:	Org.Comp.14-186
$C_4Cl_2H_{11}ISiSn$	$(CH_3)_3SiCH_2SnICl_2$	Sn:	Org.Comp.8-144
$C_4Cl_2H_{11}MnNO_2$	$Mn[NH(C_2H_4OH)_2]Cl_2$	Mn:	MVol.D4-223/4
$C_4Cl_2H_{12}MnN_2O$	$Mn[(CH_3)_2N(O)(CH_2)_2NH_2]Cl_2$	Mn:	MVol.D3-43
$C_4Cl_2H_{12}MnN_4O_2$	$Mn(OCCH_3NHNH_2)_2Cl_2$	Mn:	MVol.D5-169
$C_4Cl_2H_{12}MnN_4O_4$	$Mn(OCHNH_2)_4Cl_2$	Mn:	MVol.D5-90/1
$C_4Cl_2H_{12}NO_7Rh$	$Rh(O_2CCH_2Cl)_2(OH)(H_2O)_2(NH_3)$	Rh:	SVol.B2-61
$C_4Cl_2H_{12}N_2O_6U$	$[U(NH_2CH_2COO)_2(H_2O)_2]Cl_2 \cdot 2 H_2O$	U:	SVol.C13-148/9
$C_4Cl_2H_{12}O_2PSb$	$(CH_3)_4SbOP(O)Cl_2$	Sb:	Org.Comp.3-77,79
$C_4Cl_2H_{14}MnN_2$	$Mn(C_2H_5NH_2)_2Cl_2$	Mn:	MVol.D3-22
$C_4Cl_2H_{14}MnN_4$	$Mn(H_2N(CH_3)N(CH_2)_2N(CH_3)NH_2)Cl_2$	Mn:	MVol.D3-72
$C_4Cl_2H_{14}N_2Pt$	$(C_2H_5NH_2)_2PtCl_2$		
	Cytostatic activity	Pt:	SVol.A1-330/2
$C_4Cl_2H_{14}N_{10}Rh^+$	$[Rh((NH_2CNH)_2NH)_2Cl_2]^+$	Rh:	SVol.B2-221
$C_4Cl_2H_{14}N_{11}O_3Rh$	$[Rh((NH_2CNH)_2NH)_2Cl_2]NO_3 \cdot 4 H_2O$	Rh:	SVol.B2-221
$C_4Cl_2H_{16}IN_4O_4Rh$	$[Rh(NH_2C_2H_4NH_2)_2ClI]ClO_4$	Rh:	SVol.B2-179
$C_4Cl_2H_{16}MnN_8O_4$	$Mn(OC(NH_2)_2)_4Cl_2$	Mn:	MVol.D5-136/7
$C_4Cl_2H_{16}MnN_8O_{12}$	$Mn(OC(NH_2)_2)_4(ClO_4)_2 \cdot 2 H_2O$	Mn:	MVol.D5-140
$C_4Cl_2H_{16}N_4O_2Os$	$[OsO_2(C_2H_4(NH_2)_2)_2]Cl_2$	Os:	SVol.1-251
$C_4Cl_2H_{16}N_4O_{10}Os$	$[OsO_2(C_2H_4(NH_2)_2)_2](ClO_4)_2$	Os:	SVol.1-251
$C_4Cl_2H_{16}N_4Pd$	$Pd(NH_2C_2H_4NH_2)_2Cl_2$		
	Catalytic properties	Pt:	SVol.A1-305
$C_4Cl_2H_{16}N_4Rh^+$	$[Rh(NH_2C_2H_4NH_2)_2Cl_2]^+$	Rh:	SVol.B2-177, 179/83
$C_4Cl_2H_{16}N_4Sc^+$	$[Sc(C_2H_8N_2)_2Cl_2]^+$	Sc:	MVol.D1-22
$C_4Cl_2H_{16}N_5O_3Rh$	$[Rh(NH_2C_2H_4NH_2)_2Cl_2]NO_3$	Rh:	SVol.B2-177, 179/81
–	$[Rh(NH_2C_2H_4NH_2)_2Cl_2]NO_3 \cdot H_2O$	Rh:	SVol.B2-181
$C_4Cl_2H_{16}N_5O_6Rh$	$[Rh(NH_2C_2H_4NH_2)_2(NO_2)Cl]ClO_4$	Rh:	SVol.B2-176
–	$[Rh(NH_2C_2H_4NH_2)_2(ONO)Cl]ClO_4$	Rh:	SVol.B2-176
$C_4Cl_2H_{16}N_7O_4Rh$	$[Rh(NH_2C_2H_4NH_2)_2(N_3)Cl]ClO_4$	Rh:	SVol.B2-176
$C_4Cl_2H_{18}IN_4O_9Rh$	$[Rh(NH_2C_2H_4NH_2)_2(OH_2)I](ClO_4)_2$	Rh:	SVol.B2-173
$C_4Cl_2H_{18}N_4Os$	$[OsH_2(C_2H_4(NH_2)_2)_2]Cl_2 \cdot H_2O$	Os:	SVol.1-251
$C_4Cl_2H_{19}N_4O_9Rh$	$[Rh(NH_2C_2H_4NH_2)_2H(OH_2)](ClO_4)_2$	Rh:	SVol.B2-167
$C_4Cl_2H_{19}N_4O_{10}Rh$	$[Rh(NH_2C_2H_4NH_2)_2(OH)(OH_2)](ClO_4)_2$	Rh:	SVol.B2-168
$C_4Cl_2H_{19}N_5Rh$	$[Rh(NH_2C_2H_4NH_2)_2(NH_3)Cl]Cl$	Rh:	SVol.B2-174
$C_4Cl_2H_{20}N_4Rh^+$	$[Rh(NH_2CH_3)_4Cl_2]^+$	Rh:	SVol.B2-158

$C_4Cl_3H_{12}O_2Pr$	$PrCl_3 \cdot 2 C_2H_5OH \cdot H_2O$	Sc:	MVol.D3-19
$C_4Cl_3H_{12}O_2Sc$	$ScCl_3 \cdot 2 C_2H_5OH$	Sc:	MVol.D3-18
$C_4Cl_3H_{12}O_2Sm$	$SmCl_3 \cdot 2 C_2H_5OH \cdot H_2O$	Sc:	MVol.D3-19
$C_4Cl_3H_{12}PtSb$	$(CH_3)_4SbPtCl_3$	Sb:	Org.Comp.3-96
$C_4Cl_3H_{12}SbSn$	$(CH_3)_4SbSnCl_3$	Sb:	Org.Comp.3-94
$C_4Cl_3H_{13}LaN_3O_{12}$	$La((NH_2C_2H_4)_2NH)(ClO_4)_3$	Sc:	MVol.D1-27/8
$C_4Cl_3H_{13}N_3NdO_{12}$	$Nd((NH_2C_2H_4)_2NH)(ClO_4)_3$	Sc:	MVol.D1-27/8
$C_4Cl_3H_{13}N_3O_{12}Pr$	$Pr((NH_2C_2H_4)_2NH)(ClO_4)_3$	Sc:	MVol.D1-27/8
$C_4Cl_3H_{13}N_3O_{12}Sm$	$Sm((NH_2C_2H_4)_2NH)(ClO_4)_3$	Sc:	MVol.D1-27/8
$C_4Cl_3H_{13}N_3O_{12}Yb$	$Yb((NH_2C_2H_4)_2NH)(ClO_4)_3$	Sc:	MVol.D1-27/8
$C_4Cl_3H_{13}N_3Rh$	$Rh((NH_2C_2H_4)_2NH)Cl_3$	Rh:	SVol.B2-194
$C_4Cl_3H_{14}N_2ORhS$	$RhCl_3((CH_3)_2SO)(C_2H_8N_2)$	Rh:	SVol.B1-147
$C_4Cl_3H_{14}N_{10}Rh$	$[Rh((NH_2CNH)_2NH)_2Cl_2]Cl$	Rh:	SVol.B2-221
–	$[Rh((NH_2CNH)_2NH)_2Cl_2]Cl \cdot 2 H_2O$	Rh:	SVol.B2-221
$C_4Cl_3H_{14}O_3RhS_2$	$[RhCl_3((CH_3)_2SO)_2(H_2O)]$	Rh:	SVol.B1-144
$C_4Cl_3H_{15}NO_2RhS_2$	$RhCl_3[(CH_3)_2SO]_2(NH_3)$	Rh:	SVol.B1-146
$C_4Cl_3H_{16}HoN_4O_{12}$	$Ho(C_2H_8N_2)_2(ClO_4)_3$	Sc:	MVol.D1-19/21
$C_4Cl_3H_{16}HoO_4$	$HoCl_3 \cdot 4 CH_3OH$	Sc:	MVol.D3-16
$C_4Cl_3H_{16}LaN_4O_{12}$	$La(C_2H_8N_2)_2(ClO_4)_3$	Sc:	MVol.D1-19/21
$C_4Cl_3H_{16}LaO_4$	$LaCl_3 \cdot 4 CH_3OH$	Sc:	MVol.D3-16
$C_4Cl_3H_{16}LuN_4O_{12}$	$Lu(C_2H_8N_2)_2(ClO_4)_3$	Sc:	MVol.D1-19/21
$C_4Cl_3H_{16}N_4NdO_{12}$	$Nd(C_2H_8N_2)_2(ClO_4)_3$	Sc:	MVol.D1-19/21
$C_4Cl_3H_{16}N_4O_4Rh$	$[Rh(NH_2C_2H_4NH_2)_2Cl_2]ClO_4$	Rh:	SVol.B2-182
$C_4Cl_3H_{16}N_4O_{12}Pr$	$Pr(C_2H_8N_2)_2(ClO_4)_3$	Sc:	MVol.D1-19/21
$C_4Cl_3H_{16}N_4O_{12}Sm$	$Sm(C_2H_8N_2)_2(ClO_4)_3$	Sc:	MVol.D1-19/21
$C_4Cl_3H_{16}N_4O_{12}Tb$	$Tb(C_2H_8N_2)_2(ClO_4)_3$	Sc:	MVol.D1-19/21
$C_4Cl_3H_{16}N_4O_{12}Y$	$Y(C_2H_8N_2)_2(ClO_4)_3$	Sc:	MVol.D1-19/21
$C_4Cl_3H_{16}N_4O_{12}Yb$	$Yb(C_2H_8N_2)_2(ClO_4)_3$	Sc:	MVol.D1-19/21
$C_4Cl_3H_{16}N_4Os$	$[Os(C_2H_4(NH_2)_2)_2Cl_2]Cl \cdot H_2O$	Os:	SVol.1-251
$C_4Cl_3H_{16}N_4Rh$	$[Rh(NH_2C_2H_4NH_2)_2Cl_2]Cl$	Rh:	SVol.B2-177, 180/1, 183
–	$[Rh(NH_2C_2H_4NH_2)_2Cl_2]Cl \cdot H_2O$	Rh:	SVol.B2-177/82
$C_4Cl_3H_{16}N_4Sc$	$Sc(C_2H_8N_2)_2Cl_3 = [Sc(C_2H_8N_2)_2Cl_2]Cl$	Sc:	MVol.D1-22
$C_4Cl_3H_{16}N_6Os$	$[Os(C_4H_4N_2)(NH_3)_4Cl]Cl_2 \cdot 0.5 H_2O$	Os:	SVol.1-247
$C_4Cl_3H_{16}N_8NdO_4$	$NdCl_3 \cdot 4 OC(NH_2)_2$	Sc:	MVol.D2-220
$C_4Cl_3H_{16}N_8O_4Pr$	$PrCl_3 \cdot 4 OC(NH_2)_2$	Sc:	MVol.D2-220
$C_4Cl_3H_{16}N_8O_4Sc$	$ScCl_3 \cdot 4 OC(NH_2)_2$	Sc:	MVol.D2-219
$C_4Cl_3H_{16}N_8O_4Sm$	$SmCl_3 \cdot 4 OC(NH_2)_2$	Sc:	MVol.D2-220
$C_4Cl_3H_{16}N_8O_{16}Pr$	$Pr(ClO_4)_3 \cdot 4 OC(NH_2)_2$	Sc:	MVol.D2-220
$C_4Cl_3H_{16}NdO_4$	$NdCl_3 \cdot 4 CH_3OH$	Sc:	MVol.D3-16
$C_4Cl_3H_{16}O_4Pr$	$PrCl_3 \cdot 4 CH_3OH$	Sc:	MVol.D3-16
$C_4Cl_3H_{16}O_4Sc$	$ScCl_3 \cdot 4 CH_3OH$	Sc:	MVol.D3-15/6
$C_4Cl_3H_{16}O_4Sm$	$SmCl_3 \cdot 4 CH_3OH$	Sc:	MVol.D3-16
$C_4Cl_3H_{16}O_4Tb$	$TbCl_3 \cdot 4 CH_3OH$	Sc:	MVol.D3-16
$C_4Cl_3H_{16}O_4Y$	$YCl_3 \cdot 4 CH_3OH$	Sc:	MVol.D3-16
$C_4Cl_3H_{16}O_4Yb$	$YbCl_3 \cdot 4 CH_3OH$	Sc:	MVol.D3-16
$C_4Cl_3H_{17}N_7RhS_2$	$[Rh(NH_2NHC(S)NH_2)_2(NH_2C_2H_5)Cl]Cl_2 \cdot 2 H_2O$		
		Rh:	SVol.B3-41
$C_4Cl_3H_{18}N_4O_9Rh$	$[Rh(NH_2C_2H_4NH_2)_2(OH_2)Cl](ClO_4)_2$	Rh:	SVol.B2-172
$C_4Cl_3H_{19}N_7O_{12}Rh$	$[Rh(NH_3)_5(N(CHCH)_2N)](ClO_4)_3$	Rh:	SVol.B2-140
$C_4Cl_3H_{19}N_7Os$	$[Os(NH_3)_5(C_4H_4N_2)]Cl_3 \cdot 2 H_2O$	Os:	SVol.1-248

C_4CuH_2K	$[(CHC)_2Cu]K$	Cu:	Org.Comp.3-182
C_4CuH_3	$CHCCCH_2Cu \cdot x LiMgBrCl_2$	Cu:	Org.Comp.1-299
–	$CH_2CHCCCu$	Cu:	Org.Comp.3-7/8, 10, 19, 43, 109/10, 144, 147, 150
C_4CuH_3O	$CH_3COCCCu$	Cu:	Org.Comp.3-19, 110
–	OC_4H_3Cu	Cu:	Org.Comp.1-46
–	$OC_4H_3Cu \cdot x C_5H_5N$	Cu:	Org.Comp.1-405/6
–	$OC_4H_3Cu \cdot x C_9H_7N$	Cu:	Org.Comp.1-406
–	$OC_4H_3Cu \cdot x C_9H_7N \cdot y LiI$	Cu:	Org.Comp.1-406
–	$OC_4H_3Cu \cdot x LiBr$	Cu:	Org.Comp.1-57, 122
–	$OC_4H_3Cu \cdot x MgBrHal$	Cu:	Org.Comp.1-212, 293
$C_4CuH_3O_2$	CH_3O_2CCCCu	Cu:	Org.Comp.3-19, 70, 110
C_4CuH_3S	SC_4H_3Cu	Cu:	Org.Comp.1-46/7
–	$SC_4H_3Cu \cdot x C_5H_5N$	Cu:	Org.Comp.1-406
–	$SC_4H_3Cu \cdot x C_5H_5N \cdot y CuBr \cdot z LiBr$ . . .	Cu:	Org.Comp.1-407
–	$SC_4H_3Cu \cdot x C_5H_5N \cdot y LiI$	Cu:	Org.Comp.1-406
–	$SC_4H_3Cu \cdot x C_9H_7N$	Cu:	Org.Comp.1-408
–	$SC_4H_3Cu \cdot x C_9H_7N \cdot y LiI$	Cu:	Org.Comp.1-408
–	$SC_4H_3Cu \cdot x C_9H_7N \cdot y MgI_2$	Cu:	Org.Comp.1-408
–	$SC_4H_3Cu \cdot x LiBr$	Cu:	Org.Comp.1-57, 122
–	$SC_4H_3Cu \cdot x LiCl$	Cu:	Org.Comp.1-57, 84/5
–	$SC_4H_3Cu \cdot x MgBrHal$	Cu:	Org.Comp.1-293
–	$SC_4H_3Cu \cdot x MgBr_2$	Cu:	Org.Comp.1-212, 282
C_4CuH_3Se	$SeC_4H_3Cu \cdot x C_5H_5N \cdot y LiI$	Cu:	Org.Comp.1-409
–	$SeC_4H_3Cu \cdot x LiI$	Cu:	Org.Comp.1-58
C_4CuH_3Te	$TeC_4H_3Cu \cdot x C_5H_5N \cdot y LiI$	Cu:	Org.Comp.1-409
C_4CuH_4	$Cu(C_2H_2)_2$	Cu:	Org.Comp.4-40/2
–	$Cu(C_2D_2)_2$	Cu:	Org.Comp.4-40, 42
–	$Cu(^{12}C_2H_2)_2$	Cu:	Org.Comp.4-42
–	$(^{12}C_2H_2)Cu(^{13}C_2H_2)$	Cu:	Org.Comp.4-42
–	$Cu(^{13}C_2H_2)_2$	Cu:	Org.Comp.4-42
$C_4CuH_4^+$	$[(C_2H_2)_2Cu]^+$	Cu:	Org.Comp.4-44
C_4CuH_4IO	$C_2H_2 \cdot CuI \cdot CH_2CO$	Cu:	Org.Comp.4-52
C_4CuH_5	C_2H_5CCCu	Cu:	Org.Comp.3-19, 83, 89, 97, 103, 126
$C_4CuH_5N_2$	$C_2H_5NCCuCN$	Cu:	Org.Comp.3-217
C_4CuH_5O	$CH_3CH(OH)CCCu$	Cu:	Org.Comp.3-10, 19, 76, 83, 86
–	CH_3OCH_2CCCu	Cu:	Org.Comp.3-19, 112, 117
–	C_2H_5OCCCu	Cu:	Org.Comp.3-19, 76
–	$C_2H_5OCCCu \cdot x LiI$	Cu:	Org.Comp.3-152, 155
–	$HO(CH_2)_2CCCu$	Cu:	Org.Comp.3-10/1, 19, 110, 122/3
$C_4CuH_5O_4$	$O_2CCH_2CuO_2CCH_3$	Cu:	Org.Comp.3-243
C_4CuH_5S	CH_3SCH_2CCCu	Cu:	Org.Comp.3-41
–	C_2H_5SCCCu	Cu:	Org.Comp.3-7, 20, 43/4
$C_4CuH_6^+$	$(CH_2CHCHCH_2Cu)^+$	Cu:	Org.Comp.4-85
–	$(CH_3CHCCH_2Cu)^+$	Cu:	Org.Comp.4-65
C_4CuH_6Li	$(CH_2CH)_2CuLi$	Cu:	Org.Comp.2-6/7, 33/173

C$_4$CuH$_6$N	(CH$_3$)$_2$NCCCu	Cu: Org.Comp.3-20	
C$_4$CuH$_7$	CH$_2$CHCH$_2$CH$_2$Cu · x MgBr$_2$	Cu: Org.Comp.1-212, 274	
–	CH$_2$CHCH$_2$CH$_2$Cu · x S(CH$_3$)$_2$ · y MgBr$_2$	Cu: Org.Comp.1-359, 362	
–	CH$_2$CCH$_3$CH$_2$Cu · x MgCII	Cu: Org.Comp.1-212	
–	CH$_2$CCH$_3$CH$_2$Cu · x Mg(Hal)$_2$	Cu: Org.Comp.1-271	
–	CH$_2$CCH$_3$CH$_2$Cu · x S(C$_4$H$_9$-n)$_2$ · y LiI	Cu: Org.Comp.1-359, 378	
–	(CH$_3$)$_2$CCHCu · x LiBr	Cu: Org.Comp.1-58	
–	(CH$_3$)$_2$CCHCu · x LiI	Cu: Org.Comp.1-58, 110, 124	
–	(CH$_3$)$_2$CCHCu · x MgBr$_2$	Cu: Org.Comp.1-268	
–	(CH$_3$)$_2$CCHCu · x MgHal$_2$	Cu: Org.Comp.1-212, 258, 288	
–	(CH$_3$)$_2$CCHCu · x S(CH$_3$)$_2$ · y MgBr$_2$	Cu: Org.Comp.1-359, 367, 372/3	
–	CH$_3$CHCHCH$_2$Cu · x LiI	Cu: Org.Comp.1-58, 85	
–	CH$_3$CHCHCH$_2$Cu · x MgBrHal	Cu: Org.Comp.1-212	
–	CH$_3$CHCHCH$_2$Cu · x MgBr$_2$	Cu: Org.Comp.1-212, 265	
–	CH$_3$CHCHCH$_2$Cu · x P(C$_4$H$_9$-n)$_3$	Cu: Org.Comp.1-447	
–	CH$_3$CHCHCH$_2$Cu · x S(C$_4$H$_9$-n)$_2$ · y LiI	Cu: Org.Comp.1-359, 378	
–	CH$_3$CHCCH$_3$Cu · x P(C$_4$H$_9$-n)$_3$	Cu: Org.Comp.1-453	
–	C$_2$H$_5$CHCHCu · x C$_2$H$_4$N$_2$(CH$_3$)$_4$ · y LiI	Cu: Org.Comp.1-393	
–	C$_2$H$_5$CHCHCu · x C$_2$H$_4$N$_2$(CH$_3$)$_4$ · y MgBr$_2$	Cu: Org.Comp.1-393	
–	C$_2$H$_5$CHCHCu · x LiBr	Cu: Org.Comp.1-58	
–	C$_2$H$_5$CHCHCu · x LiI	Cu: Org.Comp.1-110	
–	C$_2$H$_5$CHCHCu · x LiSC$_6$H$_5$	Cu: Org.Comp.1-166, 189	
–	C$_2$H$_5$CHCHCu · x MgBrCl	Cu: Org.Comp.1-269/70	
–	C$_2$H$_5$CHCHCu · x MgBrHal	Cu: Org.Comp.1-213, 295	
–	C$_2$H$_5$CHCHCu · x MgBr$_2$	Cu: Org.Comp.1-213, 261/2, 266/70, 295	
–	C$_2$H$_5$CHCHCu · x P(OC$_2$H$_5$)$_3$ · y MgBr$_2$	Cu: Org.Comp.1-429, 447	
–	C$_2$H$_5$CHCHCu · x S(CH$_3$)$_2$ · y MgBr$_2$	Cu: Org.Comp.1-359	
C$_4$CuH$_7$IN	n-C$_3$H$_7$NCCuI	Cu: Org.Comp.3-217	
C$_4$CuH$_7$O	CH$_3$OC$_3$H$_4$Cu · x LiSC$_6$H$_5$	Cu: Org.Comp.1-182	
–	CH$_3$OC$_3$H$_4$Cu · x P(C$_4$H$_9$-n)$_3$ · y LiI	Cu: Org.Comp.1-450	
–	C$_2$H$_5$OCHCHCu · x LiCN	Cu: Org.Comp.1-125, 138	
C$_4$CuH$_7$O$_2$	C$_2$H$_5$O$_2$CCH$_2$Cu · x LiBr	Cu: Org.Comp.1-58	
–	C$_2$H$_5$O$_2$CCH$_2$Cu · x LiN(C$_3$H$_7$-i)$_2$	Cu: Org.Comp.1-195, 199	
C$_4$CuH$_7$S$_2$	CH$_2$(SCH$_2$)$_2$CHCu · x LiOC(CH$_3$)(C$_3$H$_7$-n)CHCH$_2$	Cu: Org.Comp.1-163	
–	CH$_2$(SCH$_2$)$_2$CHCu · x LiOCH(C$_4$H$_9$-n)CHCH$_2$	Cu: Org.Comp.1-163	
–	CH$_2$(SCH$_2$)$_2$CHCu · x LiOCH$_2$CHCHCH$_3$	Cu: Org.Comp.1-163	
C$_4$CuH$_8$	(CH$_2$CH$_2$)$_2$Cu	Cu: Org.Comp.4-37	
–	CH$_3$CHCHCH$_3$Cu	Cu: Org.Comp.4-3	
C$_4$CuH$_8$I	(CH$_3$)$_2$CCH$_2$CuI	Cu: Org.Comp.4-10	
C$_4$CuH$_8$NO	(NHC(OCH$_3$))(CH$_3$)CHCu	Cu: Org.Comp.1-28	
–	NHC(OC$_2$H$_5$)CH$_2$Cu	Cu: Org.Comp.1-28	
C$_4$CuH$_9$	n-C$_4$H$_9$Cu	Cu: Org.Comp.1-17	
–	n-C$_4$H$_9$Cu · x Al(C$_2$H$_5$)$_3$	Cu: Org.Comp.1-328, 332	
–	n-C$_4$H$_9$Cu · x AlCl$_3$	Cu: Org.Comp.1-328/31	

C_4CuH_9	n-C_4H_9Cu · x $B(C_2H_5)_3$ · y LiI	Cu: Org.Comp.1-333, 335, 353/4
		Cu: Org.Comp.1-353/4
–	n-C_4H_9Cu · x $B(C_4H_9$-$n)_3$ · y LiI	Cu: Org.Comp.1-335, 352/4
–	n-C_4H_9Cu · x BCl_3 · y LiI	Cu: Org.Comp.1-335, 352/4
–	n-C_4H_9Cu · x BF_3 · y LiHal	Cu: Org.Comp.1-335, 339/40, 348
–	n-C_4H_9Cu · x BF_3 · y LiI	Cu: Org.Comp.1-333/49
–	n-C_4H_9Cu · x BF_3 · y MgBrI	Cu: Org.Comp.1-335, 349/51
–	s-C_4H_9Cu · x BF_3 · y MgBrI	Cu: Org.Comp.1-335, 351
–	n-C_4H_9Cu · x BF_3 · y MgClI	Cu: Org.Comp.1-335, 337
–	n-C_4H_9Cu · x BF_3 · y $MgHal_2$	Cu: Org.Comp.1-335, 340
–	n-C_4H_9Cu · x $B(OCH_3)_3$ · y LiI	Cu: Org.Comp.1-335, 354
–	n-C_4H_9Cu · x $C_2H_4N_2(CH_3)_4$ · y LiI	Cu: Org.Comp.1-388
–	n-C_4H_9Cu · x $C_2H_4N_2(CH_3)_4$ · y $LiOC_4H_9$-t	
		Cu: Org.Comp.1-388
–	n-C_4H_9Cu · x $C_2H_4N_2(CH_3)_4$ · y MgBrI ...	Cu: Org.Comp.1-388
–	n-C_4H_9Cu · x $C_2H_4N_2(CH_3)_4$ · y $MgBr_2$ ..	Cu: Org.Comp.1-388
–	n-C_4H_9Cu · x n-$C_4H_9BC_8H_{14}$ · y LiI	Cu: Org.Comp.1-335, 353/4
–	n-C_4H_9Cu · x LiBr	Cu: Org.Comp.1-59, 85, 107
–	n-C_4H_9Cu · x LiCN	Cu: Org.Comp.1-125, 131/4
–	s-C_4H_9Cu · x LiCN	Cu: Org.Comp.1-134
–	t-C_4H_9Cu · x LiCN	Cu: Org.Comp.1-125, 135
–	n-C_4H_9Cu · x LiI	Cu: Org.Comp.1-59, 91/122
–	s-C_4H_9Cu · x LiI	Cu: Org.Comp.1-59
–	t-C_4H_9Cu · x LiI	Cu: Org.Comp.1-59
–	t-C_4H_9Cu · x $LiMgBrCl_2$	Cu: Org.Comp.1-300
–	n-C_4H_9Cu · x $LiMgBr_2Cl$	Cu: Org.Comp.1-300, 304, 310, 313/5
–	t-C_4H_9Cu · x $LiMgBr_2Cl$	Cu: Org.Comp.1-300, 304/14
–	n-C_4H_9Cu · x $LiMgBr_2Hal$	Cu: Org.Comp.1-300, 307/9, 312, 314
–	s-C_4H_9Cu · x $LiMgBr_2Hal$	Cu: Org.Comp.1-300, 307/9
–	t-C_4H_9Cu · x $LiMgBr_2Hal$	Cu: Org.Comp.1-300, 307/10, 312, 314
–	n-C_4H_9Cu · x $LiMgBr_2I$	Cu: Org.Comp.1-300
–	n-C_4H_9Cu · x $LiMgBr_3$	Cu: Org.Comp.1-300, 312
–	t-C_4H_9Cu · x $LiMgBr_3$	Cu: Org.Comp.1-300, 306/7
–	t-C_4H_9Cu · x $LiMgCl_3$	Cu: Org.Comp.1-300
–	n-C_4H_9Cu · x $LiN(C_2H_5)_2$	Cu: Org.Comp.1-195, 200
–	t-C_4H_9Cu · x $LiN(C_2H_5)_2$	Cu: Org.Comp.1-201
–	n-C_4H_9Cu · x $LiN(C_3H_7$-$i)_2$	Cu: Org.Comp.1-195, 200
–	n-C_4H_9Cu · x $LiN(C_6H_5)_2$	Cu: Org.Comp.1-200
–	n-C_4H_9Cu · x $LiNC_6H_5CO_2CHCH_3CHCHC_6H_5$	
		Cu: Org.Comp.1-195, 201
–	n-C_4H_9Cu · x $LiN(C_6H_{11}$-$c)_2$	Cu: Org.Comp.1-195, 200
–	t-C_4H_9Cu · x $LiN(C_6H_{11}$-$c)_2$	Cu: Org.Comp.1-201
–	n-C_4H_9Cu · x $LiOC(CH_3)(C_3H_7$-$n)CHCH_2$..	Cu: Org.Comp.1-155
–	n-C_4H_9Cu · x $LiOCH(CHCH_2)CH_2Si(CH_3)_3$	Cu: Org.Comp.1-163
–	n-C_4H_9Cu · x $LiOCHCH_3C_2H_5$	Cu: Org.Comp.1-153

C_4CuH_9	t-C_4H_9Cu · x P(C_4H_9-n)$_3$ · y LiI.	Cu:	Org.Comp.1-445
–	n-C_4H_9Cu · x P(C_4H_9-n)$_3$ · y LiI	Cu:	Org.Comp.1-442/3
–	n-C_4H_9Cu · x P(C_4H_9-n)$_3$ · y MgBrI	Cu:	Org.Comp.1-444
–	i-C_4H_9Cu · x P(C_4H_9-n)$_3$ · y MgClI	Cu:	Org.Comp.1-445
–	t-C_4H_9Cu · x P(C_4H_9-n)$_3$ · y MgClI.	Cu:	Org.Comp.1-445
–	n-C_4H_9Cu · x P(C_4H_9-n)$_3$ · y MgClI	Cu:	Org.Comp.1-444
–	n-C_4H_9Cu · x P(C_8H_{17}-n)$_3$	Cu:	Org.Comp.1-444
–	n-C_4H_9Cu · x P(OCH$_3$)$_3$ · y MgBr$_2$	Cu:	Org.Comp.1-444
–	n-C_4H_9Cu · x P(OC$_2$H$_5$)$_3$ · y MgBr$_2$	Cu:	Org.Comp.1-444/5
–	n-C_4H_9Cu · x S(CH$_3$)$_2$ · y LiBr	Cu:	Org.Comp.1-359, 365
–	n-C_4H_9Cu · x S(CH$_3$)$_2$ · y MgBr$_2$	Cu:	Org.Comp.1-359, 363/5
–	n-C_4H_9Cu · x S(C$_3$H$_7$-i)$_2$ · y LiI.	Cu:	Org.Comp.1-359, 374/5
–	n-C_4H_9Cu · x TiCl$_4$	Cu:	Org.Comp.1-328, 332
C_4CuH_{10}	(C$_2$H$_5$)$_2$Cu	Cu:	Org.Comp.2-2
$C_4CuH_{10}Li$	CH$_3$(i-C$_3$H$_7$)CuLi.	Cu:	Org.Comp.2-175, 190
–	(C$_2$H$_5$)$_2$CuLi.	Cu:	Org.Comp.2-6/7, 33/173
–	n-C_4H_9HCuLi	Cu:	Org.Comp.1-51
–	t-C_4H_9HCuLi	Cu:	Org.Comp.1-51
$C_4CuH_{10}NO$	CH$_3$Cu · HCON(CH$_3$)$_2$	Cu:	Org.Comp.1-385
$C_4CuH_{10}O_2PS$	(CH$_3$O)$_2$P(S)CH(CH$_3$)Cu · x LiI	Cu:	Org.Comp.1-59, 113
$C_4CuH_{10}O_3P$	(CH$_3$O)$_2$P(O)CH(CH$_3$)Cu · x LiI	Cu:	Org.Comp.1-59, 113
$C_4CuH_{11}Si$	(CH$_3$)$_3$SiCH$_2$Cu	Cu:	Org.Comp.1-28
–	[(CH$_3$)$_3$SiCH$_2$Cu]$_4$	Cu:	Org.Comp.4-200, 204
–	(CH$_3$)$_3$SiCH$_2$Cu · x MgBrCl	Cu:	Org.Comp.1-214, 273
–	(CH$_3$)$_3$SiCH$_2$Cu · x MgClI	Cu:	Org.Comp.1-214
–	(CH$_3$)$_3$SiCH$_2$Cu · x P(OC$_2$H$_5$)$_3$ · y MgBrCl	Cu:	Org.Comp.1-446
–	(CH$_3$)$_3$SiCH$_2$Cu · x S(CH$_3$)$_2$ · y MgBr$_2$ . . .	Cu:	Org.Comp.1-359, 366
$C_4CuH_{12}Li_3$	(CH$_3$)$_4$CuLi$_3$	Cu:	Org.Comp.2-242
$C_4CuO_4{}^+$	[(CO)$_4$Cu]$^+$	Cu:	Org.Comp.3-212/3
C_4Cu_2	CuCCCCCu	Cu:	Org.Comp.4-153, 158/9
$C_4Cu_2F_8$	Cu(CF$_2$)$_4$Cu	Cu:	Org.Comp.4-129
–	Cu(CF$_2$)$_4$Cu · x LiSC$_6$H$_5$	Cu:	Org.Comp.4-130
$C_4Cu_2H_2$	Cu$_2$C$_2$ · C$_2$H$_2$ · H$_2$O	Cu:	Org.Comp.4-150
$C_4Cu_2H_6O_3S$	(CH$_2$CHCHCH$_2$)Cu$_2$SO$_3$	Cu:	Org.Comp.4-176
$C_4Cu_2H_8O_4S$	[CH$_2$CH$_2$Cu]$_2$SO$_4$	Cu:	Org.Comp.4-5/6, 16
$C_4Cu_2H_{12}Li_2$	(CH$_3$)$_4$Cu$_2$Li$_2$ = (CH$_3$)$_2$CuLi	Cu:	Org.Comp.2-6/7, 28, 33/173
		Cu:	Org.Comp.4-103/4, 106
$C_4Cu_2H_{12}Li_2O_2$	(CH$_3$O)$_2$(CH$_3$)$_2$Cu$_2$Li$_2$	Cu:	Org.Comp.4-104
$C_4Cu_2H_{12}Mg$	(CH$_3$)$_4$Cu$_2$Mg	Cu:	Org.Comp.2-211
		Cu:	Org.Comp.4-107
$C_4Cu_2O_6$	(CO)$_2$Cu$_2$(C$_2$O$_4$)	Cu:	Org.Comp.4-113
C_4DyHO_8	HDy(C$_2$O$_4$)$_2$ · 3 H$_2$O = [H$_5$O$_2$][Dy(C$_2$O$_4$)$_2$H$_2$O]		
		Sc:	MVol.D5-139/41
$C_4DyH_2O_4{}^+$	[Dy(C$_2$H$_2$(COO)$_2$)]$^+$	Sc:	MVol.D5-197/8, 199/202
$C_4DyH_2O_5{}^+$	[Dy(OOCC(O)CH$_2$COO)]$^+$	Sc:	MVol.D5-338/9
$C_4DyH_3O_4S$	Dy(OOCCH(S)CH$_2$COO)	Sc:	MVol.D4-57
$C_4DyH_4I_2O_4{}^+$	[Dy(CH$_2$ICOO)$_2$]$^+$	Sc:	MVol.D5-53/6
$C_4DyH_4NO_8$	NH$_4$Dy(C$_2$O$_4$)$_2$ · H$_2$O	Sc:	MVol.D5-141/5
–	NH$_4$Dy(C$_2$O$_4$)$_2$ · 1.5 H$_2$O	Sc:	MVol.D5-141/5

$C_4DyH_4N_6S_4{}^+$	$Dy[(S)(H_2N)C_2N_2S]_2{}^+$	Sc:	MVol.D4-103
$C_4DyH_4O_4{}^+$	$[Dy(CH_3CH(COO)_2)]^+$	Sc:	MVol.D5-163/5
$C_4DyH_4O_5{}^+$	$[Dy(O(CH_2COO)_2)]^+$	Sc:	MVol.D5-181/5
$-$	$[Dy(OOCCH_2CHOHCOO)]^+$	Sc:	MVol.D5-318/21
$C_4DyH_4O_6{}^+$	$[Dy(OOC(CHOH)_2COO)]^+$	Sc:	MVol.D5-328/34
$C_4DyH_5NO_4{}^+$	$Dy[NH(CH_2COO)_2]^+$	Sc:	MVol.D1-119/20
$-$	$[Dy(OOCCH_2CH(NH_2)COO)]^+$	Sc:	MVol.D1-109/10
$C_4DyH_5O_4{}^{2+}$	$[DyH(C_2H_4(COO)_2)]^{2+}$	Sc:	MVol.D5-169
$C_4DyH_5O_5{}^{2+}$	$[Dy(HOOCCH_2CHOHCOO)]^{2+}$	Sc:	MVol.D5-321/2
$C_4DyH_6O_4{}^+$	$[Dy(CH_3COO)_2]^+$	Sc:	MVol.D5-24, 26/9
$C_4DyH_6O_4S_2{}^+$	$[Dy(OOCCH_2SH)_2]^+$	Sc:	MVol.D4-53/4
$C_4DyH_6O_6{}^+$	$[Dy(HOCH_2COO)_2]^+$	Sc:	MVol.D5-222/30
$C_4DyH_6O_8{}^+$	$[Dy((HO)_2CHCOO)_2]^+$	Sc:	MVol.D5-249/50
$C_4DyH_7N_2O_2{}^{2+}$	$[Dy(CH_3C(NO)C(NOH)CH_3)]^{2+}$	Sc:	MVol.D2-115
$C_4DyH_7O_2{}^{2+}$	$[Dy(i-C_3H_7COO)]^{2+}$	Sc:	MVol.D5-83/4
$C_4DyH_7O_3{}^{2+}$	$[Dy((CH_3)_2C(OH)COO)]^{2+}$	Sc:	MVol.D5-261/6
$C_4DyH_7O_4{}^{2+}$	$[Dy(HOCH_2C(OH)(CH_3)COO)]^{2+}$	Sc:	MVol.D5-268/9
$C_4DyH_7O_5{}^{2+}$	$[Dy(HOC(CH_2OH)_2COO)]^{2+}$	Sc:	MVol.D5-268/9
$C_4DyH_7O_{11}$	$[H_5O_2][Dy(C_2O_4)_2H_2O]$ = $HDy(C_2O_4)_2 \cdot 3\ H_2O$		
		Sc:	MVol.D5-139/41
$C_4DyH_8NO_3{}^{2+}$	$Dy[CH_3CHOHCH(NH_2)COO]^{2+}$	Sc:	MVol.D1-114
$C_4DyH_{10}O_4P^{2+}$	$Dy[(C_2H_5O)_2PO_2]^{2+}$	Sc:	MVol.D4-177
$C_4DyH_{11}N_2O_6P_2$	$Dy[H((O_3PCH_2NH)_2C_2H_4)]$	Sc:	MVol.D4-151
$C_4DyH_{13}N_3{}^{3+}$	$[Dy((NH_2C_2H_4)_2NH)]^{3+}$	Sc:	MVol.D1-27/8
$C_4DyH_{16}N_4{}^{3+}$	$[Dy(C_2H_8N_2)_2]^{3+}$	Sc:	MVol.D1-19/21
C_4DyKO_8	$KDy(C_2O_4)_2 \cdot 4\ H_2O$.	Sc:	MVol.D5-141/5
C_4DyLiO_8	$LiDy(C_2O_4)_2 \cdot 5\ H_2O$	Sc:	MVol.D5-141/5
C_4DyNaO_8	$NaDy(C_2O_4)_2 \cdot 5\ H_2O$	Sc:	MVol.D5-141/5
$C_4DyO_4{}^+$	$[Dy(C_4O_4)]^+$	Sc:	MVol.D3-248/50
C_4DyO_8Rb	$RbDy(C_2O_4)_2 \cdot 3.5\ H_2O$	Sc:	MVol.D5-141/5
$C_4DyO_{12}{}^{5-}$	$[Dy(CO_3)_4]^{5-}$	Sc:	MVol.D4-338/40
C_4ErIIO_8	$HEr(C_2O_4)_2 \cdot 3\ H_2O$ = $[H_5O_2][Er(C_2O_4)_2H_2O]$		
		Sc:	MVol.D5-139/41
$C_4ErH_2O_4{}^+$	$[Er(C_2H_2(COO)_2)]^+$	Sc:	MVol.D5-197/8, 199/202
$C_4ErH_4I_2O_4{}^+$	$[Er(CH_2ICOO)_2]^+$	Sc:	MVol.D5-53/6
$C_4ErH_4NO_8$	$NH_4Er(C_2O_4)_2 \cdot H_2O$	Sc:	MVol.D5-141/5
$-$	$NH_4Er(C_2O_4)_2 \cdot x\ H_2O$	Sc:	MVol.D5-141/5
$C_4ErH_4O_4{}^+$	$[Er(CH_3CH(COO)_2)]^+$	Sc:	MVol.D5-163/5
$-$	$[Er(C_2H_4(COO)_2)]^+$	Sc:	MVol.D5-169
$C_4ErH_4O_4S^+$	$Er[S(CH_2COO)_2]^+$	Sc:	MVol.D4-63
$C_4ErH_4O_5{}^+$	$[Er(O(CH_2COO)_2)]^+$	Sc:	MVol.D5-181/5
$-$	$[Er(OOCCH_2CHOHCOO)]^+$	Sc:	MVol.D5-318/21
$C_4ErH_4O_6{}^+$	$[Er(OOC(CHOH)_2COO)]^+$	Sc:	MVol.D5-328/34
$C_4ErH_5NO_4{}^+$	$Er[NH(CH_2COO)_2]^+$	Sc:	MVol.D1-119/20
$-$	$[Er(OOCCH_2CH(NH_2)COO)]^+$	Sc:	MVol.D1-109/10
$C_4ErH_5O_4{}^{2+}$	$[ErH(C_2H_4(COO)_2)]^{2+}$	Sc:	MVol.D5-169
$C_4ErH_5O_4S^{2+}$	$[Er(OOCCH_2SCH_2COOH)]^{2+}$	Sc:	MVol.D4-63
$C_4ErH_5O_5{}^{2+}$	$[Er(HOOCCH_2CHOHCOO)]^{2+}$	Sc:	MVol.D5-321/2
$C_4ErH_6O_4{}^+$	$[Er(CH_3COO)_2]^+$	Sc:	MVol.D5-25/9
$C_4ErH_6O_4S_2{}^+$	$[Er(OOCCH_2SH)_2]^+$	Sc:	MVol.D4-53/4

$C_4ErH_6O_6^+$	$[Er(HOCH_2COO)_2]^+$	Sc:	MVol.D5-222/30
$C_4ErH_6O_8^+$	$[Er((HO)_2CHCOO)_2]^+$	Sc:	MVol.D5-249/50
$C_4ErH_7N_2O_2^{2+}$	$[Er(CH_3C(NO)C(NOH)CH_3)]^{2+}$	Sc:	MVol.D2-115
$C_4ErH_7O_2^{2+}$	$[Er(i\text{-}C_3H_7COO)]^{2+}$	Sc:	MVol.D5-83/4
$C_4ErH_7O_2S^{2+}$	$[Er(C_2H_5SCH_2COO)]^{2+}$	Sc:	MVol.D4-62/3
$C_4ErH_7O_3^{2+}$	$[Er((CH_3)_2C(OH)COO)]^{2+}$	Sc:	MVol.D5-261/6
$C_4ErH_7O_4^{2+}$	$[Er(HOCH_2C(OH)(CH_3)COO)]^{2+}$	Sc:	MVol.D5-268/9
$C_4ErH_7O_5^{2+}$	$[Er(HOC(CH_2OH)_2COO)]^{2+}$	Sc:	MVol.D5-268/9
$C_4ErH_7O_7$	$[Er(HOCH_2COO)(OCH_2COO)(H_2O)] \cdot H_2O$	Sc:	MVol.D5-239/40
$C_4ErH_7O_{11}$	$[H_5O_2][Er(C_2O_4)_2H_2O] = HEr(C_2O_4)_2 \cdot 3\,H_2O$		
		Sc:	MVol.D5-139/41
$C_4ErH_{10}O_4P^{2+}$	$Er[(C_2H_5O)_2PO_2]^{2+}$	Sc:	MVol.D4-177
$C_4ErH_{11}N_2O_6P_2$	$Er[H((O_3PCH_2NH)_2C_2H_4)]$	Sc:	MVol.D4-151
$C_4ErH_{13}N_3^{3+}$	$[Er((NH_2C_2H_4)_2NH)]^{3+}$	Sc:	MVol.D1-27/8
$C_4ErH_{14}O_{10}^+$	$[Er(HOCH_2COO)_2(H_2O)_4]^+$	Sc:	MVol.D5-237/8
$C_4ErH_{16}N_4^{3+}$	$[Er(C_2H_8N_2)_2]^{3+}$	Sc:	MVol.D1-19/21
$C_4ErH_{16}N_{11}O_{13}$	$Er(NO_3)_3 \cdot 4\,OC(NH_2)_2$	Sc:	MVol.D2-220
C_4ErKO_8	$KEr(C_2O_4)_2 \cdot x\,H_2O$	Sc:	MVol.D5-141/5
C_4ErLiO_8	$LiEr(C_2O_4)_2 \cdot x\,H_2O$	Sc:	MVol.D5-141/5
C_4ErNaO_8	$NaEr(C_2O_4)_2 \cdot x\,H_2O$	Sc:	MVol.D5-141/5
$C_4ErO_4^+$	$[Er(C_4O_4)]^+$	Sc:	MVol.D3-248/50
$C_4ErO_8^-$	$[Er(C_2O_4)_2]^-$	Sc:	MVol.D5-120/4
C_4ErO_8Rb	$RbEr(C_2O_4)_2 \cdot x\,H_2O$	Sc:	MVol.D5-141/5
$C_4ErO_{12}^{5-}$	$[Er(CO_3)_4]^{5-}$	Sc:	MVol.D4-338/40
C_4EuH_2	$Eu(CCH)_2$	Sc:	MVol.D6-140
$C_4EuH_2O_4^+$	$[Eu(C_2H_2(COO)_2)]^+$	Sc:	MVol.D5-197/8, 199/202
$C_4EuH_3O_4^{2+}$	$[EuH(C_2H_2(COO)_2)]^{2+}$	Sc:	MVol.D5-200
$C_4EuH_4I_2O_4^+$	$[Eu(CH_2ICOO)_2]^+$	Sc:	MVol.D5-53/6
$C_4EuH_4NO_8$	$NH_4Eu(C_2O_4)_2 \cdot H_2O$	Sc:	MVol.D5-141/5
$C_4EuH_4O_4^+$	$[Eu(CH_3CH(COO)_2)]^+$	Sc:	MVol.D5-163/5
$-$	$[Eu(C_2H_4(COO)_2)]^+$	Sc:	MVol.D5-169
$C_4EuH_4O_5^+$	$[Eu(O(CH_2COO)_2)]^+$	Sc:	MVol.D5-181/5
$-$	$[Eu(OOCCH_2CHOHCOO)]^+$	Sc:	MVol.D5-318/21
$C_4EuH_4O_6^+$	$[Eu(OOC(CHOH)_2COO)]^+$	Sc:	MVol.D5-328/34
$C_4EuH_5NO_4^+$	$Eu[NH(CH_2COO)_2]^+$	Sc:	MVol.D1-119/21
$-$	$[Eu(OOCCH_2CH(NH_2)COO)]^+$	Sc:	MVol.D1-109
$C_4EuH_5O_2^{2+}$	$[Eu(CH_2CCH_3COO)]^{2+}$	Sc:	MVol.D5-91/2
$-$	$[Eu(CH_3CHCHCOO)]^{2+}$	Sc:	MVol.D5-91/2
$C_4EuH_5O_4^{2+}$	$[EuH(C_2H_4(COO)_2)]^{2+}$	Sc:	MVol.D5-169
$C_4EuH_5O_5^{2+}$	$[Eu(HOOCCH_2CHOHCOO)]^{2+}$	Sc:	MVol.D5-321/2
$C_4EuH_5O_7$	$Eu(OH)(OOC(CHOH)_2COO) \cdot 5\,H_2O$	Sc:	MVol.D5-336
$C_4EuH_6O_4^+$	$[Eu(CH_3COO)_2]^+$	Sc:	MVol.D5-24
$C_4EuH_6O_4S_2^+$	$[Eu(OOCCH_2SH)_2]^+$	Sc:	MVol.D4-53/4
$C_4EuH_6O_6^+$	$[Eu(HOCH_2COO)_2]^+$	Sc:	MVol.D5-222/30
$C_4EuH_6O_8^+$	$[Eu((HO)_2CHCOO)_2]^+$	Sc:	MVol.D5-249/50
$C_4EuH_7N_2O_2^{2+}$	$[Eu(CH_3C(NO)C(NOH)CH_3)]^{2+}$	Sc:	MVol.D2-115
$C_4EuH_7N_2O_3^{2+}$	$Eu(H_2NCH_2C(O)NHCH_2COO)^{2+}$	Sc:	MVol.D4-282/3
$C_4EuH_7O_2^{2+}$	$[Eu(i\text{-}C_3H_7COO)]^{2+}$	Sc:	MVol.D5-83/4
$C_4EuH_7O_2S^{2+}$	$[Eu(C_2H_5SCH_2COO)]^{2+}$	Sc:	MVol.D4-62/3
$C_4EuH_7O_3^{2+}$	$[Eu((CH_3)_2C(OH)COO)]^{2+}$	Sc:	MVol.D5-261/6

$C_4F_4H_6N_2$	$CF_3C(NF)NHC_2H_5$	F:	PFHOrg.8-221
–	$(CH_3)_2NC(CF_3)NF$	F:	PFHOrg.8-221
$C_4F_4H_7NO_2$	$(CH_3OCF_2)_2NH$	F:	PFHOrg.9-150
$C_4F_4H_8NOP$	$CF_3NPF(OC_2H_5)CH_3$	F:	PFHOrg.8-116
$C_4F_4H_9IO_3$	$CF_3IF(OCH_3)_3$	F:	PFHOrg.SVol.3-266
$C_4F_4H_{10}O_4P_2Sn$	$(C_2H_5)_2Sn(OP(O)F_2)_2$	Sn:	Org.Comp.14-196
$C_4F_4H_{12}O_2SnTe$	$(CH_3)_3SnOTe(OCH_3)F_4$	Sn:	Org.Comp.11-153/9
$C_4F_4H_{12}O_{14}S_6Th$	$Th(SO_3F)_4 \cdot 2\ (CH_3)_2SO$	Th:	SVol.C5-87
$C_4F_4H_{16}N_8O_8U_2$	$[UO_2F_2(CO(NH_2)_2)_2]_2$	U:	SVol.A6-35, 44
$C_4F_4KNO_4$	$NO_2CF_2CFCFCOOK$	F:	PFHOrg.8-11, 25
$C_4F_4N_2$	$NCCF_2CF_2CN$	F:	PFHOrg.9-56/7, 71, 82
$C_4F_4N_2O$	$NCCF_2OCF_2CN$	F:	PFHOrg.9-57, 72
$C_4F_4N_2O_2$	$NO_2CF_2CFCFCN$	F:	PFHOrg.8-11, 25
$C_4F_4N_2O_4$	$[FC(O)]_2NN[C(O)F]_2$	F:	PFHOrg.8-68/9, 91
$C_4F_4N_2O_7$	$[NO_2CF_2C(O)]_2O$	F:	PFHOrg.8-12, 26
$C_4F_4N_2S$	$NCCF_2SCF_2CN$	F:	PFHOrg.9-57/8, 72
		F:	PFHOrg.SVol.2-214/5, 228
$C_4F_4N_4$	$NCC(NF)CF(NF_2)CN$	F:	PFHOrg.8-173, 200
$C_4F_4O_3S$	$F_4C_4(O)_2OS$	F:	PFHOrg.SVol.2-41, 47, 52
$C_4F_5HN_2O_3$	$CF_3C(CN)(OH)CF_2NO_2$	F:	PFHOrg.8-9/10, 24
$C_4F_5H_2NO_4$	$NO_2CF_2C(O)OCH_2CF_3$	F:	PFHOrg.8-36
$C_4F_5H_2NO_5$	$CF_3C(CF_2NO_2)(OH)COOH$	F:	PFHOrg.8-11, 26
$C_4F_5H_2N_3S$	$C_2F_5(NH_2)C_2N_2S$	F:	PFHOrg.SVol.2-43, 53
$C_4F_5H_3OS$	$CF_3C(O)SCH_2CHF_2$	F:	PFHOrg.SVol.2-72
$C_4F_5H_4NO$	$CF_3NCFOCH_2CH_2F$	F:	PFHOrg.9-150
$C_4F_5H_4NO_3$	$NO_2CF_2CF_2OCH_2CH_2F$	F:	PFHOrg.8-36
$C_4F_5H_4NS$	$C_2F_5C(NH)SCH_3$	F:	PFHOrg.9-82
$C_4F_5H_5N_2$	$C_2F_5C(NH)NHCH_3$	F:	PFHOrg.9-79/80
$C_4F_5H_6O_2P$	$C_2F_5P(OCH_3)_2$	F:	PFHOrg.SVol.1-122
$C_4F_5H_6Sb$	$(CH_3)_2SbC_2F_5$	Sb:	Org.Comp.1-110
$C_4F_5H_{12}N_6P_3SSi_2$	$S(NSi(CH_3)_2)_2NP_3F_5N_3$	S:	S-N Comp.3-8/9
C_4F_5N	$c-C_3F_5CN$	F:	PFHOrg.9-58/9, 73
–	CF_2CFCF_2CN	F:	PFHOrg.9-52/3, 69
C_4F_5NO	$CF_2C(NCO)CF_3$	F:	PFHOrg.9-104, 111, 115, 117, 120
–	$[C_2F_5C(O)CN]_2$	F:	PFHOrg.9-53, 69, 79, 88
C_4F_5NOS	$C_2F_5C(O)NCS$	F:	PFHOrg.9-108, 114
$C_4F_5NO_2$	$C_2F_5C(O)NCO$	F:	PFHOrg.9-105/6, 112
$C_4F_5NO_3$	$NO_2CF_2CFCFC(O)F$	F:	PFHOrg.8-11, 26
$C_4F_6FeS_6^{2-}$	$[Fe(S_2C_2(CF_3)_2)(S)_4]^{2-}$	Fe:	Org.Comp.C7-319
$C_4F_6GeH_6$	$(CF_3)_2Ge(CH_3)_2$	F:	PFHOrg.SVol.1-63
C_4F_6HN	$(CF_3)_2CHCN$	F:	PFHOrg.9-79, 121
C_4F_6HNO	$(CF_3)_2C(CN)OH$	F:	PFHOrg.9-53/4, 78
$C_4F_6HNO_2$	$(CF_3)_2C(OH)NCO$	F:	PFHOrg.9-106, 112, 115
–	$C_2F_5CFNC(O)OH$	F:	PFHOrg.9-115/6
$C_4F_6HNO_4$	$(CF_3)_2C(NO_2)COOH$	F:	PFHOrg.8-11, 37
–	$NO_2(CF_2)_3C(O)OH$	F:	PFHOrg.8-11, 25, 41
$C_4F_6HO_5Sc$	$Sc(OH)(CF_3COO)_2 \cdot 2\ H_2O$	Sc:	MVol.D5-60/1
$C_4F_6H_2N_2O$	$(CF_3)_2CNNHCHO$	F:	PFHOrg.8-125, 129

$C_4F_7H_4NO$	$(CF_3)_2C(OCH_3)NFH$	F:	PFHOrg.8-221
$C_4F_7H_5N_2O$	$NF_2CF_2CF(NF_2)OC_2H_5$	F:	PFHOrg.8-207
$C_4F_7H_6N_3S$	$C_3F_7C(SNH_4)NNH_2$	F:	PFHOrg.8-66, 90
$C_4F_7H_{12}N_7P_4SSi_2$ . . .	$S(NSi(CH_3)_2)_2NP_4F_7N_4$	S:	S-N Comp.3-9
$C_4F_7I_3O_2S$	$n-C_3F_7SO_2Cl_3$	F:	PFHOrg.SVol.3-121/2,
			127
C_4F_7N	$(CF_3)_2CFCN$	F:	PFHOrg.9-52
–	$CF_3CF_2CF_2CN$	F:	PFHOrg.9-52, 65, 68,
			78/82, 84, 88, 95
C_4F_7NO	$(CF_3)_2CFNCO$	F:	PFHOrg.9-104, 110, 115
–	$CF_3CF_2CF_2NCO$	F:	PFHOrg.9-104, 110, 115/9
$C_4F_7NOS_3$	$(CF_3S)_2CFSNCO$	F:	PFHOrg.SVol.2-102, 107
$C_4F_7NO_2$	$(CF_2)_3CFNO_2$	F:	PFHOrg.8-11, 26
$C_4F_7NO_3$	$NO_2(CF_2)_3C(O)F$	F:	PFHOrg.8-11, 25, 41
$C_4F_7NO_3S$	$(CF_3)_2C(CN)OSO_2F$	F:	PFHOrg.9-55, 71
–	$(CF_3)_2CFSO_2NCO$	F:	PFHOrg.SVol.3-140/1,
			145
C_4F_7NS	C_3F_7SCN	F:	PFHOrg.SVol.2-219, 231
–	F_6C_4NSF	F:	PFHOrg.SVol.3-12/3, 36,
			45
$C_4F_7NS_3$	$(CF_3S)_2CFSCN$	F:	PFHOrg.SVol.2-220, 231
$C_4F_7NS_4$	$(CF_3S)_2CFSSCN$	F:	PFHOrg.SVol.2-176, 187,
			206
C_4F_7NSe	C_3F_7SeCN	F:	PFHOrg.SVol.3-221, 235
$C_4F_7N_3O_4$	$NO_2(CF_2)_2NCFCF_2NO_2$	F:	PFHOrg.8-12, 27
C_4F_8HNOS	$(CF_2)_4S(O)NH$	F:	PFHOrg.SVol.3-122/3,
			128
$C_4F_8HNO_3$	$(CF_3)_2C(OH)CF_2NO_2$	F:	PFHOrg.8-11, 25
C_4F_8HNS	$(CF_2)_4SNH$	F:	PFHOrg.SVol.3-9, 30, 43
$C_4F_8H_2N_4$	$(CF_3)(CF_2NF_2)CCH_2N_3F$	F:	PFHOrg.8-223
$C_4F_8H_2N_4O$	$CF_3C(O)NHNHC(NF)NFCF_3$	F:	PFHOrg.8-68
C_4F_8LiNS	$(CF_2)_4SNLi$	F:	PFHOrg.SVol.3-9, 38, 43
$C_4F_8N_2$	$CF_2NCF_2CF_2NCF_2$	F:	PFHOrg.9-130, 140, 149
–	$(CF_3)_2NCF_2CN$	F:	PFHOrg.9-55, 71
–	$CF_3C(NF)C(NF)CF_3$	F:	PFHOrg.8-171/2, 197
–	$CF_3NCFCFNCF_3$	F:	PFHOrg.9-130/1, 141,
			149, 153/4
$C_4F_8N_2OS$	$CF_3S(NCO)NC_2F_5$	F:	PFHOrg.SVol.3-7, 28
$C_4F_8N_2O_2$	$[FC(O)]_2NN(CF_3)_2$	F:	PFHOrg.8-69, 92
–	$FC(O)(CF_3)NN(CF_3)C(O)F$	F:	PFHOrg.8-68/9, 91/2, 106
$C_4F_8N_2O_4$	$(CF_3)_2C(NO_2)CF_2NO_2$	F:	PFHOrg.8-12, 26
–	$(CF_3)_2C(NO_2)CF_2ONO$	F:	PFHOrg.8-11/2, 26
–	$(CF_3)_2C(ONO)CF_2NO_2$	F:	PFHOrg.8-11/2, 26, 37
–	$NO_2(CF_2)_4NO_2$	F:	PFHOrg.8-12, 26
$C_4F_8N_2S_2$	$CF_3S(NCS)NC_2F_5$	F:	PFHOrg.SVol.3-7
–	$[S_2N_2CCF_3C_2F_5]$	S:	S-N Comp.3-76
$C_4F_8N_4O_5$	$NO_2(CF_2)_2N(O)N(CF_2)_2NO_2$	F:	PFHOrg.8-12, 26
C_4F_8OS	$C_4F_8S(O)$	F:	PFHOrg.SVol.3-3, 20, 39
–	$O(CF_2)_4S$	F:	PFHOrg.SVol.2-41
$C_4F_8OS_2$	$C_2F_5SSCF_2C(O)F$	F:	PFHOrg.SVol.2-177

$C_4F_{10}S$	$C_4F_8SF_2$	F:	PFHOrg.SVol.3-10, 33, 39, 44
$C_4F_{10}SSe$	$C_2F_5SeSC_2F_5$	F:	PFHOrg.SVol.3-218/9
$C_4F_{10}S_2$	$CF_3S(CF_2)_2SCF_3$	F:	PFHOrg.SVol.2-212, 226
–	$C_2F_5SSC_2F_5$	F:	PFHOrg.SVol.2-177, 189, 200
$C_4F_{10}S_3$	$C_2F_5S_3C_2F_5$	F:	PFHOrg.SVol.2-180/1, 193
$C_4F_{10}S_4$	$(CF_3S)_2CFSSCF_3$	F:	PFHOrg.SVol.2-176, 188
–	$C_2F_5S_4C_2F_5$	F:	PFHOrg.SVol.2-180/1, 194
$C_4F_{10}S_5$	$C_2F_5S_5C_2F_5$	F:	PFHOrg.SVol.2-180/1
$C_4F_{10}S_6$	$C_2F_5S_6C_2F_5$	F:	PFHOrg.SVol.2-180/1
$C_4F_{10}Se$	$C_2F_5SeC_2F_5$	F:	PFHOrg.SVol.3-220
$C_4F_{10}SeTe$	$C_2F_5SeTeC_2F_5$	F:	PFHOrg.SVol.3-250
$C_4F_{10}Se_2$	$C_2F_5Se_2C_2F_5$	F:	PFHOrg.SVol.3-218/9, 231, 243
$C_4F_{10}Se_3$	$C_2F_5Se_3C_2F_5$	F:	PFHOrg.SVol.3-218/9, 231/2, 243
$C_4F_{10}Te$	$C_2F_5TeC_2F_5$	F:	PFHOrg.SVol.3-249/51
$C_4F_{10}Te_2$	$C_2F_5TeTeC_2F_5$	F:	PFHOrg.SVol.3-249/51
$C_4F_{11}I$	$C_4F_9IF_2$	F:	PFHOrg.SVol.3-255/6, 259, 264, 265
$C_4F_{11}N$	$(CF_3)_2NC_2F_5$	F:	PFHOrg.9-160, 168, 176
–	$(CF_3)_3CNF_2$	F:	PFHOrg.8-156/7, 180
–	$(C_2F_5)_2NF$	F:	PFHOrg.8-162, 185
–	$(C_3F_7)(CF_3)NF$	F:	PFHOrg.8-162, 185
–	$n-C_4F_9NF_2$	F:	PFHOrg.8-156/7, 179
$C_4F_{11}NO$	$(CF_3)_2CFNFOCF_3$	F:	PFHOrg.8-161, 184/5
–	$n-C_3F_7NFOCF_3$	F:	PFHOrg.8-161, 184
$C_4F_{11}NOS$	$(CF_3)_2CFNS(O)FCF_3$	F:	PFHOrg.SVol.3-124, 129
–	$(CF_3)_2CN(CF_3)S(O)F_2$	F:	PFHOrg.SVol.3-124, 132
$C_4F_{11}NS$	$(CF_3)_2CFNSF(CF_3)$	F:	PFHOrg.9-4/5, 18
		F:	PFHOrg.SVol.3-7/8, 27, 43
$C_4F_{11}O_4PS$	$F_2P(O)OSO_2C_4F_9$	F:	PFHOrg.SVol.3-59, 77
$C_4F_{12}Ge$	$Ge(CF_3)_4$	F:	PFHOrg.SVol.1-53, 57/63
$C_4F_{12}HNP_2$	$[(CF_3)_2P]_2NH$	F:	PFHOrg.SVol.1-101
$C_4F_{12}HgN_2$	$[(CF_3)_2N]_2Hg$	F:	PFHOrg.8-207
		F:	PFHOrg.9-40/1, 43/6
$C_4F_{12}HgN_2S_4$	$Hg[N(SCF_3)_2]_2$	F:	PFHOrg.SVol.2-102, 107
$C_4F_{12}HgP_2S_4$	$Hg[SP(S)(CF_3)_2]_2$	F:	PFHOrg.SVol.1-139
$C_4F_{12}Hg_2N_2O_2S_3$ . . .	$(CF_3SNHgCF_3)_2SO_2$	F:	PFHOrg.SVol.2-102, 108/9
$C_4F_{12}IKO_{12}S_4$	$K[I(OSO_2CF_3)_4]$	F:	PFHOrg.SVol.3-56
$C_4F_{12}IO_{12}RbS_4$	$Rb[I(OSO_2CF_3)_4]$	F:	PFHOrg.SVol.3-56, 83, 85
$C_4F_{12}Mo_2O_{12}S_4$	$Mo_2(OSO_2CF_3)_4$	F:	PFHOrg.SVol.3-61, 79, 102/4
$C_4F_{12}NOP$	$(CF_3)_2PON(CF_3)_2$	F:	PFHOrg.SVol.1-89, 93, 95
$C_4F_{12}NOSb$	$(CF_3)_2SbON(CF_3)_2$	F:	PFHOrg.SVol.1-179/81

C$_4$GaH$_{11}$S	Ga(CH$_3$)$_2$SC$_2$H$_5$	Ga: Org.Comp.1-229
C$_4$GaH$_{12}$K	K[Ga(CH$_3$)$_4$]	Ga: Org.Comp.1-314, 316/8
C$_4$GaH$_{12}$N	Ga(CH$_3$)$_2$N(CH$_3$)$_2$	Ga: Org.Comp.1-244/6
−	Ga(CH$_3$)$_2$N(CD$_3$)$_2$	Ga: Org.Comp.1-245
−	Ga(C$_2$H$_5$)$_2$NH$_2$	Ga: Org.Comp.1-246/7
C$_4$GaH$_{12}$NO	Ga(CH$_3$)$_2$OCH$_2$CH$_2$NH$_2$	Ga: Org.Comp.1-182, 184/5
C$_4$GaH$_{12}$NOS	Ga(CH$_3$)$_2$NS(CH$_3$)$_2$O	Ga: Org.Comp.1-273/4
C$_4$GaH$_{12}$Na	Na[Ga(C$_2$H$_5$)$_2$H$_2$]	Ga: Org.Comp.1-321
C$_4$GaH$_{12}$OPS	Ga(CH$_3$)$_2$OP(S)(CH$_3$)$_2$	Ga: Org.Comp.1-204, 209/10
C$_4$GaH$_{12}$O$_2$P	Ga(CH$_3$)$_2$OP(O)(CH$_3$)$_2$	Ga: Org.Comp.1-204, 209/10
−	Ga(C$_2$H$_5$)$_2$OP(O)H$_2$	Ga: Org.Comp.1-207
C$_4$GaH$_{12}$P	Ga(CH$_3$)$_2$P(CH$_3$)$_2$	Ga: Org.Comp.1-296
C$_4$GaH$_{12}$PS$_2$	Ga(CH$_3$)$_2$SP(S)(CH$_3$)$_2$	Ga: Org.Comp.1-230
C$_4$GaH$_{12}$Rb	Rb[Ga(CH$_3$)$_4$]	Ga: Org.Comp.1-314, 318
C$_4$GaH$_{14}$N	Ga(CH$_3$)$_3$ · NH$_2$CH$_3$	Ga: Org.Comp.1-41/2, 49
−	GaCH$_3$(H)$_2$ · N(CH$_3$)$_3$	Ga: Org.Comp.1-124
−	GaCH$_3$(D)$_2$ · N(CH$_3$)$_3$	Ga: Org.Comp.1-124
C$_4$GaH$_{18}$NSi$_2$	Ga(CH$_3$)$_3$ · NCH$_3$(SiH$_3$)$_2$	Ga: Org.Comp.1-43
C$_4$Ga$_2$H$_{12}$O$_4$S	(Ga(CH$_3$)$_2$)$_2$SO$_4$	Ga: Org.Comp.1-214
C$_4$GdH$_2$O$_4$$^+$	[Gd(C$_2$H$_2$(COO)$_2$)]$^+$	Sc: MVol.D5-197/8, 199/202
C$_4$GdH$_2$O$_5$$^+$	[Gd(OOCC(O)CH$_2$COO)]$^+$	Sc: MVol.D5-338/9
C$_4$GdH$_3$O$_4$S	Gd(OOCCH(S)CH$_2$COO)	Sc: MVol.D4-57
C$_4$GdH$_3$O$_6$	[Gd(OOCCH(O)CHOHCOO)]	Sc: MVol.D5-333
C$_4$GdH$_4$I$_2$O$_4$$^+$	[Gd(CH$_2$ICOO)$_2$]$^+$	Sc: MVol.D5-53/6
C$_4$GdH$_4$NO$_8$	NH$_4$Gd(C$_2$O$_4$)$_2$ · H$_2$O	Sc: MVol.D5-141/5
−	NH$_4$Gd(C$_2$O$_4$)$_2$ · 2 H$_2$O	Sc: MVol.D5-141/5
−	NH$_4$Gd(C$_2$O$_4$)$_2$ · x H$_2$O	Sc: MVol.D5-141
C$_4$GdH$_4$N$_6$S$_4$$^+$	Gd[(S)(H$_2$N)C$_2$N$_2$S]$_2$$^+$	Sc: MVol.D4-103
C$_4$GdH$_4$O$_4$$^+$	[Gd(CH$_3$CH(COO)$_2$)]$^+$	Sc: MVol.D5-163/5
C$_4$GdH$_4$O$_5$$^+$	[Gd(O(CH$_2$COO)$_2$)]$^+$	Sc: MVol.D5-181/5
−	[Gd(OOCCH$_2$CHOHCOO)]$^+$	Sc: MVol.D5-318/21
C$_4$GdH$_4$O$_6$$^+$	[Gd(OOC(CHOH)$_2$COO)]$^+$	Sc: MVol.D5-328/34
C$_4$GdH$_5$NO$_4$$^+$	Gd[NH(CH$_2$COO)$_2$]$^+$	Sc: MVol.D1-119/20
−	[Gd(OOCCH$_2$CH(NH$_2$)COO)]$^+$	Sc: MVol.D1-109
C$_4$GdH$_5$O$_4$$^{2+}$	[GdH(C$_2$H$_4$(COO)$_2$)]$^{2+}$	Sc: MVol.D5-169
C$_4$GdH$_5$O$_5$$^{2+}$	[Gd(HOOCCH$_2$CHOHCOO)]$^{2+}$	Sc: MVol.D5-321/2
C$_4$GdH$_6$O$_4$$^+$	[Gd(CH$_3$COO)$_2$]$^+$	Sc: MVol.D5-24, 26/8
C$_4$GdH$_6$O$_4$S$_2$$^+$	[Gd(OOCCH$_2$SH)$_2$]$^+$	Sc: MVol.D4-53/4
C$_4$GdH$_6$O$_6$$^+$	[Gd(HOCH$_2$COO)$_2$]$^+$	Sc: MVol.D5-222/9
C$_4$GdH$_6$O$_8$$^+$	[Gd((HO)$_2$CHCOO)$_2$]$^+$	Sc: MVol.D5-249/50
C$_4$GdH$_7$N$_2$O$_2$$^{2+}$	[Gd(CH$_3$C(NO)C(NOH)CH$_3$)]$^{2+}$	Sc: MVol.D2-115
C$_4$GdH$_7$N$_2$O$_3$$^{2+}$	Gd(H$_2$NCH$_2$C(O)NHCH$_2$COO)$^{2+}$	Sc: MVol.D4-282/3
C$_4$GdH$_7$O$_2$$^{2+}$	[Gd(i-C$_3$H$_7$COO)]$^{2+}$	Sc: MVol.D5-83/4
C$_4$GdH$_7$O$_2$S^{2+}	[Gd(C$_2$H$_5$SCH$_2$COO)]$^{2+}$	Sc: MVol.D4-62/3
C$_4$GdH$_7$O$_3$$^{2+}$	[Gd((CH$_3$)$_2$C(OH)COO)]$^{2+}$	Sc: MVol.D5-261/6
C$_4$GdH$_7$O$_4$$^{2+}$	[Gd(HOCH$_2$C(OH)(CH$_3$)COO)]$^{2+}$	Sc: MVol.D5-268/9
C$_4$GdH$_7$O$_5$$^{2+}$	[Gd(HOC(CH$_2$OH)$_2$COO)]$^{2+}$	Sc: MVol.D5-268/9
C$_4$GdH$_8$NOS^{2+}	Gd[OC(CHSH)N(CH$_3$)$_2$]$^{2+}$	Sc: MVol.D4-61
C$_4$GdH$_8$NO$_2$S^{2+}	Gd[OC(CHSH)NHC$_2$H$_4$OH]$^{2+}$	Sc: MVol.D4-61
C$_4$GdH$_8$NO$_3$$^{2+}$	Gd[CH$_3$CHOHCH(NH$_2$)COO]$^{2+}$	Sc: MVol.D1-114

$C_4GdH_{10}O_4P^{2+}$	$Gd[(C_2H_5O)_2PO_2]^{2+}$	Sc:	MVol.D4-177
$C_4GdH_{11}N_2O_6P_2$	$Gd[H((O_3PCH_2NH)_2C_2H_4)]$	Sc:	MVol.D4-151
$C_4GdH_{13}N_3^{3+}$	$[Gd((NH_2C_2H_4)_2NH)]^{3+}$	Sc:	MVol.D1-27/8
$C_4GdH_{16}N_4^{3+}$	$[Gd(C_2H_8N_2)_2]^{3+}$	Sc:	MVol.D1-19/21
$C_4GdH_{16}N_{11}O_{13}$	$Gd(NO_3)_3 \cdot 4\ OC(NH_2)_2$	Sc:	MVol.D2-220
C_4GdKO_8	$KGd(C_2O_4)_2 \cdot 3\ H_2O$	Sc:	MVol.D5-141/5
C_4GdLiO_8	$LiGd(C_2O_4)_2 \cdot 5\ H_2O$	Sc:	MVol.D5-141/5
C_4GdNaO_8	$NaGd(C_2O_4)_2 \cdot 3\ H_2O$	Sc:	MVol.D5-141/5
C_4GdO_4	$Gd(CO)_4$	Sc:	MVol.D6-166
$C_4GdO_4^+$	$[Gd(C_4O_4)]^+$	Sc:	MVol.D3-248/50
$C_4GdO_8^-$	$[Gd(C_2O_4)_2]^-$	Sc:	MVol.D5-120/4
C_4GdO_8Rb	$RbGd(C_2O_4)_2 \cdot x\ H_2O$	Sc:	MVol.D5-141/5
$C_4GdO_{12}^{5-}$	$[Gd(CO_3)_4]^{5-}$	Sc:	MVol.D4-338/40
$C_4Gd_2H_3O_7^+$	$[Gd_2(OOCCH(O)CH(O)COO)(OH)]^+$	Sc:	MVol.D5-333
$C_4GeH_{11}I_3Sn$	$(CH_3)_3GeCH_2SnI_3$	Sn:	Org.Comp.8-140
$C_4GeH_{12}O_3S$	$(CH_3)_3GeOSO_2CH_3$	S:	SVol.3-321
$C_4GeH_{20}Mo_{12}N_{12}O_{38}$			
	$4\ HNC(NH_2)_2 \cdot GeO_2 \cdot 12\ MoO_3 \cdot 2\ H_2O$		
	$= [C(NH_2)_3]_4[GeMo_{12}O_{40}]$	Mo:	SVol.B4-319
$C_4GeH_{24}Mo_{10}N_{12}O_{39}V_2$			
	$[C(NH_2)_3]_4[GeMo_{10}V_2O_{39}]$	Mo:	SVol.B4-340
$C_4GeH_{24}Mo_{12}N_{12}O_{40}$			
	$[C(NH_2)_3]_4[GeMo_{12}O_{40}] = 4\ HNC(NH_2)_2$		
	$\cdot GeO_2 \cdot 12\ MoO_3 \cdot 2\ H_2O$	Mo:	SVol.B4-319
C_4HHoO_8	$HHo(C_2O_4)_2 \cdot 3\ H_2O = [H_5O_2][Ho(C_2O_4)_2H_2O]$		
		Sc:	MVol.D5-139/41
$C_4HIN_4ORh^{3-}$	$[Rh(CN)_4(OH)I]^{3-}$	Rh:	SVol.B1-187
$C_4HKN_2O_4S$	$SN_2C_2(COOH)COOK$	S:	S-N Comp.3-93/8, 184
$C_4HK_2N_4ORh$	$K_2[Rh(CN)_4OH]$	Rh:	SVol.B1-196
$C_4HK_3O_{11}U$	$K_3[UO_2(C_2O_4)_2(OH)]$	U:	SVol.C13-210/2
	$K_3[UO_2(C_2O_4)_2(OH)] \cdot 2\ H_2O$	U:	SVol.C13-210/1
$C_4HMnN_3O_4$	$Mn[(OOC)(O)_2C_3N_3H]$	Mn:	MVol.D4-337
$C_4HN_2NaO_4S$	$SN_2C_2(COOH)COONa$	S:	S-N Comp.3-93/8, 184
$C_4HN_4O_2Tc^{3-}$	$[TcO(OH)(CN)_4]^{3-}$	Tc:	SVol.2-125
$C_4HN_4O_2TcTl_3$	$Tl_3[TcO(OH)(CN)_4]$	Tc:	SVol.2-125
$C_4HN_4RhS_6$	$HRh(S(C(S)N)_2)_2$	Rh:	SVol.B3-47
$C_4HNa_2O_{16}PU_2$	$Na_2[(UO_2)_2(C_2O_4)_2(HPO_4)] \cdot 4\ H_2O$	U:	SVol.C14-129
−	$Na_2[(UO_2)_2(C_2O_4)_2(HPO_4)] \cdot 4.5\ H_2O$	U:	SVol.A5-139
		U:	SVol.C13-234
$C_4HNa_3O_{11}U$	$Na_3[UO_2(C_2O_4)_2(OH)]$	U:	SVol.C13-210/2
−	$Na_3[UO_2(C_2O_4)_2(OH)] \cdot H_2O$	U:	SVol.C13-210/1
C_4HO_8Tb	$HTb(C_2O_4)_2 \cdot 3\ H_2O = [H_5O_2][Tb(C_2O_4)_2H_2O]$		
		Sc:	MVol.D5-139/41
C_4HO_8Tm	$HTm(C_2O_4)_2 \cdot 3\ H_2O = [H_5O_2][Tm(C_2O_4)_2H_2O]$		
		Sc:	MVol.D5-139/41
C_4HO_8Y	$HY(C_2O_4)_2 \cdot 3\ H_2O = [H_5O_2][Y(C_2O_4)_2H_2O]$		
		Sc:	MVol.D5-139/41
C_4HO_8Yb	$HYb(C_2O_4)_2 \cdot 3\ H_2O = [H_5O_2][Yb(C_2O_4)_2H_2O]$		
		Sc:	MVol.D5-139/41
$C_4H_2HgN_4S_4$	$(SN_2C_2HS)_2Hg$	S:	S-N Comp.3-93/8, 161/2

$C_4H_2N_6Rh_2$	$Rh_2(CN)_4 \cdot N_2H_2 \cdot 6 H_2O$	Rh:	SVol.B1-197
$C_4H_2NaNdO_6$	$[NaNd(OOCCH(O)CH(O)COO)] \cdot 4 H_2O$	Sc:	MVol.D5-336/7
$C_4H_2NaO_6Pr$	$[NaPr(OOCCH(O)CH(O)COO)] \cdot 4 H_2O$	Sc:	MVol.D5-336/7
$C_4H_2NaO_6Y$	$[NaY(OOCCH(O)CH(O)COO)] \cdot 5 H_2O$	Sc:	MVol.D5-336/7
$C_4H_2Na_2O_{11}U$	$Na_2[UO_2(C_2O_4)_2(H_2O)] \cdot 4 H_2O$	U:	SVol.C13-193/4, 199
$C_4H_2NdO_4^+$	$[Nd(C_2H_2(COO)_2)]^+$	Sc:	MVol.D5-197/8, 199/202
$C_4H_2O_4Pr^+$	$[Pr(C_2H_2(COO)_2)]^+$	Sc:	MVol.D5-197/8, 199/202
$C_4H_2O_4Sc^+$	$[Sc(C_2H_2(COO)_2)]^+$	Sc:	MVol.D5-199/202
$C_4H_2O_4Sm^+$	$[Sm(C_2H_2(COO)_2)]^+$	Sc:	MVol.D5-197/8, 199/202
$C_4H_2O_4Tb^+$	$[Tb(C_2H_2(COO)_2)]^+$	Sc:	MVol.D5-197/8, 199/202
$C_4H_2O_4Tm^+$	$[Tm(C_2H_2(COO)_2)]^+$	Sc:	MVol.D5-197/8, 199/202
$C_4H_2O_4Y^+$	$[Y(C_2H_2(COO)_2)]^+$	Sc:	MVol.D5-199/202
$C_4H_2O_4Yb^+$	$[Yb(C_2H_2(COO)_2)]^+$	Sc:	MVol.D5-197/8, 199/202
$C_4H_2O_5Y^+$	$[Y(OOCC(O)CH_2COO)]^+$	Sc:	MVol.D5-338/9
$C_4H_2O_6Pr^-$	$[Pr(OOCCH(O)CH(O)COO)]^-$	Sc:	MVol.D5-333
$C_4H_2O_6RbY$	$[RbY(OOCCH(O)CH(O)COO)] \cdot 4 H_2O$	Sc:	MVol.D5-336/7
$C_4H_2O_6U$	$UO_2[OOC(CH)_2COO]$	U:	SVol.C13-249/50
		U:	SVol.D1-205/8
$-$	$UO_2[OOC(CH)_2COO] \cdot 3 H_2O$	U:	SVol.C13-249/50
$-$	$UO_2[OOC(CH)_2COO] \cdot 6 H_2O$	U:	SVol.C13-249/50
$C_4H_2O_{11}Rb_2U$	$Rb_2[UO_2(C_2O_4)_2(H_2O)] \cdot H_2O$	U:	SVol.C13-193, 196/9
$C_4H_2O_{12}U_2$	$(UO_2)_2(C_2O_4)(HCOO)_2$	U:	SVol.C13-235/6
$C_4H_2O_{13}U_2$	$UO_2[UO_2(C_2O_4)_2(H_2O)] \cdot 4 H_2O$	U:	SVol.C13-185
$C_4H_2O_{14}U^{6-}$	$[U(CO_3)_4(OH)_2]^{6-}$	U:	SVol.D1-169/72
$C_4H_3HoO_4S$	$Ho(OOCCH(S)CH_2COO)$	Sc:	MVol.D4-57
$C_4H_3KO_7U$	$K[UO_2(OCH(COO)CH_2COO)] \cdot H_2O$	U:	SVol.C13-245/6
$C_4H_3K_2LaO_7$	$K_2La(OOCCH(O)CH(O)COO)(OH) \cdot 4 H_2O$	Sc:	MVol.D5-337
$C_4H_3K_2N_4ORh$	$K_2[Rh(CN)_4H(H_2O)]$	Rh:	SVol.B1-195/6
$C_4H_3K_2N_4O_3Rh$	$K_2[Rh(CN)_4(OOH)(H_2O)]$	Rh:	SVol.B1-196
$C_4H_3K_2O_7Y$	$K_2Y(OOCCH(O)CH(O)COO)(OH) \cdot 4 H_2O$	Sc:	MVol.D5-337
$C_4H_3K_3O_{12}U$	$K_3[UO_2(C_2O_4)_2(OH)(H_2O)]$	U:	SVol.C13-210/1
$-$	$K_3[UO_2(C_2O_4)_2(OH)(H_2O)] \cdot H_2O$	U:	SVol.C13-210/1
$C_4H_3LaN_3O^{3+}$	$[La(HN_2C_4(O)NH_2)]^{3+}$	Sc:	MVol.D4-276
$C_4H_3LaO_4S$	$La(OOCCH(S)CH_2COO)$	Sc:	MVol.D4-57
$C_4H_3LaO_5$	$La(OOCCH_2CH(O)COO)$	Sc:	MVol.D5-321/2
$-$	$La(OOCCH_2CH(O)COO) \cdot 3 H_2O$	Sc:	MVol.D5-325
$C_4H_3LaO_6$	$La(OOCCH(O)CHOHCOO)$	Sc:	MVol.D5-333
$-$	$La(OOCCH(O)CHOHCOO) \cdot 4 H_2O$		
	$= La(OH)(OOC(CHOH)_2COO) \cdot 3 H_2O$	Sc:	MVol.D5-336
$C_4H_3MnN_2O_2^+$	$Mn[2,4-(O)_2-1,3-C_4H_2N_2H]^+$	Mn:	MVol.D4-8/9
$C_4H_3MnO_5^-$	$[Mn(OC_2H_3(COO)_2)]^-$	Mn:	MVol.D2-167
$C_4H_3MnO_6$	$Mn(OCOCH(O)CH(OH)COO) \cdot 2 H_2O$	Mn:	MVol.D2-175
$C_4H_3MnO_6^-$	$[Mn(OCOCH(O)CH(OH)COO)]^-$	Mn:	MVol.D2-169/70
$C_4H_3MnO_{10}^{2-}$	$[Mn(C_2O_4)_2(OH)(H_2O)]^{2-}$	Mn:	MVol.D2-120/1
$C_4H_3Mn_2O_7$	$Mn[Mn(OH)(OCOCH(O)CH(O)COO)] \cdot 5 H_2O$		
		Mn:	MVol.D2-175
$C_4H_3N_3OS$	$SN_2C_2(CN)OCH_3$	S:	S-N Comp.3-93/8, 168
$C_4H_3N_3O_3S$	$SN_2C_2(COOH)CONH_2$	S:	S-N Comp.3-93/8, 187
$C_4H_3N_3S$	$SN_2C_2HCH_2CN$	S:	S-N Comp.3-93/8, 124
$C_4H_3N_3S_3Sn$	$CH_3Sn(NCS)_3$	Sn:	Org.Comp.8-191

$C_4H_3N_4ORh^{2-}$	$[Rh(CN)_4H(H_2O)]^{2-}$	Rh:	SVol.B1-195
$C_4H_3N_4OS$	$SN_2C_4H(OH)N_2H$	S:	S-N Comp.3-229, 232
$C_4H_3N_4O_3Tc^{3-}$	$[Tc(OH)_3(CN)_4]^{3-}$	Tc:	SVol.2-200
$C_4H_3N_5OS$	$SN_2C_4(NH_2)(O)N_2H$	S:	S-N Comp.3-233/4, 237
–	$SN_2C_4(OH)(NH_2)N_2$	S:	S-N Comp.3-232/4, 237
$C_4H_3N_5S$	$SN_2C_4H(NH_2)N_2$	S:	S-N Comp.3-232/5
$C_4H_3N_5S_2$	$SN_2C_4(NH_2)(S)N_2H$	S:	S-N Comp.3-233/4, 237
–	$SN_2C_4(SH)(NH_2)N_2$	S:	S-N Comp.3-232/4, 237
$C_4H_3N_7S_2$	$SN_2C_2HNNNHC_2HN_2S$	S:	S-N Comp.3-93/8, 143
$C_4H_3NdO_4S$	$Nd(OOCCH(S)CH_2COO)$	Sc:	MVol.D4-57
$C_4H_3NdO_6$	$Nd(OOCCH(O)CHOHCOO) \cdot 3 H_2O$		
	$= Nd(OH)(OOC(CHOH)_2COO) \cdot 2 H_2O$	Sc:	MVol.D5-336
$C_4H_3O_4PrS$	$Pr(OOCCH(S)CH_2COO)$	Sc:	MVol.D4-57
$C_4H_3O_4SSm$	$Sm(OOCCH(S)CH_2COO)$	Sc:	MVol.D4-57
$C_4H_3O_5Sc$	$Sc(OH)(C_2H_2(COO)_2) \cdot 0.5 H_2O$	Sc:	MVol.D5-198
–	$Sc(OH)(C_2H_2(COO)_2) \cdot H_2O$	Sc:	MVol.D5-198
–	$Sc(OOCCH_2CH(O)COO) \cdot 3 H_2O$		
	$= Sc(OH)(OOCCH_2CHOHCOO) \cdot 2 H_2O$	Sc:	MVol.D5-323
$C_4H_3O_5Y$	$Y(OOCCH_2CH(O)COO)$	Sc:	MVol.D5-321/2
–	$Y(OOCCH_2CH(O)COO) \cdot H_2O$	Sc:	MVol.D5-325
$C_4H_3O_6Pr$	$Pr(OOCCH(O)CHOHCOO) \cdot 3 H_2O$		
	$= Pr(OH)(OOC(CHOH)_2COO) \cdot 2 H_2O$	Sc:	MVol.D5-336
$C_4H_3O_6U^+$	$[UO_2H(OOCCHCHCOO)]^+$	U:	SVol.D1-208
$C_4H_3O_6Y$	$Y(OOCCH(O)CHOHCOO)$	Sc:	MVol.D5-333
–	$Y(OOCCH(O)CHOHCOO) \cdot 4 H_2O$		
	$= Y(OH)(OOC(CHOH)_2COO) \cdot 3 H_2O$	Sc:	MVol.D5-336
$C_4H_3O_7U^-$	$[UO_2(OOCCH_2CHOCOO)]^-$	U:	SVol.D1-209
$C_4H_3O_8U^-$	$[UO_2H(OOCCHOCHOCOO)]^-$	U:	SVol.D1-209/10
$C_4H_3O_{12}U^{3-}$	$[UO_2(C_2O_4)_2(OH)(H_2O)]^{3-}$	U:	SVol.C13-211/2
$C_4H_4HoI_2O_4^+$	$[Ho(CH_2ICOO)_2]^+$	Sc:	MVol.D5-53/6
$C_4H_4HoNO_8$	$NH_4Ho(C_2O_4)_2 \cdot H_2O$	Sc:	MVol.D5-141/5
$C_4H_4HoN_6S_4^+$	$Ho[(S)(H_2N)C_2N_2S]_2^+$	Sc:	MVol.D4-103
$C_4H_4HoO_4^+$	$[Ho(CH_3CH(COO)_2)]^+$	Sc:	MVol.D5-163/5
$C_4H_4HoO_5^+$	$[Ho(O(CH_2COO)_2)]^+$	Sc:	MVol.D5-181/5
–	$[Ho(OOCCH_2CHOHCOO)]^+$	Sc:	MVol.D5-318/21
$C_4H_4HoO_6^+$	$[Ho(OOC(CHOH)_2COO)]^+$	Sc:	MVol.D5-328/34
$C_4H_4I_2K_2O_8Rh_2$	$K_2[Rh_2(O_2CH)_4I_2] \cdot H_2O$	Rh:	SVol.B2-13
$C_4H_4I_2LaO_4^+$	$[La(CH_2ICOO)_2]^+$	Sc:	MVol.D5-53/6
$C_4H_4I_2LuO_4^+$	$[Lu(CH_2ICOO)_2]^+$	Sc:	MVol.D5-53/6
$C_4H_4I_2MnO_4$	$Mn(CH_2ICOO)_2 \cdot 2 H_2O$	Mn:	MVol.D2-62
$C_4H_4I_2NdO_4^+$	$[Nd(CH_2ICOO)_2]^+$	Sc:	MVol.D5-53/6
$C_4H_4I_2O_4Pr^+$	$[Pr(CH_2ICOO)_2]^+$	Sc:	MVol.D5-53/6
$C_4H_4I_2O_4Sm^+$	$[Sm(CH_2ICOO)_2]^+$	Sc:	MVol.D5-53/6
$C_4H_4I_2O_4Tb^+$	$[Tb(CH_2ICOO)_2]^+$	Sc:	MVol.D5-53/6
$C_4H_4I_2O_4Tm^+$	$[Tm(CH_2ICOO)_2]^+$	Sc:	MVol.D5-53/6
$C_4H_4I_2O_4Y^+$	$[Y(CH_2ICOO)_2]^+$	Sc:	MVol.D5-53/6
$C_4H_4I_2O_4Yb^+$	$[Yb(CH_2ICOO)_2]^+$	Sc:	MVol.D5-53/6
$C_4H_4I_2O_5U$	$UO(CH_2ICOO)_2 \cdot H_2O$	U:	SVol.C13-151/2
$C_4H_4I_3N_2RhS_2$	$[Rh((NCSCH_2)_2)I_3]$	Rh:	SVol.B3-21
$C_4H_4KMnO_7$	$K[Mn(OH)(C_4H_3O_6)] \cdot H_2O$	Mn:	MVol.D2-177

$C_4H_4KMnO_{10}$	$K[Mn(C_2O_4)_2(H_2O)_2]$	Mn:	MVol.D2-116, 119
–	$K[Mn(C_2O_4)_2(H_2O)_2] \cdot 3 H_2O$	Mn:	MVol.D2-116/20
$C_4H_4KNO_7SU$	$K[U(HCOO)_3(OH)(NCS)] \cdot 2 H_2O$	U:	SVol.C13-109
$C_4H_4KN_3O_3S$	$(O)SN_2C_2(OK)(N(C_2H_4)O)$	S:	S-N Comp.3-250/7
$C_4H_4KN_4O_2Rh$	$K[Rh(CN)_4(H_2O)_2]$	Rh:	SVol.B1-196
$C_4H_4KO_6Y$	$KY(OH)(OOCCH_2CH(O)COO)$	Sc:	MVol.D5-325
$C_4H_4KO_{10}Rh$	$K[Rh(C_2O_4)_2(H_2O)_2]$	Rh:	SVol.B2-7
–	$K[Rh(C_2O_4)_2(H_2O)_2] \cdot H_2O$	Rh:	SVol.B2-7
–	$K[Rh(C_2O_4)_2(H_2O)_2] \cdot 1.75 H_2O$	Rh:	SVol.B2-7
–	$K[Rh(C_2O_4)_2(H_2O)_2] \cdot 2 H_2O$	Rh:	SVol.B2-7
–	$K[Rh(C_2O_4)_2(H_2O)_2] \cdot 3 H_2O$	Rh:	SVol.B2-7
$C_4H_4K_2MnO_8$	$K_2Mn(HCOO)_4 \cdot n H_2O$	Mn:	MVol.D2-1/2, 26
$C_4H_4K_2O_{12}U$	$K_2[UO_2(C_2O_4)_2(H_2O)_2] \cdot H_2O$	U:	SVol.C13-193/4, 198/9
$C_4H_4K_3O_8Y$	$K_3Y(OOCCH(O)CH(O)COO)(OH)_2 \cdot 4 H_2O$.	Sc:	MVol.D5-337
$C_4H_4LaNO_8$	$NH_4La(C_2O_4)_2 \cdot 3 H_2O$	Sc:	MVol.D5-141/5
$C_4H_4LaO_4^+$	$[La(CH_3CH(COO)_2)]^+$	Sc:	MVol.D5-163/5
–	$[La(C_2H_4(COO)_2)]^+$	Sc:	MVol.D5-169
$C_4H_4LaO_4S^+$	$[La(OOCCH(SH)CH_2COO)]^+$	Sc:	MVol.D4-57
$C_4H_4LaO_5^+$	$[La(O(CH_2COO)_2)]^+$	Sc:	MVol.D5-181/5
–	$[La(OOCCH_2CHOHCOO)]^+$	Sc:	MVol.D5-318/21
$C_4H_4LaO_6^+$	$[La(OOC(CHOH)_2COO)]^+$	Sc:	MVol.D5-328/34
$C_4H_4La_2O_5^{4+}$	$[La_2(OOCCH_2CHOHCOO)]^{4+}$	Sc:	MVol.D5-321/2
$C_4H_4La_2O_6^{4+}$	$[La_2(OOC(CHOH)_2COO)]^{4+}$	Sc:	MVol.D5-333
$C_4H_4LiMnO_7$	$Li[Mn(OH)(C_4H_3O_6)] \cdot H_2O$	Mn:	MVol.D2-177
$C_4H_4LuO_4^+$	$[Lu(CH_3CH(COO)_2)]^+$	Sc:	MVol.D5-163/5
$C_4H_4LuO_5^+$	$[Lu(O(CH_2COO)_2)]^+$	Sc:	MVol.D5-181/5
–	$[Lu(OOCCH_2CHOHCOO)]^+$	Sc:	MVol.D5-318/21
$C_4H_4LuO_6^+$	$[Lu(OOC(CHOH)_2COO)]^+$	Sc:	MVol.D5-328/34
$C_4H_4MnN_2O_4S$	$Mn(1,2-C_4H_4N_2)SO_4$	Mn:	MVol.D4-3
$C_4H_4MnN_3O_2^+$	$Mn(HN_2C_3(O)(CH_3)(NO))^+$	Mn:	MVol.D5-248/9
$C_4H_4MnN_4O_2^{2-}$	$[Mn(CN)_4(H_2O)_2]^{2-}$	Mn:	MVol.D2-203, 207
$C_4H_4MnN_4O_2S_4^{2-}$ ..	$[Mn(NCS)_4(H_2O)_2]^{2-}$	Mn:	MVol.D2-282/5
$C_4H_4MnN_6^{2+}$	$Mn(6-NH_2-1,3,7,8,9-C_4HN_5H)^{2+}$	Mn:	MVol.D4-71/2
$C_4H_4MnN_{10}O_2$	$Mn(N_4CCONH_2)_2$	Mn:	MVol.D5-121/2
–	$Mn(N_4CCONH_2)_2 \cdot 2 H_2O$	Mn:	MVol.D5-121/2
$C_4H_4MnNaO_7$	$Na[Mn(OH)(C_4H_3O_6)] \cdot 4 H_2O$	Mn:	MVol.D2-177
$C_4H_4MnNaO_{10}$	$Na[Mn(C_2O_4)_2(H_2O)_2] \cdot n H_2O$	Mn:	MVol.D2-119
$C_4H_4MnNa_2O_8$	$Na_2Mn(HCOO)_4 \cdot n H_2O$	Mn:	MVol.D2-1/2, 26
$C_4H_4MnO_4$	$Mn[C_2H_4(COO)_2]$	Mn:	MVol.D2-136/7
–	$Mn[C_2H_4(COO)_2] \cdot 4 H_2O$	Mn:	MVol.D2-137
$C_4H_4MnO_5$	$Mn(HOC_2H_3(COO)_2)$	Mn:	MVol.D2-167
–	$Mn(HOC_2H_3(COO)_2) \cdot 3 H_2O$	Mn:	MVol.D2-168/9
–	$Mn(HOC_2H_3(COO)_2) \cdot 4 H_2O$	Mn:	MVol.D2-168
–	$Mn[O(CH_2COO)_2]$	Mn:	MVol.D2-138
$C_4H_4MnO_5^+$	$[Mn(HOC_2H_3(COO)_2)]^+$	Mn:	MVol.D2-167
$C_4H_4MnO_6$	$Mn(OCO(CHOH)_2COO)$	Mn:	MVol.D2-169/71
–	$Mn(OCO(CHOH)_2COO) \cdot 2 H_2O$	Mn:	MVol.D2-171/2
–	$Mn(OCO(CHOH)_2COO) \cdot 3 H_2O$	Mn:	MVol.D2-171
$C_4H_4MnO_{10}^-$	$[Mn(C_2O_4)_2(H_2O)_2]^-$	Mn:	MVol.D2-107/8, 116/8
$C_4H_4MoO_{10}U$	$UO_2[MoO_3(HOCHCH_2(COO)_2)] \cdot 2 H_2O$ ...	U:	SVol.C13-245/6

$C_4H_4O_4SY^+$	$[Y(OOCCH(SH)CH_2COO)]^+$	Sc:	MVol.D4-57
$C_4H_4O_4SYb^+$	$Yb[S(CH_2COO)_2]^+$	Sc:	MVol.D4-63
$C_4H_4O_4Sm^+$	$[Sm(CH_3CH(COO)_2)]^+$	Sc:	MVol.D5-163/5
—	$[Sm(C_2H_4(COO)_2)]^+$	Sc:	MVol.D5-169
$C_4H_4O_4Tb^+$	$[Tb(CH_3CH(COO)_2)]^+$	Sc:	MVol.D5-163/5
$C_4H_4O_4Tm^+$	$[Tm(CH_3CH(COO)_2)]^+$	Sc:	MVol.D5-163/5
$C_4H_4O_4U^{2+}$	$[U(OOC(CH_2)_2COO)]^{2+}$	U:	SVol.D1-207
$C_4H_4O_4Yb^+$	$[Yb(CH_3CH(COO)_2)]^+$	Sc:	MVol.D5-163/5
$C_4H_4O_5Pr^+$	$[Pr(O(CH_2COO)_2)]^+$	Sc:	MVol.D5-181/5
—	$[Pr(OOCCH_2CHOHCOO)]^+$	Sc:	MVol.D5-318/21
$C_4H_4O_5S_2Tc^-$	$[TcO(SCH_2COO)_2]^-$	Tc:	SVol.2-232
$C_4H_4O_5Sc^+$	$[Sc(O(CH_2COO)_2)]^+$	Sc:	MVol.D5-181/5
$C_4H_4O_5Sm^+$	$[Sm(O(CH_2COO)_2)]^+$	Sc:	MVol.D5-181/5
—	$[Sm(OOCCH_2CHOHCOO)]^+$	Sc:	MVol.D5-318/21
$C_4H_4O_5Tb^+$	$[Tb(O(CH_2COO)_2)]^+$	Sc:	MVol.D5-181/5
—	$[Tb(OOCCH_2CHOHCOO)]^+$	Sc:	MVol.D5-318/21
$C_4H_4O_5Tm^+$	$[Tm(O(CH_2COO)_2)]^+$	Sc:	MVol.D5-181/5
—	$[Tm(OOCCH_2CHOHCOO)]^+$	Sc:	MVol.D5-318/21
$C_4H_4O_5U$	$UO[(CH_2COO)_2] \cdot 3 H_2O$	U:	SVol.C13-242/4
$C_4H_4O_5Y^+$	$[Y(O(CH_2COO)_2)]^+$	Sc:	MVol.D5-181/5
—	$[Y(OOCCH_2CHOHCOO)]^+$	Sc:	MVol.D5-318/21
$C_4H_4O_5Y_2^{4+}$	$[Y_2(OOCCH_2CHOHCOO)]^{4+}$	Sc:	MVol.D5-321/2
$C_4H_4O_5Yb^+$	$[Yb(O(CH_2COO)_2)]^+$	Sc:	MVol.D5-181/5
—	$[Yb(OOCCH_2CHOHCOO)]^+$	Sc:	MVol.D5-318/21
$C_4H_4O_6Pm^+$	$[Pm(OOC(CHOH)_2COO)]^+$	Sc:	MVol.D5-328/34
$C_4H_4O_6Pr^+$	$[Pr(OOC(CHOH)_2COO)]^+$	Sc:	MVol.D5-328/34
$C_4H_4O_6SU$	$UO_2[OOCCH(SH)CH_2COO]$	U:	SVol.C13-248/9
		U:	SVol.D1-209
—	$UO_2((OOCCH_2)_2S)$	U:	SVol.D1-210
$C_4H_4O_6S_2U$	$UO_2(OOCCHSHCHSHCOO)$	U:	SVol.D1-209
$C_4H_4O_6Sc^+$	$[Sc(OOC(CHOH)_2COO)]^+$	Sc:	MVol.D5-328/34
$C_4H_4O_6Sm^+$	$[Sm(OOC(CHOH)_2COO)]^+$	Sc:	MVol.D5-328/34
$C_4H_4O_6Tb^+$	$[Tb(OOC(CHOH)_2COO)]^+$	Sc:	MVol.D5-328/34
$C_4H_4O_6Tm^+$	$[Tm(OOC(CHOH)_2COO)]^+$	Sc:	MVol.D5-328/34
$C_4H_4O_6U$	$UO_2[(CH_2COO)_2]$	U:	SVol.D1-205/7
—	$UO_2[(CH_2COO)_2] \cdot H_2O$	U:	SVol.A6-35
		U:	SVol.C13-242/4
—	$UO_2[(CH_2COO)_2] \cdot 2 H_2O$	U:	SVol.C13-242/4
—	$UO_2[(CH_2COO)_2] \cdot 3 H_2O$	U:	SVol.C13-242/4
—	$UO_2(CH_3CH(COO)_2)$	U:	SVol.D1-207
$C_4H_4O_6Y^+$	$[Y(OOC(CHOH)_2COO)]^+$	Sc:	MVol.D5-328/34
$C_4H_4O_6Y_2^{4+}$	$[Y_2(OOC(CHOH)_2COO)]^{4+}$	Sc:	MVol.D5-333
$C_4H_4O_6Yb^+$	$[Yb(OOC(CHOH)_2COO)]^+$	Sc:	MVol.D5-328/34
$C_4H_4O_7Pr^-$	$[Pr(OOCCH(O)CHOHCOO)(OH)]^-$	Sc:	MVol.D5-333
$C_4H_4O_7U$	$UO_2[HOCHCH_2(COO)_2]$	U:	SVol.D1-209
—	$UO_2[HOCHCH_2(COO)_2] \cdot H_2O$	U:	SVol.C13-245/6
—	$UO_2[HOCHCH_2(COO)_2] \cdot 3 H_2O$	U:	SVol.C13-245/6
—	$UO_2[O(CH_2COO)_2]$	U:	SVol.A6-34
		U:	SVol.D1-210
—	$[UO_2(O(CH_2COO)_2)]_n$	U:	SVol.C13-255/8

$C_4H_4O_8Rh_2$	$Rh_2(O_2CH)_4$	Rh:	SVol.B2-10
$C_4H_4O_8Rh_2^+$	$[Rh_2(O_2CH)_4]^+$	Rh:	SVol.B2-56
$C_4H_4O_8U$	$HUO_2(C_2O_4)(CH_3COO)$	U:	SVol.C13-235/6
$-$	$U(HCOO)_4$	U:	SVol.A5-94
		U:	SVol.A6-22
		U:	SVol.C13-97/100
$-$	$U(HCOO)_4$ systems		
	$U(HCOO)_4-HCOOH-H_2O$	U:	SVol.C13-100
$-$	$UO_2[OOC(CHOH)_2COO] \cdot H_2O$	U:	SVol.C13-246/8
$-$	$UO_2[OOC(CHOH)_2COO] \cdot 4 H_2O$	U:	SVol.C13-246/8
$C_4H_4O_9STh$	$Th(O(CH_2COO)_2)SO_4 \cdot 3 H_2O$	Th:	SVol.C5-77/9
$C_4H_4O_{10}Rh^-$	$[Rh(C_2O_4)_2(H_2O)_2]^-$	Rh:	SVol.B2-2/5
$C_4H_4O_{10}SrU$	$SrUO_2(HCOO)_4 \cdot 1.38 H_2O$	U:	SVol.A6-35
		U:	SVol.C13-102, 105, 107/8
$C_4H_4O_{10}U^{2-}$	$UO_2(HCOO)_4^{2-}$	U:	SVol.C13-106/7
$C_4H_4O_{12}Rb_2U$	$Rb_2[UO_2(C_2O_4)_2(H_2O)_2]$	U:	SVol.C13-193, 196, 198/9
$C_4H_4O_{12}Tl_2U$	$Tl_2[UO_2(C_2O_4)_2(H_2O)_2]$	U:	SVol.C13-193/9
$C_4H_4O_{14}Rh_2^{4-}$	$[Rh_2(CO_3)_4(H_2O)_2]^{4-}$	Rh:	SVol.B1-183/4
$C_4H_5HoNO_4^+$	$Ho[NH(CH_2COO)_2]^+$	Sc:	MVol.D1-119/21
$-$	$[Ho(OOCCH_2CH(NH_2)COO)]^+$	Sc:	MVol.D1-109/10
$C_4H_5HoO_5^{2+}$	$[Ho(HOOCCH_2CHOHCOO)]^{2+}$	Sc:	MVol.D5-321/2
$C_4H_5KN_2O_3S$	$(O)SN_2C_2(OK)(OC_2H_5)$	S:	S-N Comp.3-250/7
$C_4H_5KO_9U$	$K[U(HCOO)_4(OH)] \cdot 3 H_2O$	U:	SVol.C13-104
$C_4H_5KO_9Y_2$	$KY_2(OOCCH(O)CH(O)COO)(OH)_3 \cdot 5 H_2O$	Sc:	MVol.D5-337
$C_4H_5K_2NO_4$	$K_2(HN(CH_2COO)_2)$ systems		
	$K_2(HN(CH_2COO)_2)-Mn(HN(CH_2COO)_2)-H_2O$		
		Mn:	MVol.D5-6
$C_4H_5K_4O_9Y$	$K_4Y(OOCCH(O)CH(O)COO)(OH)_3 \cdot 3 H_2O$	Sc:	MVol.D5-337
$C_4H_5LaNO_4^+$	$La[NH(CH_2COO)_2]^+$	Sc:	MVol.D1-119/20
$-$	$[La(OOCCH_2CH(NH_2)COO)]^+$	Sc:	MVol.D1-109
$C_4H_5LaO_4^{2+}$	$[LaH(C_2H_4(COO)_2)]^{2+}$	Sc:	MVol.D5-169
$C_4H_5LaO_5^{2+}$	$[La(HOOCCH_2CHOHCOO)]^{2+}$	Sc:	MVol.D5-321/2
$C_4H_5LaO_6$	$La(OH)(OOCCH_2CHOHCOO) \cdot 4 H_2O$	Sc:	MVol.D5-324
$C_4H_5LaO_6^{2+}$	$[La(HOOC(CHOH)_2COO)]^{2+}$	Sc:	MVol.D5-333
$C_4H_5LaO_7$	$La(OH)(OOC(CHOH)_2COO) \cdot 3 H_2O$		
	$= La(OOCCH(O)CHOHCOO) \cdot 4 H_2O$ ...	Sc:	MVol.D5-336
$C_4H_5Li_2NO_4$	$Li_2(HN(CH_2COO)_2)$ systems		
	$Li_2(HN(CH_2COO)_2)-Mn(HN(CH_2COO)_2)-H_2O$		
		Mn:	MVol.D5-6
$C_4H_5LuNO_4^+$	$Lu[NH(CH_2COO)_2]^+$	Sc:	MVol.D1-119/21
$-$	$[Lu(OOCCH_2CH(NH_2)COO)]^+$	Sc:	MVol.D1-109/10
$C_4H_5LuO_4^{2+}$	$[LuH(C_2H_4(COO)_2)]^{2+}$	Sc:	MVol.D5-169
$C_4H_5LuO_5^{2+}$	$[Lu(HOOCCH_2CHOHCOO)]^{2+}$	Sc:	MVol.D5-321/2
$C_4H_5MnNO_4$	$Mn(HN(CH_2COO)_2)$	Mn:	MVol.D5-4, 5
$-$	$Mn(HN(CH_2COO)_2) \cdot 1.5 H_2O$	Mn:	MVol.D5-5
$-$	$Mn(HN(CH_2COO)_2) \cdot 3.5 H_2O$	Mn:	MVol.D5-5
$-$	$Mn(HN(CH_2COO)_2) \cdot 6 H_2O$	Mn:	MVol.D5-5

C$_4$H$_6$MgO$_4$.........	Mg(CH$_3$COO)$_2$ systems		
	Mg(CH$_3$COO)$_2$-H$_3$BO$_3$-H$_2$O...........	B:	B Comp.SVol.1/1-188
		B:	B Comp.SVol.3/2-105
	Mg(CH$_3$COO)$_2$-Mn(CH$_3$COO)$_2$-H$_2$O	Mn:	MVol.D2-53
C$_4$H$_6$MnN$_2$O$_2$	Mn(N$_2$(COCH$_3$)$_2$) · H$_2$O	Mn:	MVol.D5-196
–	Mn(N$_2$(COCH$_3$)$_2$) · 2 H$_2$O	Mn:	MVol.D5-196
C$_4$H$_6$MnNa$_2$O$_8$	Na$_2$[Mn(OH)$_2$(OCO(CHOH)$_2$COO)] · 4 H$_2$O	Mn:	MVol.D2-173
C$_4$H$_6$MnO$_4$.........	Mn(CH$_3$COO)$_2$......................	Mn:	MVol.A1-165
		Mn:	MVol.D2-29/34
–	Mn(CH$_3$COO)$_2$ · H$_2$O	Mn:	MVol.D2-29, 34
–	Mn(CH$_3$COO)$_2$ · 2 H$_2$O	Mn:	MVol.D2-29, 34, 47
–	Mn(CH$_3$COO)$_2$ · 4 H$_2$O	Mn:	MVol.D2-29, 34/46
–	Mn(CH$_3$COO)$_2$ · 9 H$_2$O	Mn:	MVol.D2-43
–	Mn(CH$_3$COO)$_2$ solutions		
	Mn(CH$_3$COO)$_2$-CH$_3$COOH-H$_2$O........	Mn:	MVol.D2-30/1, 47/50
	Mn(CH$_3$COO)$_2$-H$_2$O	Mn:	MVol.D2-30/1, 47/50
–	Mn(CH$_3$COO)$_2$ systems		
	Mn(CH$_3$COO)$_2$-CH$_3$COOH-H$_2$O........	Mn:	MVol.D2-50/2
	Mn(CH$_3$COO)$_2$-H$_2$O	Mn:	MVol.D2-34
	Mn(CH$_3$COO)$_2$-Mg(CH$_3$COO)$_2$-H$_2$O	Mn:	MVol.D2-53
	Mn(CH$_3$COO)$_2$-Zn(CH$_3$COO)$_2$-H$_2$O	Mn:	MVol.D2-53/4
C$_4$H$_6$MnO$_5$.........	[Mn(H$_2$O)(C$_2$H$_4$(COO)$_2$)]	Mn:	MVol.D2-137
C$_4$H$_6$MnO$_6$.........	Mn(CH$_2$(OH)COO)$_2$	Mn:	MVol.D2-146/7
–	Mn(CH$_2$(OH)COO)$_2$ · 2 H$_2$O.............	Mn:	MVol.D2-147/8
–	[MnO$_2$(CH$_3$CH(OH)CH(O)COO)]	Mn:	MVol.D2-155
C$_4$H$_6$MnO$_6$$^-$	[MnO$_2$(CH$_3$CH(OH)CH(O)COO)]$^-$	Mn:	MVol.D2-155
C$_4$H$_6$NO$_5$U$^+$	[UO$_2$(OOCCH$_2$NHCOCH$_3$)]$^+$	U:	SVol.D1-216
C$_4$H$_6$NO$_6$U$^+$	[UO$_2$H((OOCCH$_2$)$_2$NH)]$^+$.............	U:	SVol.D1-220
–	[UO$_2$H(OOCCH$_2$CH(NH$_2$)COO)]$^+$	U:	SVol.D1-218
C$_4$H$_6$NO$_6$Y........	[(NH$_4$)Y(OOCCH(O)CH(O)COO)] · 3 H$_2$O ..	Sc:	MVol.D5-336/7
C$_4$H$_6$N$_2$NdO$_4$$^+$.....	Nd[H$_2$NN(CH$_2$COO)$_2$]'	Sc:	MVol.D1-243
C$_4$H$_6$N$_2$OS........	SN$_2$C$_2$HCHOHCH$_3$...................	S:	S-N Comp.3-93/8, 123
–	SN$_2$C$_2$OHC$_2$H$_5$	S:	S-N Comp.3-93/8, 150/1
C$_4$H$_6$N$_2$OS$_2$	SN$_2$C$_2$OHSC$_2$H$_5$	S:	S-N Comp.3-93/8, 154
C$_4$H$_6$N$_2$OS$_3$	(O)SN$_2$C$_2$(SCH$_3$)$_2$	S:	S-N Comp.3-250/7
C$_4$H$_6$N$_2$O$_2$S........	SN$_2$C$_2$(OCH$_3$)$_2$..................	S:	S-N Comp.3-93/8, 159
–	SN$_2$C$_2$OHOC$_2$H$_5$	S:	S-N Comp.3-93/8, 155/6
C$_4$H$_6$N$_2$O$_2$Sn	(CH$_3$)$_2$Sn(NCO)$_2$	Sn:	Org.Comp.8-169/70
C$_4$H$_6$N$_2$O$_3$S........	2 CH$_3$CN · SO$_3$	S:	SVol.3-316
–	(O)SN$_2$C$_2$(OCH$_3$)$_2$	S:	S-N Comp.3-250/7
–	[OS(O)N(COCH$_3$)NCCH$_3$]	S:	S-N Comp.3-309/12
C$_4$H$_6$N$_2$S	SN$_2$C$_2$(CH$_3$)$_2$....................	S:	S-N Comp.3-93/8, 124/5
C$_4$H$_6$N$_2$S$_2$Sn	(CH$_3$)$_2$Sn(NCS)$_2$	Sn:	Org.Comp.8-184/6
C$_4$H$_6$N$_2$Sn	(CH$_3$)$_2$Sn(CN)$_2$	Sn:	Org.Comp.8-156/7
C$_4$H$_6$N$_4$NiOS$_4$	[Ni(S$_2$N$_2$CHC(OC$_2$H$_5$)S$_2$N$_2$)]	S:	S-N Comp.2-293, 296
C$_4$H$_6$N$_4$NiO$_2$S$_4$	[Ni((S$_2$N$_2$CHOCH$_2$)$_2$)]	S:	S-N Comp.2-291, 293, 296
C$_4$H$_6$N$_6$O$_2$S........	SN$_2$C$_2$(CONHNH$_2$)$_2$...................	S:	S-N Comp.3-93/8, 188
C$_4$H$_6$NaO$_9$Y........	Na[Y(HCOO)$_4$H$_2$O]	Sc:	MVol.D5-18/9

$C_4H_6NdO_4^+$	$[Nd(CH_3COO)_2]^+$	Sc:	MVol.D5-23/9
$C_4H_6NdO_4S_2^+$	$[Nd(OOCCH_2SH)_2]^+$	Sc:	MVol.D4-53/4
$C_4H_6NdO_6^+$	$[Nd(HOCH_2COO)_2]^+$	Sc:	MVol.D5-222/30
$C_4H_6NdO_8^+$	$[Nd((HO)_2CHCOO)_2]^+$	Sc:	MVol.D5-249/50
$C_4H_6O_4Pr^+$	$[Pr(CH_3COO)_2]^+$	Sc:	MVol.D5-23, 25/8
$C_4H_6O_4PrS_2^+$	$[Pr(OOCCH_2SH)_2]^+$	Sc:	MVol.D4-53/4
$C_4H_6O_4S_2Sm^+$	$[Sm(OOCCH_2SH)_2]^+$	Sc:	MVol.D4-53/4
$C_4H_6O_4S_2Tb^+$	$[Tb(OOCCH_2SH)_2]^+$	Sc:	MVol.D4-53/4
$C_4H_6O_4S_2Tm^+$	$[Tm(OOCCH_2SH)_2]^+$	Sc:	MVol.D4-53/4
$C_4H_6O_4S_2Y^+$	$[Y(OOCCH_2SH)_2]^+$	Sc:	MVol.D4-53/4
$C_4H_6O_4S_2Yb^+$	$[Yb(OOCCH_2SH)_2]^+$	Sc:	MVol.D4-53/4
$C_4H_6O_4Sc^+$	$Sc(CH_3COO)_2^+$	Sc:	MVol.D5-21
$C_4H_6O_4Sm$	$Sm(CH_3COO)_2$	Sc:	MVol.D5-22
$C_4H_6O_4Sm^+$	$[Sm(CH_3COO)_2]^+$	Sc:	MVol.D5-24, 26/9
$C_4H_6O_4Sn$	$(CH_3)_2SnOOCCOO$	Sn:	Org.Comp.14-103, 105
$C_4H_6O_4Tb^+$	$[Tb(CH_3COO)_2]^+$	Sc:	MVol.D5-24
$C_4H_6O_4Tm^+$	$[Tm(CH_3COO)_2]^+$	Sc:	MVol.D5-25
$C_4H_6O_4U^+$	$[U(CH_3COO)_2]^+$	U:	SVol.D1-200
$C_4H_6O_4U^{2+}$	$[U(CH_3COO)_2]^{2+}$	U:	SVol.D1-200
$C_4H_6O_4Y^+$	$[Y(CH_3COO)_2]^+$	Sc:	MVol.D5-23, 27/8
$C_4H_6O_4Yb$	$Yb(CH_3COO)_2$	Sc:	MVol.D5-22
$C_4H_6O_4Yb^+$	$[Yb(CH_3COO)_2]^+$	Sc:	MVol.D5-25/9
$C_4H_6O_4Zn$	$Zn(CH_3COO)_2$ systems		
	$Zn(CH_3COO)_2-Mn(CH_3COO)_2-H_2O$	Mn:	MVol.D2-53/4
$C_4H_6O_5U$	$UO(CH_3COO)_2$	U:	SVol.C13-111, 113
–	$UO(CH_3COO)_2 \cdot H_2O$	U:	SVol.C13-111, 113
–	$UO(CH_3COO)_2 \cdot 1.5\ H_2O$	U:	SVol.C13-111, 113
–	$UO(CH_3COO)_2 \cdot 2\ H_2O$	U:	SVol.C13-111, 113
–	$UO(CH_3COO)_2 \cdot 2.5\ H_2O$	U:	SVol.C13-112/3
$C_4H_6O_6Pr^+$	$[Pr(HOCH_2COO)_2]^+$	Sc:	MVol.D5-222/7
$C_4H_6O_6S_2U$	$UO_2(HSCH_2COO)_2$	U:	SVol.C13-154
		U:	SVol.D1-202
$C_4H_6O_6Sc^+$	$[Sc(HOCH_2COO)_2]^+$	Sc:	MVol.D5-222
$C_4H_6O_6Sm^+$	$[Sm(HOCH_2COO)_2]^+$	Sc:	MVol.D5-222/30
$C_4H_6O_6Tb^+$	$[Tb(HOCH_2COO)_2]^+$	Sc:	MVol.D5-222/30
$C_4H_6O_6Tm^+$	$[Tm(HOCH_2COO)_2]^+$	Sc:	MVol.D5-222/30
$C_4H_6O_6U$	$UO_2(CH_3COO)_2$	U:	SVol.A5-150, 154
		U:	SVol.A6-181
		U:	SVol.C13-116/20
		U:	SVol.D1-33, 200
–	$UO_2(CH_3COO)_2 \cdot H_2O$	U:	SVol.A6-55
–	$UO_2(CH_3COO)_2 \cdot 2\ H_2O$	U:	SVol.A5-137, 194
		U:	SVol.A6-36, 50, 55/7, 61/4, 181
		U:	SVol.C13-117, 120/1
–	$UO_2(CH_3COO)_2 \cdot 2\ D_2O$	U:	SVol.C13-121
–	$UO_2(CD_3COO)_2 \cdot 2\ H_2O$	U:	SVol.C13-121
–	$UO_2(CD_3COO)_2 \cdot 2\ D_2O$	U:	SVol.C13-121
–	$UO_2(CH_3COO)_2 \cdot 3\ H_2O$	U:	SVol.C13-117, 121
–	$UO_2(CH_3COO)_2 \cdot 6\ H_2O$	U:	SVol.C13-118, 121

$C_4H_7MnO_2{}^+$	$[Mn(C_3H_7COO)]^+$	Mn:	MVol.D2-68
$C_4H_7MnO_3{}^+$	$[Mn(HOC(CH_3)_2COO)]^+$	Mn:	MVol.D2-153
$C_4H_7MnO_3{}^{2+}$	$[Mn(HOC(CH_3)_2COO)]^{2+}$	Mn:	MVol.D2-153
$C_4H_7MnO_7{}^{2-}$	$[MnO_3(CH_3CH(OH)CH(OH)COO)]^{2-}$	Mn:	MVol.D2-155
$C_4H_7NO_3S$	$[OS(O)NHC(O)C(CH_3)_2]$.	S:	S-N Comp.3-288/90
$C_4H_7NO_6SU$	$UO_2(NH_2CH_2COO)(HSCH_2COO)$.	U:	SVol.D1-265
$C_4H_7NO_7U$	$[UO_2(HN(CH_2COO)_2)(H_2O)]_n$	U:	SVol.C13-259/60
$C_4H_7NO_8U$	$UO_2(NH_2CH_2COO)((HOCH_2COO)O)$	U:	SVol.D1-264
$C_4H_7N_2NdO_2{}^{2+}$	$[Nd(CH_3C(NO)C(NOH)CH_3)]^{2+}$	Sc:	MVol.D2-115
$C_4H_7N_2NdO_3{}^{2+}$	$Nd(H_2NCH_2C(O)NHCH_2COO)^{2+}$	Sc:	MVol.D4-282/3
–	$Nd[OOCCH(NH_2)CH_2C(O)NH_2]^{2+}$	Sc:	MVol.D1-114
$C_4H_7N_2O_2Pr^{2+}$	$[Pr(CH_3C(NO)C(NOH)CH_3)]^{2+}$	Sc:	MVol.D2-115
$C_4H_7N_2O_2Sm^{2+}$	$[Sm(CH_3C(NO)C(NOH)CH_3)]^{2+}$	Sc:	MVol.D2-115
$C_4H_7N_2O_2Tb^{2+}$	$[Tb(CH_3C(NO)C(NOH)CH_3)]^{2+}$	Sc:	MVol.D2-115
$C_4H_7N_2O_2Tm^{2+}$	$[Tm(CH_3C(NO)C(NOH)CH_3)]^{2+}$	Sc:	MVol.D2-115
$C_4H_7N_2O_2Y^{2+}$	$[Y(CH_3C(NO)C(NOH)CH_3)]^{2+}$	Sc:	MVol.D2-115
$C_4H_7N_2O_2Yb^{2+}$	$[Yb(CH_3C(NO)C(NOH)CH_3)]^{2+}$	Sc:	MVol.D2-115
$C_4H_7N_2O_3PS_2$	$SN_2C_2OP(S)H(OCH_3)_2$.	S:	S-N Comp.3-93/8, 160
$C_4H_7N_2O_3Pr^{2+}$	$Pr[OOCCH(NH_2)CH_2C(O)NH_2]^{2+}$	Sc:	MVol.D1-114
$C_4H_7N_2O_3Sc^{2+}$	$Sc[OOCCH(NH_2)CH_2C(O)NH_2]^{2+}$	Sc:	MVol.D1-114
$C_4H_7N_2O_3Y^{2+}$	$Y[OOCCH(NH_2)CH_2C(O)NH_2]^{2+}$	Sc:	MVol.D1-114
$C_4H_7N_2O_3Yb^{2+}$	$Yb(H_2NCH_2C(O)NHCH_2COO)^{2+}$	Sc:	MVol.D4-282/3
$C_4H_7N_2O_4PS$	$SN_2C_2HOP(O)(OCH_3)_2$	S:	S-N Comp.3-93/8, 160
$C_4H_7N_2O_4Sm^{2+}$	$Sm[H_2NN(CH_2COO)_2H]^{2+}$	Sc:	MVol.D1-243
$C_4H_7N_2O_4Y^{2+}$	$Y[H_2NN(CH_2COO)_2H]^{2+}$	Sc:	MVol.D1-243
$C_4H_7N_2O_4Yb^{2+}$	$Yb[H_2NN(CH_2COO)_2H]^{2+}$	Sc:	MVol.D1-243
$C_4H_7N_2O_5U^+$	$[UO_2(OOCCH(NH_2)CH_2CONH_2)]^+$	U:	SVol.D1-218
–	$[UO_2(OOCCH_2NHCOCH_2NH_2)]^+$	U:	SVol.D1-216
$C_4H_7N_3OS$	$SN_2C_2NH_2OC_2H_5$	S:	S-N Comp.3-93/8, 146
$C_4H_7N_3O_2S$	$(O)SN_2C_2(OC_2H_5)(NH_2)$.	S:	S-N Comp.3-250/7
$C_4H_7N_3O_2S_2$	$[C(O)NHC(SCH_3)NS(O)NCH_3]$	S:	S-N Comp.4-23/7
$C_4H_7N_3S$	$SN_2C_2HCH(NH_2)CH_3$	S:	S-N Comp.3-93/8, 123
$C_4H_7NdO_2{}^{2+}$	$[Nd(n-C_3H_7COO)]^{2+}$	Sc:	MVol.D5-80
–	$[Nd(i-C_3H_7COO)]^{2+}$	Sc:	MVol.D5-83/4
$C_4H_7NdO_2S^{2+}$	$[Nd(C_2H_5SCH_2COO)]^{2+}$	Sc:	MVol.D4-62/3
$C_4H_7NdO_3{}^{2+}$	$[Nd((CH_3)_2C(OH)COO)]^{2+}$	Sc:	MVol.D5-261/6
$C_4H_7NdO_4{}^{2+}$	$[Nd(HOCH_2C(OH)(CH_3)COO)]^{2+}$	Sc:	MVol.D5-268/9
$C_4H_7NdO_5{}^{2+}$	$[Nd(HOC(CH_2OH)_2COO)]^{2+}$	Sc:	MVol.D5-268/9
$C_4H_7O_2Pr^{2+}$	$[Pr(n-C_3H_7COO)]^{2+}$	Sc:	MVol.D5-80
–	$[Pr(i-C_3H_7COO)]^{2+}$	Sc:	MVol.D5-83/4
$C_4H_7O_2PrS^{2+}$	$[Pr(SC_2H_4COOCH_3)]^{2+}$	Sc:	MVol.D4-59
$C_4H_7O_2SSm^{2+}$	$[Sm(C_2H_5SCH_2COO)]^{2+}$	Sc:	MVol.D4-62/3
–	$[Sm(SC_2H_4COOCH_3)]^{2+}$	Sc:	MVol.D4-59
$C_4H_7O_2STb^{2+}$	$[Tb(C_2H_5SCH_2COO)]^{2+}$	Sc:	MVol.D4-62/3
$C_4H_7O_2SY^{2+}$	$[Y(C_2H_5SCH_2COO)]^{2+}$	Sc:	MVol.D4-62/3
$C_4H_7O_2SYb^{2+}$	$[Yb(C_2H_5SCH_2COO)]^{2+}$	Sc:	MVol.D4-62/3
$C_4H_7O_2Sc^{2+}$	$[Sc(n-C_3H_7COO)]^{2+}$	Sc:	MVol.D5-80
$C_4H_7O_2Sm^{2+}$	$[Sm(i-C_3H_7COO)]^{2+}$	Sc:	MVol.D5-83/4
$C_4H_7O_2Tb^{2+}$	$[Tb(i-C_3H_7COO)]^{2+}$	Sc:	MVol.D5-83/4
$C_4H_7O_2Tm^{2+}$	$[Tm(i-C_3H_7COO)]^{2+}$	Sc:	MVol.D5-83/4

$C_4H_7O_2Y^{2+}$	$[Y(i\text{-}C_3H_7COO)]^{2+}$	Sc:	MVol.D5-83/4
$C_4H_7O_2Yb^{2+}$	$[Yb(i\text{-}C_3H_7COO)]^{2+}$	Sc:	MVol.D5-83/4
$C_4H_7O_3Pm^{2+}$	$[Pm((CH_3)_2C(OH)COO)]^{2+}$	Sc:	MVol.D5-261/6
$C_4H_7O_3Pr^{2+}$	$[Pr((CH_3)_2C(OH)COO)]^{2+}$	Sc:	MVol.D5-261/6
$C_4H_7O_3Sm^{2+}$	$[Sm((CH_3)_2C(OH)COO)]^{2+}$	Sc:	MVol.D5-261/6
$C_4H_7O_3Tb^{2+}$	$[Tb((CH_3)_2C(OH)COO)]^{2+}$	Sc:	MVol.D5-261/6
$C_4H_7O_3Tm^{2+}$	$[Tm((CH_3)_2C(OH)COO)]^{2+}$	Sc:	MVol.D5-261/6
$C_4H_7O_3U^{2+}$	$[U((CH_3)_2COHCOO)]^{2+}$	U:	SVol.D1-203
$C_4H_7O_3Y^{2+}$	$[Y((CH_3)_2C(OH)COO)]^{2+}$	Sc:	MVol.D5-261/6
$C_4H_7O_3Yb^{2+}$	$[Yb((CH_3)_2C(OH)COO)]^{2+}$	Sc:	MVol.D5-261/6
$C_4H_7O_4Pr^{2+}$	$[Pr(HOCH_2C(OH)(CH_3)COO)]^{2+}$	Sc:	MVol.D5-268/9
$C_4H_7O_4Sm^{2+}$	$[Sm(HOCH_2C(OH)(CH_3)COO)]^{2+}$	Sc:	MVol.D5-268/9
$C_4H_7O_4Tb^{2+}$	$[Tb(HOCH_2C(OH)(CH_3)COO)]^{2+}$	Sc:	MVol.D5-268/9
$C_4H_7O_4Tm^{2+}$	$[Tm(HOCH_2C(OH)(CH_3)COO)]^{2+}$	Sc:	MVol.D5-268/9
$C_4H_7O_4U^{+}$	$[UO_2(C_2H_5CH_2COO)]^{+}$	U:	SVol.D1-198, 201
–	$[UO_2(i\text{-}C_3H_7COO)]^{+}$	U:	SVol.D1-198, 202
$C_4H_7O_4Y^{2+}$	$[Y(HOCH_2C(OH)(CH_3)COO)]^{2+}$	Sc:	MVol.D5-268/9
$C_4H_7O_4Yb^{2+}$	$[Yb(HOCH_2C(OH)(CH_3)COO)]^{2+}$	Sc:	MVol.D5-268/9
$C_4H_7O_5Pr^{2+}$	$[Pr(HOC(CH_2OH)_2COO)]^{2+}$	Sc:	MVol.D5-268/9
$C_4H_7O_5Sc$	$Sc(OH)(CH_3COO)_2$	Sc:	MVol.D5-21
–	$Sc(OH)(CH_3COO)_2 \cdot H_2O$	Sc:	MVol.D5-22
–	$Sc(OH)(CH_3COO)_2 \cdot 2\,H_2O$	Sc:	MVol.D5-22
–	$Sc(OH)(CH_3COO)_2 \cdot x\,H_2O$	Sc:	MVol.D5-22
$C_4H_7O_5Sm^{2+}$	$[Sm(HOC(CH_2OH)_2COO)]^{2+}$	Sc:	MVol.D5-268/9
$C_4H_7O_5Tb^{2+}$	$[Tb(HOC(CH_2OH)_2COO)]^{2+}$	Sc:	MVol.D5-268/9
$C_4H_7O_5Tm^{2+}$	$[Tm(HOC(CH_2OH)_2COO)]^{2+}$	Sc:	MVol.D5-268/9
$C_4H_7O_5U^{+}$	$[UO_2((CH_3)_2COHCOO)]^{+}$	U:	SVol.D1-203
–	$[UO_2(CH_3CHOHCH_2COO)]^{+}$	U:	SVol.D1-198, 203
–	$[UO_2(C_2H_5CHOHCOO)]^{+}$	U:	SVol.D1-203
–	$[UO_2(HOCH_2CH_2CH_2COO)]^{+}$	U:	SVol.D1-198, 204
$C_4H_7O_5Y^{2+}$	$[Y(HOC(CH_2OH)_2COO)]^{2+}$	Sc:	MVol.D5-268/9
$C_4H_7O_5Yb^{2+}$	$[Yb(HOC(CH_2OH)_2COO)]^{2+}$	Sc:	MVol.D5-268/9
$C_4H_7O_7U^{-}$	$[UO_2(CH_3COO)_2(OH)]^{-}$	U:	SVol.D1-200
		U:	SVol.D2-377
$C_4H_7O_{11}Tb$	$[H_5O_2][Tb(C_2O_4)_2H_2O] = HTb(C_2O_4)_2 \cdot 3\,H_2O$		
		Sc:	MVol.D5-139/41
$C_4H_7O_{11}Tm$	$[H_5O_2][Tm(C_2O_4)_2H_2O] = HTm(C_2O_4)_2 \cdot 3\,H_2O$		
		Sc:	MVol.D5-139/41
$C_4H_7O_{11}Y$	$[H_5O_2][Y(C_2O_4)_2H_2O] = HY(C_2O_4)_2 \cdot 3\,H_2O$		
		Sc:	MVol.D5-139/41
$C_4H_7O_{11}Yb$	$[H_5O_2][Yb(C_2O_4)_2H_2O] = HYb(C_2O_4)_2 \cdot 3\,H_2O$		
		Sc:	MVol.D5-139/41
$C_4H_7O_{15}P_4Pr^{8-}$	$[Pr(OH)(OC(PO_3)_2CH_3)_2]^{8-}$	Sc:	MVol.D4-147
$C_4H_8K_2O_6Os$	$K_2[OsO_2(O_2C_2H_4)_2]$	Os:	SVol.1-188
$C_4H_8K_4O_{20}U_2$	$K_4[(UO_2)_2(OO)_2(C_2O_4)_2(H_2O)_4] \cdot 2\,H_2O$ ...	U:	SVol.C13-213/5
$C_4H_8LaNOS^{2+}$	$La[OC(CHSH)N(CH_3)_2]^{2+}$	Sc:	MVol.D4-61
$C_4H_8LaNO_2S^{2+}$	$La[OC(CHSH)NHC_2H_4OH]^{2+}$	Sc:	MVol.D4-61
$C_4H_8LaNO_3^{2+}$	$La[CH_3CHOHCH(NH_2)COO]^{2+}$	Sc:	MVol.D1-114
$C_4H_8LaN_2O_4^{+}$	$La(NH_2CH_2COO)_2^{+}$	Sc:	MVol.D1-103
$C_4H_8LaN_3O_{11}$	$La(NO_3)_3 \cdot C_4H_8O_2 \cdot 2\,H_2O$	Sc:	MVol.D3-273, 275

$C_4H_8N_2S$	$S(NCH_2CH_2)_2$	S:	S-N Comp.4-124
–	$SN_2C_2H_2(CH_3)_2$	S:	S-N Comp.3-91/3
$C_4H_8N_4NiO_2S_4$	$[Ni((S_2N_2CHOCH_3)_2)]$	S:	S-N Comp.2-291, 293, 296
–	$[Ni(S_2N_2CHOHCH(OC_2H_5)S_2N_2)]$	S:	S-N Comp.2-291, 293, 296
$C_4H_8N_4O_3S_4$	$S_3N_2NSO_2NC_4H_8O$	S:	S-N Comp.2-46
$C_4H_8N_4O_4S$	$SN_2C_2(COONH_4)_2$	S:	S-N Comp.3-93/8, 185
$C_4H_8N_4O_4S_4U$	$U(NCS)_4(H_2O)_4 \cdot 1.5\ C_{12}H_{24}O_6$ $\cdot\ CH_3CO(i\text{-}C_4H_9) \cdot 3\ H_2O$	U:	SVol.C13-35/6
$C_4H_8Na_4O_{20}U_2$	$Na_4[(UO_2)_2(OO)_2(C_2O_4)_2(H_2O)_4] \cdot 2\ H_2O$	U:	SVol.C13-213/5
$C_4H_8OS_4Tc^-$	$[TcO(SC_2H_4S)_2]^-$	Tc:	SVol.2-141/3, 229
$C_4H_8O_2SSn$	$(CH_3)_2Sn(SCH_2C(O)O)$	Sn:	Org.Comp.10-230
$C_4H_8O_3Sn$	$(CH_3)_2SnO_2CCH_2O$	Sn:	Org.Comp.14-107
$C_4H_8O_4Sn$	$(CH_3)_2Sn(OOCH)_2$	Sn:	Org.Comp.14-82, 84
$C_4H_8O_5Os$	$OsO(O_2C_2H_4)_2$	Os:	SVol.1-185, 187
$C_4H_8O_5Rh_2^{2+}$	$[Rh_2(O_2CCH_3)_2(H_2O)]^{2+}$	Rh:	SVol.B2-16
$C_4H_8O_5S$	$SO_3 \cdot C_4H_8O_2$	S:	SVol.3-315
$C_4H_8O_6Os^{2-}$	$[OsO_2(O_2(CH_2)_2)_2]^{2-}$	Os:	SVol.1-186
$C_4H_8O_7S$	$SO_3 \cdot 2\ CH_3COOH$	S:	SVol.3-315
$C_4H_8O_8S_2$	$2\ SO_3 \cdot C_4H_8O_2$	S:	SVol.3-315
$C_4H_8O_8Sc_2$	$Sc_2(C_2H_4(COO)_2)(OH)_4 \cdot H_2O$	Sc:	MVol.D5-170/1
$C_4H_8O_8U^{2-}$	$[UO_2(CH_3COO)_2(OH)_2]^{2-}$	U:	SVol.D1-200
		U:	SVol.D2-377
$C_4H_8O_{10}Rh_2$	$Rh_2(O_2CH)_4(H_2O)_2$	Rh:	SVol.B2-10
$C_4H_8O_{10}Rh_2^+$	$[Rh_2(O_2CH)_4(H_2O)_2]^+$	Rh:	SVol.B2-56
$C_4H_8O_{11}STh$	$[Th(O(CH_2COO)_2)SO_4(H_2O)_2] \cdot H_2O$	Th:	SVol.C5-78/9
$C_4H_8O_{11}S_3$	$3\ SO_3 \cdot C_4H_8O_2$	S:	SVol.3-315
$C_4H_8O_{12}U_2$	$[(UO_2)_2(O)(HOCH_2COO)_2(H_2O)]_n$	U:	SVol.C13-144/5
$C_4H_8O_{15}P_4Tc^{5-}$	$[TcO(CH_3C(PO_3)_2OH)_2]^{5-}$	Tc:	SVol.2-210/1
$C_4H_8S_2Sn$	$(CH_3)_2Sn(SCH)_2$	Sn:	Org.Comp.10-32
$C_4H_9HgMnN_7$	$Mn(CN)_2 \cdot Hg(CN)_2 \cdot 3\ NH_3$	Mn:	MVol.C7-229
C_4H_9ISn	$(CH_3)_2(CH_2CH)SnI$	Sn:	Org.Comp.8-53
$C_4H_9I_3Sn$	$C_4H_9SnI_3$	Sn:	Org.Comp.8-139
$C_4H_9MnN_2O_2^+$	$Mn[H_2NC_2H_4CH(NH_2)COO]^+$	Mn:	MVol.D4-288/9
–	$Mn(NH_2C_2H_4NHCH_2COO)^+$	Mn:	MVol.D5-24/5
$C_4H_9MnN_2O_4^+$	$Mn(NH_2CH_2COO)(NH_3CH_2COO)^+$	Mn:	MVol.D4-248/50
$C_4H_9MnN_3O_2S_2$	$Mn(NH_2OH)(NCS)_2 \cdot C_2H_5OH$	Mn:	MVol.D3-76
C_4H_9NOSn	$(CH_3)_3SnNCO$	Sn:	Org.Comp.8-158/9
$C_4H_9NO_3Os$	$OsO_3(NC(CH_3)_3)$	Os:	SVol.1-240
$C_4H_9NO_6U$	$UO_2(HN(CH_2COO)_2)(H_2O)_2$	U:	SVol.C13-259/60
$C_4H_9NO_{11}Rh$	$[Rh(NH_3)(H_2O)_3(C_2O_4)]C_2O_4$	Rh:	SVol.B2-154
C_4H_9NSSn	$(CH_3)_3SnNCS$	Sn:	Org.Comp.8-172/5
C_4H_9NSeSn	$(CH_3)_3SnNCSe$	Sn:	Org.Comp.8-192
C_4H_9NSn	$(CH_3)_3SnCN$	Sn:	Org.Comp.8-148/51
$C_4H_9N_2NdO_5$	$Nd(NH_2CH_2COO)_2(OH) \cdot 4\ H_2O$	Sc:	MVol.D1-105
$C_4H_9N_2O_5Pr$	$Pr(NH_2CH_2COO)_2(OH) \cdot 3\ H_2O$	Sc:	MVol.D1-105
$C_4H_9N_2O_5Sc$	$Sc(NH_2CH_2COO)_2(OH) \cdot n\ H_2O$	Sc:	MVol.D1-105
$C_4H_9N_5NiOS_4$	$[Ni(S_2N_2CH_2N(CH_2OCH_3)CH_2S_2N_2)]$	S:	S-N Comp.2-291, 294, 296

$C_4H_9N_5NiS_4$	$[Ni((S_2N_2CH(CH_3))_2NH)]$	S:	S–N Comp.2–291, 293, 296
–	$[Ni((S_2N_2CH_2)_2NC_2H_5)]$	S:	S–N Comp.2–291, 294, 296
$C_4H_9NaO_{13}U_2$	$Na[(UO_2)_2(O)(HOCH_2COO)_2(OH)(H_2O)]$	U:	SVol.C13–144/5
$C_4H_9Na_3S_3Sn$	$C_4H_9Sn(SNa)_3$	Sn:	Org.Comp.10–190
$C_4H_9O_3Sb$	$(CH_3)_3SbCO_3$	Sb:	Org.Comp.4–111
$C_4H_9O_4Sm$	$Sm(OCH_2)_2(OCH_2CH_2OH)$	Sc:	MVol.D3–31/2
$C_4H_9O_6PU$	$UO_2[(n-C_4H_9O)PO_3] \cdot H_2O$	U:	SVol.C14–124/5
–	$UO_2[(n-C_4H_9O)PO_3] \cdot 2 H_2O$	U:	SVol.C14–124/5
–	$UO_2(C_4H_9OPO_3)$	U:	SVol.D2–228
–	$UO_2(C_4H_9OPO_3) \cdot 2 H_2O$	U:	SVol.D2–204
C_4H_9Sb	$(CH_3)_2SbCHCH_2$	Sb:	Org.Comp.1–112
$C_4H_{10}HoO_4P^{2+}$	$Ho[(C_2H_5O)_2PO_2]^{2+}$	Sc:	MVol.D4–177
$C_4H_{10}ISb$	$(C_2H_5)_2SbI$	Sb:	Org.Comp.2–25
$C_4H_{10}I_2MnN_2O_4$	$Mn(NH_3CH_2COO)_2I_2$	Mn:	MVol.D4–254/5, 261
–	$Mn(NH_3CH_2COO)_2I_2 \cdot 2 H_2O$	Mn:	MVol.D4–254, 261
$C_4H_{10}I_2O_6Sn$	$(C_2H_5)_2Sn(OIO_2)_2$	Sn:	Org.Comp.14–186
$C_4H_{10}I_2Sn$	$(C_2H_5)_2SnI_2$	Sn:	Org.Comp.8–96/102
$C_4H_{10}LaN_2O_4P_2^+$ ...	$La(C_2H_4(NHCH_2PO_2)_2)^+$	Sc:	MVol.D4–142
$C_4H_{10}LaO_4P^{2+}$	$La[(C_2H_5O)_2PO_2]^{2+}$	Sc:	MVol.D4–177
$C_4H_{10}LiSb$	$(C_2H_5)_2SbLi \cdot C_4H_8O_2$	Sb:	Org.Comp.2–57
$C_4H_{10}LuO_4P^{2+}$	$Lu[(C_2H_5O)_2PO_2]^{2+}$	Sc:	MVol.D4–177
$C_4H_{10}MnN_2O_4^{2+}$	$Mn(NH_3CH_2COO)_2^{2+}$	Mn:	MVol.D4–248/50
$C_4H_{10}MnN_2O_6$	$(NH_4)_2[Mn(C_4H_2O_6)] \cdot 6 H_2O$	Mn:	MVol.D2–177
$C_4H_{10}MnN_2O_8S$	$Mn(NH_3CH_2COO)_2SO_4$	Mn:	MVol.D4–262
$C_4H_{10}MnN_4O_2^{2+}$	$Mn(C_2H_4(C(O)NHNH_2)_2)^{2+}$	Mn:	MVol.D5–204
$C_4H_{10}MnN_4O_6S$	$Mn(CH_3CH(C(O)NHNH_2)_2)SO_4$	Mn:	MVol.D5–203
–	$Mn(C_2H_4(C(O)NHNH_2)_2)SO_4 \cdot H_2O$	Mn:	MVol.D5–204/5
–	$Mn(C_2H_4(C(O)NHNH_2)_2)SO_4 \cdot 3 H_2O$	Mn:	MVol.D5–204/5
$C_4H_{10}MnN_4O_{10}$	$Mn(NH_3CH_2COO)_2(NO_3)_2$	Mn:	MVol.D4–252
$C_4H_{10}MnN_6O_8S$	$Mn(HN(CONH_2)_2)_2SO_4 \cdot H_2O$	Mn:	MVol.D5–156/7
$C_4H_{10}MnN_8O_{10}$	$Mn(HN(CONH_2)_2)_2(NO_3)_2 \cdot 2 H_2O$	Mn:	MVol.D5–155/6
$C_4H_{10}MnO_6$	$[Mn(CH_3COO)_2(H_2O)_2]$	Mn:	MVol.D2–30/1
$C_4H_{10}MnO_8$	$[Mn(CH_2(OH)COO)_2(H_2O)_2]$	Mn:	MVol.D2–147/8
$C_4H_{10}NO_2SU^+$	$[UO_2(SC_2H_4N(CH_3)_2 \cdot HCl)]^+$	U:	SVol.D1–251
$C_4H_{10}N_2OS$	$OS(NCH_3CH_2)_2$	S:	S–N Comp.3–264/6
$C_4H_{10}N_2O_6Sn$	$(C_2H_5)_2Sn(ONO_2)_2$	Sn:	Org.Comp.14–187
$C_4H_{10}N_2O_6U^{2+}$	$[UO_2H_2(OOCCH_2NH_2)_2]^{2+}$	U:	SVol.D1–216
$C_4H_{10}N_2O_8Rh_2$	$Rh_2(O_2CH)_4(NH_3)_2$	Rh:	SVol.B2–10
$C_4H_{10}N_2O_8SU$	$UO_2SO_4 \cdot 2 CH_3CONH_2$	U:	SVol.C10–163
$C_4H_{10}N_2O_9U$	$UO_2(NO_3)_2 \cdot (C_2H_5)_2O \cdot 3 H_2O$	U:	SVol.C7–166/8
		U:	SVol.D2–40, 49
–	$UO_2(NO_3)_2 \cdot i-C_4H_9OH \cdot 2 H_2O$	U:	SVol.C7–163/5
$C_4H_{10}N_2O_{11}U$	$(NH_4)_2[UO_2(C_2O_4)_2(H_2O)]$	U:	SVol.C13–194
–	$(NH_4)_2[UO_2(C_2O_4)_2(H_2O)] \cdot H_2O$	U:	SVol.C13–193/4, 198/9
–	$(NH_4)_2[UO_2(C_2O_4)_2(H_2O)] \cdot 2 H_2O$	U:	SVol.C13–193, 195/9
$C_4H_{10}N_2O_{14}S_2U_2$ ...	$[NH_3C_2H_4NH_3][(UO_2)_2(C_2O_4)(SO_3)_2]$	U:	SVol.C13–223/5
$C_4H_{10}N_4O_{10}U$	$(N_2H_5)_2[UO_2(C_2O_4)_2] \cdot 2 H_2O$	U:	SVol.C13–195/7
$C_4H_{10}N_6O_4Rh^{3+}$	$[Rh((NH_2CO)_2NH)_2]^{3+}$	Rh:	SVol.B2–321

$C_4H_{14}Mo_4N_2O_{15}Tm_2$ $2\,(CH_3)_2NH \cdot Tm_2O_3 \cdot 4\,MoO_3 \cdot 9\,H_2O$

         $= (CH_3)_2NH_2Tm(MoO_4)_2 \cdot 4\,H_2O$ ...... Mo: SVol.B4-277/8

$C_4H_{14}NNaO_4OsS$ ... $Na[Os(SNC_4H_{10})(OH)_4]$ .... Os: SVol.1-298

$C_4H_{14}NOSb$ .... $(CH_3)_4SbONH_2$ .... Sb: Org.Comp.3-114/5

$C_4H_{14}N_6O_9U$ .... $[C(NH_2)_3]_2[UO_2(OO)(C_2O_4)(H_2O)]$ .... U: SVol.C13-214/5

$C_4H_{14}N_{10}O_3OsS$ .... $[OsO_2(H_2NC(NH)NHC(NH)NH_2)_2]SO \cdot H_2O$ Os: SVol.1-271

$C_4H_{14}N_{13}O_7Rh$ ..... $[Rh((NH_2CNH)_2NH)_2(NO_2)_2]NO_3$ .... Rh: SVol.B2-221

$C_4H_{14}O_2PSb$ .... $(CH_3)_4SbOP(O)H_2$ .... Sb: Org.Comp.3-75/7

$C_4H_{14}O_4P_2Sn$ ...... $(C_2H_5)_2Sn(OP(O)H_2)_2$ .... Sn: Org.Comp.14-195, 196

$C_4H_{14}O_6Os$ .... $Os(OCH_3)_4(OH)_2$ .... Os: SVol.1-184

$C_4H_{14}O_{10}Sc^+$ ...... $[Sc(HOCH_2COO)_2(H_2O)_4]^+$ .... Sc: MVol.D5-232

$C_4H_{15}NS_2$ .... $H_2S \cdot [(CH_3)_4N]SH$ .... S: SVol.4a/b-368

$C_4H_{16}HoN_4^{3+}$ ...... $[Ho(C_2H_8N_2)_2]^{3+}$ .... Sc: MVol.D1-19/21

$C_4H_{16}IN_{10}ORh^{2+}$ ... $[Rh((NH_2CNH)_2NH)_2I(H_2O)]^{2+}$ .... Rh: SVol.B2-221

$C_4H_{16}I_2MnN_8O_4$ .... $Mn(OC(NH_2)_2)_4I_2 \cdot 2\,H_2O$ .... Mn: MVol.D5-142/3

$C_4H_{16}I_2MnN_{10}O_8$.... $[Mn(OH)_2((NHC(NH_2))_2NH)_2](IO_3)_2 \cdot 2\,H_2O$ Mn: MVol.D5-162/4

$C_4H_{16}I_2N_2Sn$ .... $(C_2H_5)_2SnI_2 \cdot 2\,NH_3$ .... Sn: Org.Comp.8-101

$C_4H_{16}I_2N_4O_2Os$ ..... $[OsO_2(C_2H_4(NH_2)_2)_2]I_2$ .... Os: SVol.1-251

$C_4H_{16}I_2N_4Rh^+$ ..... $[Rh(NH_2C_2H_4NH_2)_2I_2]^+$ .... Rh: SVol.B2-179/83

$C_4H_{16}I_3N_4Rh$ ...... $[Rh(NH_2C_2H_4NH_2)_2I_2]I$ .... Rh: SVol.B2-178/9

$C_4H_{16}I_4N_8O_4Th$ ..... $ThI_4 \cdot 4\,(NH_2)_2CO \cdot 4\,H_2O$ .... Th: SVol.E-37, 40

$C_4H_{16}LaN_4^{3+}$ ...... $[La(C_2H_8N_2)_2]^{3+}$ .... Sc: MVol.D1-19/21

$C_4H_{16}LuN_4^{3+}$ ...... $[Lu(C_2H_8N_2)_2]^{3+}$ .... Sc: MVol.D1-19/21

$C_4H_{16}LuN_{11}O_{13}$.... $Lu(NO_3)_3 \cdot 4\,OC(NH_2)_2$ .... Sc: MVol.D2-220

$C_4H_{16}MnN_4^{2+}$ ...... $[Mn(C_2H_8N_2)_2]^{2+}$ .... Mn: MVol.D3-37/9

$C_4H_{16}MnN_8O_8S$ ..... $Mn(OC(NH_2)_2)_4SO_4$ .... Mn: MVol.D5-144

− .... $Mn(OC(NH_2)_2)_4SO_4 \cdot 2\,H_2O$ .... Mn: MVol.D5-144

$C_4H_{16}MnN_{10}O_6S$.... $[Mn(OH)_2((NHC(NH_2))_2NH)_2]SO_4 \cdot 4.5\,H_2O$ Mn: MVol.D5-162/4

$C_4H_{16}MnN_{10}O_{10}$ .... $Mn(OC(NH_2)_2)_4(NO_3)_2$ .... Mn: MVol.D5-134/5

− .... $Mn(OC(NH_2)_2)_4(NO_3)_2 \cdot 2\,H_2O$ .... Mn: MVol.D5-135

$C_4H_{16}MnN_{12}O_8$ ..... $[Mn(OH)_2((NHC(NH_2))_2NH)_2](NO_3)_2$ ... Mn: MVol.D5-162/4

$C_4H_{16}Mo_2N_2O_7$ ..... $(C_2H_5NH_3)_2O \cdot 2\,MoO_3 \cdot 2\,H_2O$ ...... Mo: SVol.B4-131, 138

$C_4H_{16}Mo_3N_2O_{10}$ .... $[(CH_3)_2NH_2]_2O \cdot 3\,MoO_3$ .... Mo: SVol.B4-135

− .... $[(CH_3)_2NH_2]_2O \cdot 3\,MoO_3 \cdot H_2O$ ...... Mo: SVol.B4-14, 40, 43, 130

− .... $(C_2H_5NH_3)_2O \cdot 3\,MoO_3$ .... Mo: SVol.B4-138

$C_4H_{16}Mo_4N_2O_{13}$ .... $[(CH_3)_2NH_2]_2O \cdot 4\,MoO_3$ .... Mo: SVol.B4-135

− .... $(C_2H_5NH_3)_2O \cdot 4\,MoO_3 \cdot 2\,H_2O$ .... Mo: SVol.B4-138

$C_4H_{16}Mo_{8.41}N_2O_{26.23}$

       $(C_2H_5NH_3)_2O \cdot 8.41\,MoO_3 \cdot x\,H_2O$ ...... Mo: SVol.B4-131, 138

$C_4H_{16}N_2O_{10}U$ ...... $C_2H_4(NH_3)_2[U(OH)_2(CO_3)_2(H_2O)_2] \cdot H_2O$ .. U: SVol.C13-7/8

$C_4H_{16}N_4NaO_6RhS_4$.. $Na[Rh(NH_2C_2H_4NH_2)_2(S_2O_3)_2]$ .... Rh: SVol.B2-183

$C_4H_{16}N_4Na_3O_9RhS_6$ $Na_3[Rh(S_2O_3)_3(C_2H_8N_2)_2]$ .... Rh: SVol.B2-185

$C_4H_{16}N_4Nd^{3+}$ ...... $[Nd(C_2H_8N_2)_2]^{3+}$ .... Sc: MVol.D1-19/21

$C_4H_{16}N_4O_2Os^{2+}$ .... $[OsO_2(C_2H_4(NH_2)_2)_2]^{2+}$ .... Os: SVol.1-251

$C_4H_{16}N_4O_2Tc^+$ .... $[TcO_2((NH_2)_2C_2H_4)_2]^+$ .... Tc: SVol.2-214

$C_4H_{16}N_4O_{12}Rh_2$ .... $(NH_4)_4[Rh_2(CO_3)_4] \cdot n\,H_2O$ .... Rh: SVol.B1-184

$C_4H_{16}N_4Pr^{3+}$ ...... $[Pr(C_2H_8N_2)_2]^{3+}$ .... Sc: MVol.D1-19/21

$C_4H_{16}N_4Rh^+$ .... $[Rh(NH_2C_2H_4NH_2)_2]^+$ .... Rh: SVol.B2-161/2

$C_4H_{16}N_4Sm^{3+}$ ..... $[Sm(C_2H_8N_2)_2]^{3+}$ .... Sc: MVol.D1-19/21

$C_4H_{16}N_4Tb^{3+}$ ...... $[Tb(C_2H_8N_2)_2]^{3+}$ .... Sc: MVol.D1-19/21

$C_4H_{16}N_4Y^{3+}$	$[Y(C_2H_8N_2)_2]^{3+}$	Sc:	MVol.D1-19/21
$C_4H_{16}N_4Yb^{3+}$	$[Yb(C_2H_8N_2)_2]^{3+}$	Sc:	MVol.D1-19/21
$C_4H_{16}N_5O_4Rh^+$	$[Rh(NH_2C_2H_4NH_2)_2(NO_2)(O_2)]^+$	Rh:	SVol.B2-172
$C_4H_{16}N_6O_7Rh$	$[Rh(NH_2C_2H_4NH_2)_2(NO_2)(O_2)]NO_3$	Rh:	SVol.B2-172
$C_4H_{16}N_6O_{14}U$	$[(C_2H_5)NH_3]_2UO_2(NO_3)_4$	U:	SVol.C7-191, 195
$C_4H_{16}N_6O_{16}S_4U_2$	$(NH_4)_4[(UO_2)_2(C_2O_4)(NCS)_2(SO_4)_2]$	U:	SVol.C13-233/4
—	$(NH_4)_4[(UO_2)_2(C_2O_4)(NCS)_2(SO_4)_2] \cdot 6\ H_2O$		
		U:	SVol.C13-233/4
$C_4H_{16}N_7O_7Rh$	$[Rh(NH_2C_2H_4NH_2)_2(NO_2)_2]NO_3$	Rh:	SVol.B2-175/6
$C_4H_{16}N_8O_2OsS_4^{2+}$	$[OsO_2(SC(NH_2)_2)_4]^{2+}$	Os:	SVol.1-288
$C_4H_{16}N_8O_2S_4Tc^+$	$[TcO_2(SC(NH_2)_2)_4]^+$	Tc:	SVol.2-228
$C_4H_{16}N_8O_6OsS_5$	$[OsO_2(SC(NH_2)_2)_4]SO_4$	Os:	SVol.1-289
$C_4H_{16}N_8O_{10}SU$	$UO_2SO_4 \cdot 4\ (NH_2)_2CO$	U:	SVol.C10-163
$C_4H_{16}N_8O_{12}S_2Th$	$Th(SO_4)_2 \cdot 4\ (NH_2)_2CO$	Th:	SVol.E-38, 41/2
—	$Th(SO_4)_2 \cdot 4\ (NH_2)_2CO \cdot H_2O$	Th:	SVol.E-38, 42
$C_4H_{16}N_8O_{12}S_2U$	$U(SO_4)_2 \cdot 4\ CO(NH_2)_2$	U:	SVol.C10-149
—	$U(SO_4)_2 \cdot 4\ CO(NH_2)_2 \cdot 4\ H_2O$	U:	SVol.C10-149
$C_4H_{16}N_8O_{18}Th$	$[(CH_3)_2NH_2]_2Th(NO_3)_6$	Th:	SVol.C3-108/9
$C_4H_{16}N_{10}Rh^+$	$[Rh(NH_2C_2H_4NH_2)_2(N_3)_2]^+$	Rh:	SVol.B2-176
$C_4H_{16}N_{11}O_{13}Pr$	$Pr(NO_3)_3 \cdot 4\ OC(NH_2)_2$	Sc:	MVol.D2-220
$C_4H_{16}N_{11}O_{13}Sc$	$Sc(NO_3)_3 \cdot 4\ OC(NH_2)_2$	Sc:	MVol.D2-219
$C_4H_{16}N_{11}O_{13}Yb$	$Yb(NO_3)_3 \cdot 4\ OC(NH_2)_2$	Sc:	MVol.D2-220
$C_4H_{16}N_{12}O_{16}Th$	$Th(NO_3)_4 \cdot 4\ (NH_2)_2CO \cdot 4\ H_2O$	Th:	SVol.E-38, 41
$C_4H_{16}N_{13}Rh$	$[Rh(NH_2C_2H_4NH_2)_2(N_3)_2]N_3$	Rh:	SVol.B2-176
$C_4H_{17}IN_4ORh^+$	$[Rh(NH_2C_2H_4NH_2)_2(OH)I]^+$	Rh:	SVol.B2-173
$C_4H_{17}IN_4Rh^+$	$[Rh(NH_2C_2H_4NH_2)_2HI]^+$	Rh:	SVol.B2-167
$C_4H_{17}MnN_{10}O_6P$	$[Mn(OH)_2((NHC(NH_2))_2NH)]HPO_4 \cdot 1.5\ H_2O$		
		Mn:	MVol.D5-162/4
$C_4H_{17}N_6NiS_2$	$[Ni(S_2N_2H)(C_2H_8N_2)_2]$	S:	S-N Comp.2-290
$C_4H_{17}N_{10}O_8Sc$	$N_2H_5[Sc(H_2NNHCOO)_4] \cdot 3\ H_2O$	Sc:	MVol.D1-242
$C_4H_{17}N_{11}O_4Os$	$[Os(N)(O)(H_2NC(NH)NHC(NH)NH_2)_2](OH)_3$	Os:	SVol.1-271, 284
$C_4H_{18}IN_4ORh^{2+}$	$[Rh(NH_2C_2H_4NH_2)_2(OH_2)I]^{2+}$	Rh:	SVol.B2-173
$C_4H_{18}N_4ORh^+$	$[Rh(NH_2C_2H_4NH_2)_2H(OH)]^+$	Rh:	SVol.B2-168
$C_4H_{18}N_4O_2Rh^+$	$[Rh(NH_2C_2H_4NH_2)_2(OH)_2]^+$	Rh:	SVol.B2-169
$C_4H_{18}N_4O_3Rh^+$	$[Rh(NH_2C_2H_4NH_2)_2(OH)(OOH)]^+$	Rh:	SVol.B2-171
$C_4H_{18}N_4O_3Rh^{2+}$	$[Rh(NH_2C_2H_4NH_2)_2(H_2O)(O_2)]^{2+}$	Rh:	SVol.B2-172
$C_4H_{18}N_4O_{10}OsS_2$	$[OsO_2(C_2H_4(NH_2)_2)_2](HSO_4)_2$	Os:	SVol.1-251
$C_4H_{18}N_4O_{14}S_3U$	$(NH_4)_2U(SO_4)_3 \cdot 2\ CH_3CONH_2$	U:	SVol.C10-188/90
—	$(NH_4)_2U(SO_4)_3 \cdot 2\ CH_3CONH_2 \cdot 4\ H_2O$	U:	SVol.C10-188
$C_4H_{18}N_4Os^{2+}$	$[OsH_2(C_2H_4(NH_2)_2)_2]^{2+}$	Os:	SVol.1-251
$C_4H_{18}N_4Rh^+$	$[Rh(NH_2C_2H_4NH_2)_2H_2]^+$	Rh:	SVol.B2-167
$C_4H_{18}N_7ORh^{2+}$	$[Rh(NH_2C_2H_4NH_2)_2(N_3)(OH_2)]^{2+}$	Rh:	SVol.B2-176
$C_4H_{18}N_{10}O_2Rh^{3+}$	$[Rh((NH_2CNH)_2NH)_2(H_2O)_2]^{3+}$	Rh:	SVol.B2-222
$C_4H_{18}N_{10}O_{13}U$	$[UO_2(CO(NH_2)_2)_4(H_2O)](NO_3)_2$	U:	SVol.A6-38
$C_4H_{19}IN_5O_6RhS_2$	$[Rh(NH_2C_2H_4NH_2)_2(NH_3)I]S_2O_6$	Rh:	SVol.B2-175
$C_4H_{19}I_3N_5Rh$	$[Rh(NH_2C_2H_4NH_2)_2(NH_3)I]I_2$	Rh:	SVol.B2-175
$C_4H_{19}Mo_3NO_{19}V_3$	$[(CH_3)_4N]H_7V_3Mo_3O_{19} \cdot 14\ H_2O$	Mo:	SVol.B4-327
$C_4H_{19}NS_4$	$3\ H_2S \cdot [(CH_3)_4N]SH$	S:	SVol.4a/b-368, 495
$C_4H_{19}N_4ORh^{2+}$	$[Rh(NH_2C_2H_4NH_2)_2H(OH_2)]^{2+}$	Rh:	SVol.B2-162, 167/8
$C_4H_{19}N_4O_2Rh^{2+}$	$[Rh(NH_2C_2H_4NH_2)_2(OH)(OH_2)]^{2+}$	Rh:	SVol.B2-168

$C_4H_{19}N_4O_8RhS_2$	$[Rh(NH_2C_2H_4NH_2)_2(OH)(OH_2)]S_2O_6$	Rh:	SVol.B2-168
$C_4H_{19}N_7Rh^{3+}$	$[Rh(NH_3)_5(N(CHCH)_2N)]^{3+}$	Rh:	SVol.B2-140
$C_4H_{19}N_{16}Rh_3$	$Rh_3(CN)_4 \cdot 3 N_2H_4 \cdot N_2H_3 \cdot 2 N_2H_2 \cdot 9 H_2O$		
		Rh:	SVol.B1-198
$C_4H_{20}MnN_{10}S_4$	$Mn(NH_3)_4(NCS)_2 \cdot 2 NH_4SCN$	Mn:	MVol.D3-19
$C_4H_{20}N_4O_2Rh^{3+}$	$[Rh(NH_2C_2H_4NH_2)_2(OH_2)_2]^{3+}$	Rh:	SVol.B2-169
$C_4H_{20}N_4O_{16}S_4U$	$(C_2H_4(NH_2)_2H_2)_2U(SO_4)_4 \cdot 2 H_2O$	U:	SVol.C10-188/90
$C_4H_{20}N_5ORh^{2+}$	$[Rh(NH_2C_2H_4NH_2)_2(NH_3)(OH)]^{2+}$	Rh:	SVol.B2-174
$C_4H_{20}N_5Rh^{2+}$	$[Rh(NH_2C_2H_4NH_2)_2H(NH_3)]^{2+}$	Rh:	SVol.B2-168
$C_4H_{20}N_6Rh^{3+}$	$[Rh(NH_3)_5(NCC(CH_3)CH_2)]^{3+}$	Rh:	SVol.B2-138/9
$C_4H_{20}NdO_6^{3+}$	$[Nd(H_2O)_2(CH_3OH)_4]^{3+}$	Sc:	MVol.D3-17
$C_4H_{21}N_8O_{10}Rh$	$[Rh(NH_2C_2H_4NH_2)_2(NH_3)(OH_2)](NO_3)_3$	Rh:	SVol.B2-173/4
$C_4H_{22}I_2N_4Sn$	$(C_2H_5)_2SnI_2 \cdot 4 NH_3$	Sn:	Org.Comp.8-101
$C_4H_{24}Mo_{12}N_{12}O_{40}Ti$	$(C(NH_2)_3)_4[TiMo_{12}O_{40}] \cdot n H_2O$	Mo:	SVol.B4-295
$C_4H_{24}Mo_{12}N_{12}O_{40}Zr$	$(C(NH_2)_3)_4[ZrMo_{12}O_{40}] \cdot n H_2O$	Mo:	SVol.B4-299/300
$C_4H_{24}N_8Rh^{3+}$	$[Rh(NH_2C_2H_4NH_2)_2(N_2H_4)_2]^{3+}$	Rh:	SVol.B2-174
$C_4H_{25}MnN_{11}O_8S$	$Mn(OC(NH_2)_2)_4SO_4 \cdot 3 NH_3$	Mn:	MVol.D5-144
$C_4H_{28}N_7Na_2O_9RhS_6$	$Na_2(NH_4)_3[Rh(S_2O_3)_3(C_2H_8N_2)_2] \cdot 3.5 H_2O$	Rh:	SVol.B2-185
$C_4H_{34}N_{12}RhRu^{5+}$ ...	$[Rh(NH_3)_5(N(CHCH)_2N)Ru(NH_3)_5]^{5+}$	Rh:	SVol.B2-140
$C_4H_{34}N_{12}RhRu^{6+}$ ...	$[Rh(NH_3)_5(N(CHCH)_2N)Ru(NH_3)_5]^{6+}$	Rh:	SVol.B2-140
$C_4HgMn_{0.115}N_4S_4Zn_{0.885}$			
	$HgMn_{0.115}Zn_{0.885}(SCN)_4$	Mn:	MVol.C7-236
$C_4HgMnN_4S_2Se_2$	$MnHg(SeCN)_2(SCN)_2 \cdot 2 (CH_3)_2CO$	Mn:	MVol.D2-297
$C_4HgMnN_4S_4$	$HgMn(SCN)_4$	Mn:	MVol.C7-236
−	$HgMn(SCN)_4$ solid solutions		
	$HgMn(SCN)_4-HgZn(SCN)_4$	Mn:	MVol.C7-236
$C_4HgN_4S_4Zn$	$HgZn(SCN)_4$ solid solutions		
	$HgZn(SCN)_4-HgMn(SCN)_4$	Mn:	MVol.C7-236
C_4HoO_4	$Ho(CO)_4$	Sc:	MVol.D6-166
$C_4HoO_4^+$	$[Ho(C_4O_4)]^+$	Sc:	MVol.D3-248/50
$C_4HoO_{12}^{5-}$	$[Ho(CO_3)_4]^{5-}$	Sc:	MVol.D4-338/40
$C_4I_2MnN_4S_4$	$Mn(SCN)_4I_2$	Mn:	MVol.C7-239
C_4KO_8Sc	$K[Sc(C_2O_4)_2]$	Sc:	MVol.D5-116/7
−	$K[Sc(C_2O_4)_2] \cdot 2 H_2O$	Sc:	MVol.D5-116/7
C_4KO_8Sm	$KSm(C_2O_4)_2 \cdot 3.5 H_2O$	Sc:	MVol.D5-141/5
C_4KO_8Tb	$KTb(C_2O_4)_2 \cdot 3.5 H_2O$	Sc:	MVol.D5-141/5
C_4KO_8Y	$KY(C_2O_4)_2 \cdot 5 H_2O$	Sc:	MVol.D5-141/5
C_4KO_8Yb	$KYb(C_2O_4)_2 \cdot 4 H_2O$	Sc:	MVol.D5-141/5
$C_4K_2MnN_4S_4$	$K_2[Mn(NCS)_4] \cdot 6 H_2O$	Mn:	MVol.D2-290
$C_4K_2MnO_8$	$K_2[Mn(C_2O_4)_2]$	Mn:	MVol.D2-104
−	$K_2[Mn(C_2O_4)_2] \cdot 2 H_2O$	Mn:	MVol.D2-104/5
$C_4K_2N_2O_4S$	$SN_2C_2(COOK)_2$	S:	S-N Comp.3-93/8, 184
$C_4K_2N_4O_2Os$	$K_2[OsO_2(CN)_4]$	Os:	SVol.1-178
$C_4K_2N_4O_2U$	$K_2UO_2(CN)_4$	U:	SVol.C13-28/9
$C_4K_2N_4Pd$	$K_2Pd(CN)_4$		
	Catalytic properties	Pt:	SVol.A1-305
$C_4K_2N_4PdS_4$	$K_2Pd(SCN)_4$		
	Catalytic properties	Pt:	SVol.A1-305
$C_4K_2O_{10}Os$	$K_2[OsO_2(C_2O_4)_2]$	Os:	SVol.1-203

$C_4N_4O_4Sn$	$Sn(NCO)_4$	Sn:	Org.Comp.8-171
$C_4N_4O_6U^{2-}$	$[UO_2(CNO)_4]^{2-}$	U:	SVol.D1-170
$C_4N_4OsS_4$	$Os(SCN)_4$	Os:	SVol.1-168/9
$C_4N_4Rh^{3-}$	$[Rh(CN)_4]^{3-}$	Rh:	SVol.B1-187
C_4N_4S	$SN_2C_2(CN)_2$	S:	S-N Comp.3-93/8, 169/70
$C_4N_4S_4U$	$U(NCS)_4 \cdot x H_2O$	U:	SVol.C13-35
C_4NaNdO_8	$NaNd(C_2O_4)_2$	Sc:	MVol.D5-143
–	$NaNd(C_2O_4)_2 \cdot 5 H_2O$	Sc:	MVol.D5-141/5
C_4NaO_8Sc	$NaSc(C_2O_4)_2$	Sc:	MVol.D5-116/7
–	$NaSc(C_2O_4)_2 \cdot 2 H_2O$	Sc:	MVol.D5-116/7
–	$NaSc(C_2O_4)_2 \cdot 4 H_2O$	Sc:	MVol.D5-116/7
–	$NaSc(C_2O_4)_2 \cdot x H_2O$	Sc:	MVol.D5-116/7
C_4NaO_8Sm	$NaSm(C_2O_4)_2 \cdot 4 H_2O$	Sc:	MVol.D5-141/5
C_4NaO_8Tb	$NaTb(C_2O_4)_2 \cdot 3 H_2O$	Sc:	MVol.D5-141/5
C_4NaO_8Y	$NaY(C_2O_4)_2 \cdot 5 H_2O$	Sc:	MVol.D5-141/5
C_4NaO_8Yb	$NaYb(C_2O_4)_2 \cdot 4 H_2O$	Sc:	MVol.D5-141/5
$C_4Na_2O_{10}U$	$Na_2[UO_2(C_2O_4)_2]$	U:	SVol.C13-193/4, 198
–	$Na_2[UO_2(C_2O_4)_2] \cdot x H_2O$ (x = 1 to 5)	U:	SVol.C13-194/8
$C_4Na_2O_{11}SU$	$Na_2[U(C_2O_4)_2(SO_3)] \cdot 5.5 H_2O$	U:	SVol.C10-143
$C_4Na_4O_{12}Rh_2$	$Na_4[Rh_2(CO_3)_4] \cdot 2.5 H_2O$	Rh:	SVol.B1-184
$C_4Na_4O_{14}S_2U$	$Na_4[U(C_2O_4)_2(SO_3)_2] \cdot 2 H_2O$	U:	SVol.C10-143
		U:	SVol.C13-223/5
$C_4Na_6O_{17}S_3Th$	$Na_6Th(SO_3)_3(C_2O_4)_2 \cdot 6 H_2O$	Th:	SVol.C5-62
$C_4Na_6O_{17}S_3U$	$Na_6[U(C_2O_4)_2(SO_3)_3] \cdot 5 H_2O$	U:	SVol.C10-143
		U:	SVol.C13-223/5
$C_4Na_8O_{20}S_4Th$	$Na_8Th(SO_3)_4(C_2O_4)_2 \cdot 6 H_2O$	Th:	SVol.C5-62
$C_4Na_8O_{20}S_4U$	$Na_8[U(C_2O_4)_2(SO_3)_4] \cdot 4 H_2O$	U:	SVol.C10-143
		U:	SVol.C13-223/5
$C_4Na_{10}O_{23}S_5Th$	$Na_{10}Th(SO_3)_5(C_2O_4)_2 \cdot 6 H_2O$	Th:	SVol.C5-62
$C_4Na_{10}O_{23}S_5U$	$Na_{10}[U(C_2O_4)_2(SO_3)_5] \cdot 7.5 H_2O$	U:	SVol.C10-143
		U:	SVol.C13-223/5
$C_4Na_{12}O_{26}S_6Th$	$Na_{12}Th(SO_3)_6(C_2O_4)_2 \cdot 8 H_2O$	Th:	SVol.C5-62
$C_4Na_{12}O_{26}S_6U$	$Na_{12}[U(C_2O_4)_2(SO_3)_6] \cdot x H_2O$	U:	SVol.C10-143
		U:	SVol.C13-223/5
$C_4Na_{14}O_{29}S_7Th$	$Na_{14}Th(SO_3)_7(C_2O_4)_2 \cdot 5 H_2O$	Th:	SVol.C5-62
$C_4Na_{18}O_{35}S_9Th$	$Na_{18}Th(SO_3)_9(C_2O_4)_2 \cdot 6 H_2O$	Th:	SVol.C5-62
C_4NdO_4	$Nd(CO)_4$	Sc:	MVol.D6-166
$C_4NdO_4^+$	$[Nd(C_4O_4)]^+$	Sc:	MVol.D3-248/50
$C_4NdO_8^-$	$[Nd(C_2O_4)_2]^-$	Sc:	MVol.D5-120/4
C_4NdO_8Rb	$RbNd(C_2O_4)_2 \cdot x H_2O$	Sc:	MVol.D5-141/5
$C_4NdO_{12}^{5-}$	$[Nd(CO_3)_4]^{5-}$	Sc:	MVol.D4-338/40
C_4O_4Pr	$Pr(CO)_4$	Sc:	MVol.D6-166
$C_4O_4Pr^+$	$[Pr(C_4O_4)]^+$	Sc:	MVol.D3-248/50
$C_4O_4Si_4$	$Si_4O_4C_4$	Si:	SVol.B3-522
$C_4O_4Sm^+$	$[Sm(C_4O_4)]^+$	Sc:	MVol.D3-248/50
$C_4O_4Tb^+$	$[Tb(C_4O_4)]^+$	Sc:	MVol.D3-248/50
C_4O_4Tc	$[Tc(CO)_4]_n$	Tc:	SVol.2-130
$C_4O_4Tm^+$	$[Tm(C_4O_4)]^+$	Sc:	MVol.D3-248/50
C_4O_4U	$U(CO)_4$	U:	SVol.E2-171/2
$C_4O_4Y^+$	$[Y(C_4O_4)]^+$	Sc:	MVol.D3-248/50

$C_4O_{14}U^{6-}$ . . . . $[UO_2(CO_3)_4]^{6-}$
                    Sorption on anion exchangers . . . . . . U: SVol.D3-328

$C_4O_{16}S_2U_2$ . . . . $U_2(C_2O_4)_2(SO_4)_2 \cdot 6 H_2O$
                $= U(C_2O_4)(SO_4) \cdot 3 H_2O$ . . . . U: SVol.C13-225/7

C_4U . . . . UC_4 . . . . U: SVol.A6-180
                                                               U: SVol.C12-54/5

$C_{4.02}Cl_2H_{4.69}MnN_{0.67}$
                $Mn(CH_3C_5H_4N)_{0.67}Cl_2$ . . . . Mn: MVol.D3-121

$C_{4.5}Cl_5H_6N_3U$ . . . . . . $UCl_5 \cdot 1.5 NC_3H_3NH$ . . . . U: SVol.A6-23

$C_{4.5}H_9O_{7.5}S_2Th$ . . . . $Th(SO_3)_2 \cdot 1.5 (CH_3)_2CO \cdot 1.5 H_2O$ . . . . . . Th: SVol.C5-58

$C_5CdH_{15}Sb$ . . . . $Cd(CH_3)_2[Sb(CH_3)_3]$ . . . . Sb: Org.Comp.1-15

$C_5CeClH_7N_3O_4S$ . . . . $Ce[H_2NC(S)NHN(CH_2COO)_2]Cl \cdot 2 H_2O$ . . . Sc: MVol.D4-89/93

$C_5CeCl_3H_5$ . . . . $Ce(C_5H_5)Cl_3$ . . . . Sc: MVol.D6-230

$C_5CeCl_3H_{25}N_5$ . . . . . . $CeCl_3 \cdot 5 CH_3NH_2$ . . . . Sc: MVol.D1-15/6

$C_5CeH_4NO_2^{2+}$ . . . . . . $[Ce(NC_5H_3(OH)O)]^{2+}$ . . . . Sc: MVol.D2-6

$C_5CeH_4O_4^+$ . . . . $[Ce(OOCC(CH_2)CH_2COO)]^+$ . . . . Sc: MVol.D5-206/7

$-$ . . . . $[Ce(OOCCH_3CCHCOO)]^+$ . . . . Sc: MVol.D5-206

$C_5CeH_5OS^{2+}$ . . . . . . $Ce[(SCH_2)C_4H_3O]^{2+}$ . . . . Sc: MVol.D4-51

$C_5CeH_5O_7$ . . . . $Ce(OCH(CHOHCOO)_2) \cdot H_2O$
                $= Ce(OH)(OOC(CHOH)_3COO)$ . . . . Sc: MVol.D5-339/40

$C_5CeH_6NO_3^{2+}$ . . . . . . $[Ce(ONC(COCH_3)_2)]^{2+}$ . . . . Sc: MVol.D2-102

$C_5CeH_6NO_7P^-$ . . . . . $Ce[(OOCCH_2)_2NCH_2PO_3]^-$ . . . . Sc: MVol.D4-148

$C_5CeH_6O_7^+$ . . . . $[Ce(OOC(CHOH)_3COO)]^+$ . . . . Sc: MVol.D5-339/40

$C_5CeH_7NO_4^+$ . . . . . . $[Ce(CH_3N(CH_2COO)_2)]^+$ . . . . Sc: MVol.D1-127

$C_5CeH_7NO_7P$ . . . . . . $Ce[H((OOCCH_2)_2NCH_2PO_3)]$ . . . . Sc: MVol.D4-148

$C_5CeH_7NO_8P^{2-}$ . . . . $[Ce((OOCCH_2)_2NCH_2PO_3)(OH)]^{2-}$ . . . . Sc: MVol.D4-148

$C_5CeH_7N_2O_3^{2+}$ . . . . . $[Ce(CH_3C(NO)C(NOH)COCH_3)]^{2+}$ . . . . Sc: MVol.D2-116/7

$C_5CeH_7O_2^{2+}$ . . . . $[Ce(CH_3COCHCOCH_3)]^{2+}$ . . . . Sc: MVol.D3-76/82

$C_5CeH_7O_4^{3+}$ . . . . $[Ce(C_2H_5CH(COO)_2H)]^{3+}$ . . . . Sc: MVol.D5-164/5

$C_5CeH_7O_8$ . . . . $Ce(OH)(OOC(CHOH)_3COO)$
                $= Ce(OCH(CHOHCOO)_2) \cdot H_2O$ . . . . Sc: MVol.D5-339/40

$C_5CeH_8NO_2^{2+}$ . . . . . . $Ce(HNC_4H_7COO)^{2+}$ . . . . Sc: MVol.D1-215

$C_5CeH_8NO_3^{2+}$ . . . . . . $Ce[HNC_4H_6(OH)COO]^{2+}$ . . . . Sc: MVol.D1-216

$C_5CeH_8NO_4^{2+}$ . . . . . . $Ce[HOOCC_2H_4CH(NH_2)COO]^{2+}$ . . . . Sc: MVol.D1-110

$C_5CeH_8N_3O_5S$ . . . . . . $Ce[H_2NC(S)NHN(CH_2COO)_2]OH \cdot 2 H_2O$ . . Sc: MVol.D4-89/93

$C_5CeH_8O_4^{4+}$ . . . . $[Ce(C_2H_5CH(COOH)_2)]^{4+}$ . . . . Sc: MVol.D5-164/5

$C_5CeH_9N_2O_3^{2+}$ . . . . . $Ce[OOCCH(NH_2)C_2H_4C(O)NH_2]^{2+}$ . . . . Sc: MVol.D1-115

$C_5CeH_9O_3^{2+}$ . . . . $[Ce(C_2H_5C(OH)(CH_3)COO)]^{2+}$ . . . . Sc: MVol.D5-270

$-$ . . . . $[Ce(i-C_3H_7CH(OH)COO)]^{2+}$ . . . . Sc: MVol.D5-271

$C_5CeH_9O_4^{2+}$ . . . . $[Ce(H_3CCH(OH)C(OH)(CH_3)COO)]^{2+}$ . . . . . . Sc: MVol.D5-272

$C_5CeH_{10}NO_2^{2+}$ . . . . . $Ce[(CH_3)_2CHCH(NH_2)COO]^{2+}$ . . . . Sc: MVol.D1-115

$C_5CeH_{10}NO_2S^{2+}$ . . . . $Ce[CH_3SC_2H_4CH(NH_2)COO]^{2+}$ . . . . Sc: MVol.D1-115

$C_5CeH_{20}I_3N_{10}O_5$ . . . . $CeI_3 \cdot 5 OC(NH_2)_2$ . . . . Sc: MVol.D2-221

$C_5CeH_{20}I_{11}N_{10}O_5$ . . . . . $CeI_3 \cdot 5 OC(NH_2)_2 \cdot 4 I_2 \cdot 10 H_2O$ . . . . Sc: MVol.D2-221

$C_5CeH_{20}N_{13}O_{14}$ . . . . $Ce(NO_3)_3 \cdot 5 OC(NH_2)_2$ . . . . Sc: MVol.D2-220

$C_5CeO_5^+$ . . . . $[Ce(C_5O_5)]^+$ . . . . Sc: MVol.D3-250/1

$C_5CeO_{15}^{6-}$ . . . . $[Ce(CO_3)_5]^{6-}$ . . . . Sc: MVol.D4-338/40

$C_5Ce_4Cl_{12}H_{20}N_{10}S_5$ $Ce_4Cl_{12} \cdot 5 (NH_2)_2CS \cdot 24 H_2O$ . . . . Sc: MVol.D4-77/8

$C_5ClCoH_{18}N_{11}Rh$ . . . $[Co(NH_3)_6][Rh(CN)_5Cl]$ . . . . Rh: SVol.B1-199

C_5ClO_5Tc	$Tc(CO)_5Cl$	Tc:	SVol.2-129
$C_5Cl_2CoF_{12}NO_2P_2$ . .	$[Co(CO)(NO)(ClP(CF_3)_2)_2]$.	F:	PFHOrg.SVol.1-130
$C_5Cl_2Cu_2H_8$.	$[CH_2C(CH_3)CHCH_2]Cu_2Cl_2$	Cu:	Org.Comp.4-176
$C_5Cl_2DyH_5$	$Dy(C_5H_5)Cl_2 \cdot 3\ C_4H_8O$	Sc:	MVol.D6-226/9
$C_5Cl_2ErH_5$	$Er(C_5H_5)Cl_2 \cdot 3\ C_4H_8O$	Sc:	MVol.D6-226/9
$C_5Cl_2EuH_5$.	$Eu(C_5H_5)Cl_2 \cdot n\ C_4H_8O$ (n = 2, 3)	Sc:	MVol.D6-226/9
$C_5Cl_2FH_3S_2$	$SC_4H_3(SCFCl_2)$	F:	PFHOrg.SVol.2-166
$C_5Cl_2FH_4N_3O_2S$	$NC_5(Cl)_2(F)(NH_2)SO_2NH_2$	F:	PFHOrg.SVol.3-138/9, 143
$C_5Cl_2FH_7N_2S$.	$i\text{-}C_3H_7N(CN)SCFCl_2$	F:	PFHOrg.SVol.2-144
$C_5Cl_2F_2HNS$	$Cl_2(SH)F_2C_5N$	F:	PFHOrg.SVol.2-64, 67, 81
$C_5Cl_2F_2H_2N_2O_2S$. . . .	$NC_5(Cl)_2(F)_2SO_2NH_2$	F:	PFHOrg.SVol.3-138/9, 143
$C_5Cl_2F_2H_5NO_3S$	$CF_2ClCCl(NO_2)C(O)SC_2H_5$	F:	PFHOrg.8-38
$C_5Cl_2F_2H_5NO_4$	$CF_2ClCCl(NO_2)C(O)OC_2H_5$	F:	PFHOrg.8-38
$C_5Cl_2F_3H_2N_3OS$	$N_2SC_2(CF_3)NHC(O)CHCl_2$	F:	PFHOrg.SVol.2-54
$C_5Cl_2F_3H_6IO_2S$	$FSO_2CF_2(CH_2CHCl)_2I$	F:	PFHOrg.SVol.3-179
$C_5Cl_2F_3NOS$	$Cl(CF_3)C_3NSC(O)Cl$.	F:	PFHOrg.SVol.2-42, 52
$C_5Cl_2F_5H_4NO_2$	$ClCF_2CCl(CF_3)NHC(O)OCH_3$.	F:	PFHOrg.9-116/7
$C_5Cl_2F_8O_3S$	$CF_2CF(CF_2ClCFClCF_2)OSO_2$	F:	PFHOrg.SVol.3-125/6, 130, 135
$C_5Cl_2F_9N$.	$CF_3CCl_2NC(CF_3)_2$	F:	PFHOrg.9-131/2, 142
$C_5Cl_2F_{10}S_5$	$CF_3SCCl_2SSCF(SCF_3)_2$	F:	PFHOrg.SVol.2-177, 189
$C_5Cl_2GdH_5$	$Gd(C_5H_5)Cl_2 \cdot 3\ C_4H_8O$	Sc:	MVol.D6-226/9
$C_5Cl_2H_5Ho$	$Ho(C_5H_5)Cl_2 \cdot 3\ C_4H_8O$	Sc:	MVol.D6-226/9
$C_5Cl_2H_5La$.	$La(C_5H_5)Cl_2 \cdot 3\ C_4H_8O$	Sc:	MVol.D6-226/9
$C_5Cl_2H_5Lu$.	$Lu(C_5H_5)Cl_2 \cdot 3\ C_4H_8O$	Sc:	MVol.D6-226/9
$C_5Cl_2H_5MnN$	$Mn(C_5H_5N)Cl_2$.	Mn:	MVol.D3-102/3
$C_5Cl_2H_5MnNO$.	$Mn(C_5H_5NO)Cl_2$	Mn:	MVol.D3-150/1
–	$Mn(C_5H_5NO)Cl_2 \cdot H_2O$.	Mn:	MVol.D3-151
–	$Mn(NHC_5H_4O)Cl_2$	Mn:	MVol.D3-133
$C_5Cl_2H_5Sb$.	$C_5H_5SbCl_2$.	Sb:	Org.Comp.2-94/5
$C_5Cl_2H_5Sm$	$Sm(C_5H_5)Cl_2 \cdot 3\ C_4H_8O$.	Sc:	MVol.D6-226/9
$C_5Cl_2H_5Ti^+$	$[C_5H_5TiCl_2]^+$	Ti:	Org.Verb.2-116/7
$C_5Cl_2H_5Tm$	$Tm(C_5H_5)Cl_2 \cdot n\ C_4H_8O$ (n = 3, 4)	Sc:	MVol.D6-226/9
$C_5Cl_2H_5Y$.	$Y(C_5H_5)Cl_2 \cdot 3\ C_4H_8O$	Sc:	MVol.D6-226/9
$C_5Cl_2H_5Yb$.	$Yb(C_5H_5)Cl_2 \cdot 3\ C_4H_8O$	Sc:	MVol.D6-226/9
$C_5Cl_2H_6MnN_2O$	$Mn(NC_3H_3NCOCH_3)Cl_2$.	Mn:	MVol.D3-280
$C_5Cl_2H_6MnN_2O_2$	$Mn[2,4\text{-}(O)_2\text{-}5\text{-}CH_3\text{-}1,3\text{-}C_4HN_2H_2]Cl_2 \cdot 2\ H_2O$	Mn:	MVol.D4-9
$C_5Cl_2H_7MnNO$.	$Mn(ONC_3H(CH_3)_2)Cl_2 \cdot 0.5\ H_2O$	Mn:	MVol.D4-213/5
$C_5Cl_2H_9MnNO$.	$Mn(CH_3N(CH_2)_3C(O))Cl_2$.	Mn:	MVol.D3-79
$C_5Cl_2H_9Sb$.	$C_5H_9SbCl_2$.	Sb:	Org.Comp.2-93
$C_5Cl_2H_{10}N_4S_2$	$Cl_2S_2N_3C(N(C_2H_5)_2)$	S:	S-N Comp.4-6/9
$C_5Cl_2H_{10}O_2Sn$.	$(CH_3)_3SnOOCCHCl_2$	Sn:	Org.Comp.11-92, 101/2
$C_5Cl_2H_{11}MnN$	$Mn(C_5H_{10}NH)Cl_2$.	Mn:	MVol.D3-82
$C_5Cl_2H_{11}MnNO_2S$ . . .	$MnCl_2 \cdot CH_3SC_2H_4CH(NH_2)COOH \cdot 3\ H_2O$	Mn:	MVol.D4-294
$C_5Cl_2H_{12}MnN_2O$	$Mn(i\text{-}C_4H_9C(O)NHNH_2)Cl_2 \cdot H_2O$	Mn:	MVol.D5-172/3
$C_5Cl_2H_{12}SSn$.	$C_4H_9SnCl_2(SCH_3)$	Sn:	Org.Comp.10-253
$C_5Cl_2H_{14}PdSb_2$	$PdCl_2[((CH_3)_2Sb)_2CH_2]$	Sb:	Org.Comp.1-166

$C_5Cl_3H_{25}N_5O_8Rh$ ...	$[Rh(NH_2CH_3)_5Cl](ClO_4)_2$	Rh:	SVol.B2-156
$C_5Cl_3H_{25}N_5Pr$	$PrCl_3 \cdot 5\ CH_3NH_2$	Sc:	MVol.D1-15/6
$C_5Cl_3H_{25}N_5Rh$	$[Rh(NH_2CH_3)_5Cl]Cl_2$	Rh:	SVol.B2-156/7
$C_5Cl_3H_{25}N_5Sm$	$SmCl_3 \cdot 5\ CH_3NH_2$	Sc:	MVol.D1-15/6
$C_5Cl_3H_{27}N_5O_{13}Rh$...	$[Rh(NH_2CH_3)_5(H_2O)](ClO_4)_3$	Rh:	SVol.B2-156
$C_5Cl_4CsH_7O_2Os$	$Cs[Os(CH_3COCHCOCH_3)Cl_4]$	Os:	SVol.1-205
C_5Cl_4CuN	$NC_5Cl_4Cu \cdot x\ LiBr$	Cu:	Org.Comp.1-114, 116/7
−	$NC_5Cl_4Cu \cdot x\ LiCl$	Cu:	Org.Comp.1-59, 91, 109
−	$NC_5Cl_4Cu \cdot x\ LiHal$	Cu:	Org.Comp.1-120
−	$NC_5Cl_4Cu \cdot x\ LiI$	Cu:	Org.Comp.1-59, 91, 112, 117
−	$NC_5Cl_4Cu \cdot x\ MgBrCl$	Cu:	Org.Comp.1-214, 287/9
−	$NC_5Cl_4Cu \cdot x\ MgClHal$	Cu:	Org.Comp.1-279
−	$NC_5Cl_4Cu \cdot x\ MgCII$	Cu:	Org.Comp.1-214, 290/1
−	$NC_5Cl_4Cu \cdot x\ MgCl_2$	Cu:	Org.Comp.1-214, 262, 268, 290/2
$C_5Cl_4FNO_2S$	$NC_5(Cl)_3(F)SO_2Cl$	F:	PFHOrg.SVol.3-160, 169
C_5Cl_4FNS	$Cl_3(SCl)FC_5N$	F:	PFHOrg.SVol.2-120
$C_5Cl_4F_2HN_3OS$	$(CF_2Cl)C_2N_2S(NHC(O)CCl_3)$	F:	PFHOrg.SVol.2-43/4, 49
$C_5Cl_4GaH_{14}Sb$	$(CH_3)_3C_2H_5SbGaCl_4$	Sb:	Org.Comp.3-157
$C_5Cl_4H_5MnN$	$(ClC_5H_4NH)MnCl_3$		
	Spectra	Mn:	MVol.C10-40
$C_5Cl_4H_5NOTh$	$ThCl_4 \cdot C_5H_5NO$	Th:	SVol.E-68
$C_5Cl_4H_7KO_2Os$	$K[Os(CH_3COCHCOCH_3)Cl_4]$	Os:	SVol.1-206
$C_5Cl_4H_{14}MnN_3O$	$Mn[((CH_3)_3NCH_2C(O)NHNH_2)Cl]Cl_3$	Mn:	MVol.D5-171
$C_5Cl_4H_{19}N_6O_{12}Rh$...	$[Rh(NH_3)_5(NC_5H_4Cl)](ClO_4)_3$	Rh:	SVol.B2-140
$C_5Cl_4H_{20}N_{10}O_5Th$...	$ThCl_4 \cdot 5\ (NH_2)_2CO \cdot 6\ H_2O$	Th:	SVol.E-37, 40
$C_5Cl_5Cu_4H_{10}N$	$2\ C_2H_2 \cdot 4\ CuCl \cdot (CH_3NH_3)Cl$	Cu:	Org.Comp.4-51
$C_5Cl_6F_6N_2O_2P_2$	$[Cl_3PNC(O)CF_2]_2CF_2$	F:	PFHOrg.9-40, 42, 45
$C_5Cl_6H_3N_3OS$	$(CH_3O)SN_3C_2(CCl_3)_2$	S:	S-N Comp.4-31/2
$C_5Cl_7H_{10}N_4S_2Sb$	$[ClS_2N_3C(N(C_2H_5)_2)]SbCl_6$	S:	S-N Comp.4-6
$C_5Cl_7H_{37}N_{16}O_5Rh_4$ ..	$Rh_4Cl_7(O_2CCH_3)_{2.5}(N_2H_3)_{2.5}(N_2H_4)_{5.5}$	Rh:	SVol.B2-60/1
$C_5Cl_{12}Dy_4H_{20}N_{10}S_5$	$Dy_4Cl_{12} \cdot 5\ (NH_2)_2CS \cdot 20\ H_2O$	Sc:	MVol.D4-77/8
$C_5Cl_{12}Er_4H_{20}N_{10}S_5$..	$Er_4Cl_{12} \cdot 5\ (NH_2)_2CS \cdot 20\ H_2O$.	Sc:	MVol.D4-77/8
$C_5Cl_{12}Eu_4H_{20}N_{10}S_5$	$Eu_4Cl_{12} \cdot 5\ (NH_2)_2CS \cdot 22\ H_2O$	Sc:	MVol.D4-77/8
$C_5Cl_{12}Gd_4H_{20}N_{10}S_5$	$Gd_4Cl_{12} \cdot 5\ (NH_2)_2CS \cdot 22\ H_2O$	Sc:	MVol.D4-77/8
$C_5Cl_{12}H_{20}Ho_4N_{10}S_5$	$Ho_4Cl_{12} \cdot 5\ (NH_2)_2CS \cdot 20\ H_2O$	Sc:	MVol.D4-77/8
$C_5Cl_{12}H_{20}La_4N_{10}S_5$	$La_4Cl_{12} \cdot 5\ (NH_2)_2CS \cdot 24\ H_2O$	Sc:	MVol.D4-77/8
$C_5Cl_{12}H_{20}Lu_4N_{10}S_5$	$Lu_4Cl_{12} \cdot 5\ (NH_2)_2CS \cdot 20\ H_2O$	Sc:	MVol.D4-77/8
$C_5Cl_{12}H_{20}N_{10}Nd_4S_5$	$Nd_4Cl_{12} \cdot 5\ (NH_2)_2CS \cdot 20\ H_2O$	Sc:	MVol.D4-77/8
$C_5Cl_{12}H_{20}N_{10}Pr_4S_5$..	$Pr_4Cl_{12} \cdot 5\ (NH_2)_2CS \cdot 20\ H_2O$	Sc:	MVol.D4-77/8
$C_5Cl_{12}H_{20}N_{10}S_5Sm_4$	$Sm_4Cl_{12} \cdot 5\ (NH_2)_2CS \cdot 22\ H_2O$	Sc:	MVol.D4-77/8
$C_5Cl_{12}H_{20}N_{10}S_5Tb_4$	$Tb_4Cl_{12} \cdot 5\ (NH_2)_2CS \cdot 20\ H_2O$	Sc:	MVol.D4-77/8
$C_5Cl_{12}H_{20}N_{10}S_5Tm_4$	$Tm_4Cl_{12} \cdot 5\ (NH_2)_2CS \cdot 20\ H_2O$	Sc:	MVol.D4-77/8
$C_5Cl_{12}H_{20}N_{10}S_5Y_4$ ..	$Y_4Cl_{12} \cdot 5\ (NH_2)_2CS \cdot 20\ H_2O$	Sc:	MVol.D4-77/8
$C_5Cl_{12}H_{20}N_{10}S_5Yb_4$	$Yb_4Cl_{12} \cdot 5\ (NH_2)_2CS \cdot 20\ H_2O$	Sc:	MVol.D4-77/8
$C_5CoF_{14}NO_2P_2$	$[Co(CO)(NO)(FP(CF_3)_2)_2]$	F:	PFHOrg.SVol.1-130
$C_5CoH_5N_2S_2$	$(C_5H_5)CoS_2N_2$	S:	S-N Comp.2-191
$C_5CoH_{18}IN_{11}Rh$	$[Co(NH_3)_6][Rh(CN)_5I]$	Rh:	SVol.B1-199
$C_5Co_2H_{36}N_{12}O_{15}U$ ..	$[Co(NH_3)_6]_2[U(CO_3)_5] \cdot 4\ H_2O$	U:	SVol.C13-8/9

$C_5Co_2H_{36}N_{12}O_{15}U$ ..	$[Co(NH_3)_6]_2[U(CO_3)_5]$ · 5 H_2O	U:	SVol.C13-8/9
—	$[Co(NH_3)_6]_2[U(CO_3)_5]$ · x H_2O	U:	SVol.C13-8/9
$C_5Co_2H_{38}N_{12}O_{16}U$ ..	$[Co(NH_3)_6]_2[U(CO_3)_5(H_2O)]$ · x H_2O	U:	SVol.C13-8/9
$C_5Cr_2H_4NO_7Th$	$ThCr_2O_7$ · 0.5 $C_{10}H_8N_2$	Th:	SVol.E-17, 20
$C_5CsH_6NO_6SU$	$Cs[UO_2(CH_3COO)_2(NCS)]$ · 3 H_2O	U:	SVol.C13-141/2
$C_5CsH_7I_4O_2Os$	$Cs[Os(CH_3COCHCOCH_3)I_4]$	Os:	SVol.1-205
$C_5Cs_3HN_5Rh$	$Cs_3[Rh(CN)_5H]$	Rh:	SVol.B1-195
$C_5Cs_3N_5O_2S_5U$	$Cs_3UO_2(NCS)_5$	U:	SVol.A6-34
		U:	SVol.C13-45/6
$C_5CuF_3H_6O_2$	$(CH_3CHCH_2Cu)OOCCF_3$	Cu:	Org.Comp.4-7
$C_5CuF_3H_6O_3$	$(HOCH_2CHCH_2Cu)OOCCF_3$	Cu:	Org.Comp.4-9
$C_5CuF_3H_6O_3S$	$(CH_2CHCHCH_2)CuO_3SCF_3$	Cu:	Org.Comp.4-85
$C_5CuF_3H_7NO_3S$	$(CH_2CH_2CuNCCH_3)(O_3SCF_3)$	Cu:	Org.Comp.4-25
C_5CuF_4N	NC_5F_4Cu · x $MgBrCl$	Cu:	Org.Comp.1-215, 287
$C_5CuF_6HO_5$	$(CO)CuO_2CCF_3$ · CF_3CO_2H	Cu:	Org.Comp.3-191
C_5CuF_{11}	$n-C_5F_{11}Cu$	Cu:	Org.Comp.1-23
C_5CuH_3	$CH_3CCCCCu$	Cu:	Org.Comp.3-20, 86, 110
C_5CuH_4N	NC_5H_4Cu	Cu:	Org.Comp.1-46
—	NC_5H_4Cu · x $P(C_4H_9-n)_3$ · y LiI	Cu:	Org.Comp.1-459
—	NC_5H_4Cu · x $S(C_4H_9-n)_2$ · y LiI	Cu:	Org.Comp.1-360, 377
C_5CuH_5	$c-C_3H_5CCCu$	Cu:	Org.Comp.3-21, 93
—	$c-C_3H_5CCCu$ · x $LiBr$	Cu:	Org.Comp.3-152
—	$CH_2C(CH_3)CCCu$	Cu:	Org.Comp.3-21, 83, 97,
			116/24, 157
—	$CH_2C(CH_3)CCCu$ · x $LiBr$	Cu:	Org.Comp.3-152
—	$CH_3CHCHCCCu$	Cu:	Org.Comp.3-20, 110
C_5CuH_5O	$CH_3OCHCHCCCu$	Cu:	Org.Comp.3-11, 21,
			76/98, 119
—	$HOCH_2CHCHCCCu$	Cu:	Org.Comp.3-21, 150
—	$OC_4H_2(CH_3)Cu$ · x $LiBr$	Cu:	Org.Comp.1-59
—	$OC_4H_2(CH_3)Cu$ · x $MgBrHal$	Cu:	Org.Comp.1-215, 293
$C_5CuH_5O_2$	$CH_3CO_2CH_2CCCu$	Cu:	Org.Comp.3-21
—	$C_2H_5O_2CCCCu$	Cu:	Org.Comp.3-8, 11, 21,
			98, 101, 110, 127/8, 140
—	$HO_2C(CH_2)_2CCCu$	Cu:	Org.Comp.3-16, 21
$C_5CuH_6^+$	$(c-C_5H_6Cu)^+$	Cu:	Org.Comp.4-87
C_5CuH_6N	$CH_3NC_4H_3Cu$ · x $C_2H_4N_2(CH_3)_4$		
	· y C_5H_5N · z $LiBr$	Cu:	Org.Comp.1-409/10
—	$CH_3NC_4H_3Cu$ · x $C_2H_4N_2(CH_3)_4$		
	· y C_5H_5N · z LiI.	Cu:	Org.Comp.1-411
—	$CH_3NC_4H_3Cu$ · x $C_2H_4N_2(CH_3)_4$ · y $LiBr$..	Cu:	Org.Comp.1-409
—	$CH_3NC_4H_3Cu$ · x $C_2H_4N_2(CH_3)_4$ · y $LiCl$..	Cu:	Org.Comp.1-409
C_5CuH_7	$CH_2CHCHCHCH_2Cu$ · x LiI	Cu:	Org.Comp.1-60
—	$CH_2CHCHCHCH_2Cu$ · x $LiSC_6H_5$	Cu:	Org.Comp.1-166, 182/3
—	$CH_2CHCHCHCH_2Cu$ · x $P(CH_3)_3$	Cu:	Org.Comp.1-447
—	$(CH_3)_2CCCHCu$ · x $LiBr$.............	Cu:	Org.Comp.1-60, 94, 121/2
—	$n-C_3H_7CCCu$.	Cu:	Org.Comp.3-10/1, 21, 44,
			79/130
—	$i-C_3H_7CCCu$	Cu:	Org.Comp.3-21, 93, 98
—	$n-C_3H_7CCCu$ · x LiI.	Cu:	Org.Comp.3-152

C_5CuH_9 . . . . i-C_3H_7CHCHCu · x MgBrHal . . . . Cu: Org.Comp.1-215, 258, 261, 295

− . . . . i-C_3H_7CHCHCu · x MgBr$_2$ . . . . Cu: Org.Comp.1-262, 269/70, 295

− . . . . i-C_3H_7CHCHCu · x MgIBr . . . . Cu: Org.Comp.1-269

C_5CuH_9N . . . . t-C_4H_9CNCu . . . . Cu: Org.Comp.3-242

− . . . . NC(C_4H_9-t)Cu (radical) . . . . Cu: Org.Comp.3-242

C_5CuH_9OS . . . . C_2H_5CHC(S(O)CH$_3$)Cu · x MgBrHal . . . . . . Cu: Org.Comp.1-215

− . . . . C_2H_5S(O)C(CHCH$_3$)Cu · x LiI . . . . Cu: Org.Comp.1-60

$C_5CuH_9O_2$ . . . . (CO)CuOC$_4H_9$-t . . . . Cu: Org.Comp.3-191

$C_5CuH_9O_2S$ . . . . (CH$_3$)$_2$CC(SO$_2$CH$_3$)Cu · x MgBr$_2$ . . . . Cu: Org.Comp.1-215

− . . . . C_2H_5CHC(SO$_2$CH$_3$)Cu · x MgBrHal . . . . . . Cu: Org.Comp.1-215

C_5CuH_9Si . . . . (CH$_3$)$_3$SiCCCu . . . . Cu: Org.Comp.3-8, 10/1, 22, 44, 77/86, 156

− . . . . (CH$_3$)$_3$SiCCCu · x LiBr . . . . Cu: Org.Comp.3-152

− . . . . (CH$_3$)$_3$SiCCCu · x LiI . . . . Cu: Org.Comp.3-152

C_5CuH_{11} . . . . (C_2H_5)$_2$CHCu · x MgHal$_2$ . . . . Cu: Org.Comp.1-216

− . . . . C_2H_5CH(CH$_3$)CH$_2$Cu · x LiMgBr$_2$Cl . . . . . . Cu: Org.Comp.1-300, 314

− . . . . t-C_4H_9CH$_2$Cu · x C_5H_5N · y MgBrCl . . . . . Cu: Org.Comp.1-389

− . . . . t-C_4H_9CH$_2$Cu · x MgBrCl . . . . Cu: Org.Comp.1-216, 258

− . . . . t-C_4H_9CH$_2$Cu · x MgBr$_2$ . . . . Cu: Org.Comp.1-216

− . . . . t-C_4H_9CH$_2$Cu · x MgCl$_2$ . . . . Cu: Org.Comp.1-278

− . . . . t-C_4H_9CH$_2$Cu · x MgHal$_2$ . . . . Cu: Org.Comp.1-216

− . . . . t-C_4H_9CH$_2$Cu · x MgIHal . . . . Cu: Org.Comp.1-216, 258

− . . . . i-C_5H_{11}Cu · x LiCN . . . . Cu: Org.Comp.1-126, 135/6

− . . . . n-C_5H_{11}Cu · x LiMgBrCl$_2$ . . . . Cu: Org.Comp.1-300

− . . . . n-C_5H_{11}Cu · x LiMgBr$_2$Cl . . . . Cu: Org.Comp.1-300, 312, 314

− . . . . n-C_5H_{11}Cu · x LiMgBr$_3$ . . . . Cu: Org.Comp.1-300

− . . . . n-C_5H_{11}Cu · x LiSC$_6H_5$ . . . . Cu: Org.Comp.1-180

− . . . . n-C_5H_{11}Cu · x MgBrCl . . . . Cu: Org.Comp.1-216

− . . . . n-C_5H_{11}Cu · x MgBrHal . . . . Cu: Org.Comp.1-216, 281, 284

− . . . . n-C_5H_{11}Cu · x MgBr$_2$ . . . . Cu: Org.Comp.1-216, 284

− . . . . n-C_5H_{11}Cu · x P(OC$_2H_5$)$_3$ · y MgBr$_2$ . . . . . Cu: Org.Comp.1-445/6

− . . . . n-C_5H_{11}Cu · x S(CH$_3$)$_2$ · y MgBr$_2$ . . . . Cu: Org.Comp.1-359, 363

$C_5CuH_{11}N_2$ . . . . (CH$_3$)$_2$NNC(CH$_3$)CH$_2$Cu · x LiSC$_6H_5$ . . . . . . Cu: Org.Comp.1-166, 181

$C_5CuH_{11}Si$ . . . . (CH$_3$)$_3$SiC(CH$_2$)Cu · x LiI . . . . Cu: Org.Comp.1-92

− . . . . (CH$_3$)$_3$SiC(CH$_2$)Cu · x S(CH$_3$)$_2$ · y MgBr$_2$ Cu: Org.Comp.1-359, 372/3

$C_5CuH_{12}Li$ . . . . CH$_3$(n-C_4H_9)CuLi . . . . Cu: Org.Comp.2-175, 190, 206

− . . . . CH$_3$(t-C_4H_9)CuLi . . . . Cu: Org.Comp.2-175, 190, 206

$C_5CuH_{12}O_2PS$ . . . . . . (C$_2H_5$O)$_2$P(S)CH$_2$Cu · x LiI . . . . Cu: Org.Comp.1-60, 111, 113/7

$C_5CuH_{12}O_3P$ . . . . (C$_2H_5$O)$_2$P(O)CH$_2$Cu · x LiBr . . . . Cu: Org.Comp.1-61, 113

− . . . . (C$_2H_5$O)$_2$P(O)CH$_2$Cu · x LiCl . . . . Cu: Org.Comp.1-61, 113

− . . . . (C$_2H_5$O)$_2$P(O)CH$_2$Cu · x LiI . . . . Cu: Org.Comp.1-61, 100/18

$C_5CuH_{14}P$ . . . . (CH$_3$)$_3$PCH$_2$CuCH$_3$ . . . . Cu: Org.Comp.2-3

$C_5CuH_{15}Li_4$ . . . . (CH$_3$)$_5$CuLi$_4$ . . . . Cu: Org.Comp.2-243

$C_5F_5H_8O_2P$	$C_2F_5PH(O)OC_3H_7\text{-}i$	F:	PFHOrg.SVol.1-123
$C_5F_5H_9O_3SSn$	$(CH_3)_3SnOSO_2C_2F_5$	Sn:	Org.Comp.11-148
$C_5F_6FeH_5KP_2$	$K[C_5H_5Fe(PF_3)_2]$	Fe:	Org.Comp.B11-1/2
$C_5F_6FeH_6OP_2$	$(CH_2)_3CFe(PF_3)_2CO$	Fe:	Org.Comp.B6-27, 28, 40
–	$C_4H_6Fe(PF_3)_2CO$	Fe:	Org.Comp.B6-27, 28, 40
$C_5F_6FeH_6P_2$	$C_5H_5Fe(PF_3)_2H$	Fe:	Org.Comp.B6-18
		Fe:	Org.Comp.B11-17, 26
–	$C_5H_5Fe(PF_3)_2D$	Fe:	Org.Comp.B11-17, 26
C_5F_6HNO	$(CF_2)_3C(OH)CN$	F:	PFHOrg.9-59, 73, 90/1
$C_5F_6HN_3OS$	$CF_3C_2N_2S(NHC(O)CF_3)$	F:	PFHOrg.SVol.2-43/4, 49
$C_5F_6H_2N_2O_2S_3$	$NSC_3H(SCF_3)NHSO_2CF_3$	F:	PFHOrg.SVol.2-165
$C_5F_6H_2O_4$	$HOCO(CF_2)_3COOH$	F:	PFHOrg.9-150
$C_5F_6H_3N$	$(CF_3)_2C(CH_3)CN$	F:	PFHOrg.9-82
$C_5F_6H_3NO$	$(CF_3)_2C(OCH_3)CN$	F:	PFHOrg.9-82
–	$CF_3CHCHC(O)NHCF_3$	F:	PFHOrg.9-151
$C_5F_6H_5N$	$(CF_3)_2NCH_2CHCH_2$	F:	PFHOrg.9-47
$C_5F_6H_5NO$	$(CF_3)_2H_4C_3ONH$	F:	PFHOrg.SVol.2-9
–	$CF_3CH_2CH_2C(O)NHCF_3$	F:	PFHOrg.9-151
$C_5F_6H_5NO_2$	$CF_3CF(NF_2)COOC_2H_5$	F:	PFHOrg.8-207
–	$CF_3NHC(O)OC_2H_4CF_3$	F:	PFHOrg.9-116
$C_5F_6H_5N_3O$	$(CF_3)_2CNN(CH_3)C(O)NH_2$	F:	PFHOrg.8-126
–	$(CF_3)_2CNNHC(O)NHCH_3$	F:	PFHOrg.8-126, 130
$C_5F_6H_6N_2$	$(CF_3)_2CN(C_2H_5)NH$	F:	PFHOrg.8-221
–	$(CF_3)_2CNN(CH_3)_2$	F:	PFHOrg.8-221
$C_5F_6H_6N_2S$	$(CF_3)_2CNSN(CH_3)_2$	F:	PFHOrg.9-35
–	$CF_3SNC(CF_3)N(CH_3)_2$	F:	PFHOrg.SVol.2-113
$C_5F_6H_6N_4O_2$	$H_2NNHCO(CF_2)_3CONHNH_2$	F:	PFHOrg.8-66
$C_5F_6H_6OS$	$(CF_3)_2C(OCH_3)SCH_3$	F:	PFHOrg.SVol.2-7
$C_5F_6H_6O_2Sn$	$(CH_3)_2(CF_3)SnOOCCF_3$	Sn:	Org.Comp.13-237
$C_5F_6H_6S_2$	$(CF_3)_2CHSSC_2H_5$	F:	PFHOrg.SVol.2-7
$C_5F_6H_7NS$	$(CF_3)_2SNC_3H_7\text{-}i$	F:	PFHOrg.SVol.3-44
$C_5F_6H_7N_3O_2$	$(CF_3)_2C(OCH_3)NHNHC(O)NH_2$	F:	PFHOrg.8-122
$C_5F_6H_7OP$	$(CF_3)_2PO(C_3H_7\text{-}i)$	F:	PFHOrg.SVol.1-122
$C_5F_6H_7O_2P_2Rh$	$Rh(PF_3)_2(CH(COCH_3)_2)$	Rh:	SVol.B3-74
$C_5F_6H_7PS$	$(CF_3)_2PS(C_3H_7\text{-}i)$	F:	PFHOrg.SVol.1-122
$C_5F_6H_9NSi$	$(CF_3)_2NSi(CH_3)_3$	F:	PFHOrg.8-207
$C_5F_6H_9PSi$	$(CH_3)_3SiP(CF_3)_2$	F:	PFHOrg.SVol.1-86
$C_5F_6H_9PSn$	$(CH_3)_3SnP(CF_3)_2$	F:	PFHOrg.SVol.1-86
$C_5F_6H_{13}Mn_2N_3$	$2\,MnF_2 \cdot 2\,NH_4F \cdot C_5H_5N$	Mn:	MVol.D3-105
C_5F_6INO	$NC(CF_2)_3C(O)I$	F:	PFHOrg.9-54/5
$C_5F_6N_2$	$NC(CF_2)_3CN$	F:	PFHOrg.9-56/7, 71, 82, 88, 95
$C_5F_6N_2O$	$CF_3CF(CN)OCF_2CN$	F:	PFHOrg.9-57/8
$C_5F_6N_2O_2$	$OCN(CF_2)_3NCO$	F:	PFHOrg.9-106, 112
$C_5F_6SSe_2$	$SC(SeCCF_3)_2$	F:	PFHOrg.SVol.3-209/10
$C_5F_6S_2$	$(CF_3)_2C_2S_2C$	F:	PFHOrg.SVol.2-19
$C_5F_6S_3$	$(CF_3)_2C_2S_2C(S)$	F:	PFHOrg.SVol.2-18
$C_5F_6Se_3$	$SeC(SeCCF_3)_2$	F:	PFHOrg.SVol.3-209/10
$C_5F_7H_3N_2O$	$CF_3NCFC(OCH_3)NCF_3$	F:	PFHOrg.9-153
$C_5F_7H_4NO$	$C_3F_7C(NH)OCH_3$	F:	PFHOrg.9-81

$C_5F_7H_4NO_5S_2$	$CF_3CH(SO_2F)C(O)NHSO_2CH_2CF_3$	F:	PFHOrg.SVol.3-174
$C_5F_7H_4NS$	$C_3F_7C(NH)SCH_3$	F:	PFHOrg.9-82
$C_5F_7H_5N_2$	$C_3F_7C(NH)NHCH_3$	F:	PFHOrg.9-79/80
$C_5F_7H_6NO$	$(CF_3)_2C(OC_2H_5)NFH$	F:	PFHOrg.8-221
$C_5F_7H_6N_3$	$[N(C_2H_5)NHC(CF_3)(CF_2NF_2)]$	F:	PFHOrg.8-223
C_5F_7N	$(CF_3)_2CCFCN$	F:	PFHOrg.9-52/3, 69
C_5F_7NO	$(CF_2)_3CFNCO$	F:	PFHOrg.9-104/5, 111
–	$[C_3F_7C(O)CN]_2$	F:	PFHOrg.9-53, 69, 79, 88
–	$NC(CF_2)_3C(O)F$	F:	PFHOrg.9-54/5, 70
C_5F_7NOS	$C_3F_7C(O)NCS$	F:	PFHOrg.9-108, 114
C_5F_7NS	$(CF_3)_2CCFNCS$	F:	PFHOrg.9-108, 114, 121
$C_5F_8H_3NS$	$(CF_2)_4SNCH_3$	F:	PFHOrg.SVol.3-44
$C_5F_8H_5N_3O$	$CF_3NHCF_2OCH_2CH_2NNCF_3$	F:	PFHOrg.9-150/1
$C_5F_8N_2$	$n-C_3F_7C(CN)NF$	F:	PFHOrg.8-170/1
$C_5F_8N_2S$	$(CF_2)_4SNCN$	F:	PFHOrg.SVol.3-9, 31
$C_5F_9FeH_6P_3$	$C_5H_6Fe(PF_3)_3$	Fe:	Org.Comp.B6-4, 10, 18
$C_5F_9FeH_8P_3$	$(CH_2CHC(CH_3)CH_2)Fe(PF_3)_3$	Fe:	Org.Comp.B6-4, 7, 16
–	$(CH_2CHCHCHCH_3)Fe(PF_3)_3$	Fe:	Org.Comp.B6-4, 6, 16
$C_5F_9HN_2$	$(CF_3)_2CNNCHCF_3$	F:	PFHOrg.8-125, 128
$C_5F_9HO_2$	$C_4F_9C(O)OH$	F:	PFHOrg.9-150
$C_5F_9HO_2S$	$CF_3C(O)SC(CF_3)_2OH$	F:	PFHOrg.SVol.2-73
$C_5F_9HS_3$	$(CF_3S)_2CCHSCF_3$	F:	PFHOrg.SVol.2-71
$C_5F_9H_2N$	$(CF_3)_2NCHCHCF_3$	F:	PFHOrg.9-178
$C_5F_9H_2NO$	$(CF_3)_2NC(O)CH_2CF_3$	F:	PFHOrg.9-178
–	$C_4F_9C(O)NH_2$	F:	PFHOrg.9-155
$C_5F_9H_2NOS$	$CF_3SNC(CF_3)OCH_2CF_3$	F:	PFHOrg.SVol.2-111/2
$C_5F_9H_2O_2S_2$	$(CF_3)C(OH)SSC(OH)(CF_3)_2$	F:	PFHOrg.SVol.2-179, 191
$C_5F_9H_3N_2O$	$C_4F_9C(O)NHNH_2$	F:	PFHOrg.8-65/6, 90
$C_5F_9H_3N_2OS$	$CF_3S(O)(NCH_3)NC(CF_3)_2$	F:	PFHOrg.SVol.3-134/5
$C_5F_9H_5N_2$	$(CF_3)_2NN(CF_3)C_2H_5$	F:	PFHOrg.8-105
$C_5F_9H_6O_2P$	$(CF_3)_3P(OCH_3)_2$	F:	PFHOrg.SVol.1-126
$C_5F_9IO_4$	$CF_3I[OC(O)CF_3]_2$	F:	PFHOrg.SVol.3-254, 258, 263
C_5F_9N	$(CF_3)_2NCCCF_3$	F:	PFHOrg.9-161/2, 171, 178
C_5F_9NO	$(CF_3)_3CNCO$	F:	PFHOrg.9-104/5, 111, 117/20
–	$CF_3CF_2OCF_2CF_2CN$	F:	PFHOrg.9-52/3, 71
–	$CF_3C(O)NC(CF_3)_2$	F:	PFHOrg.9-131/2, 142
–	$n-C_4F_9NCO$	F:	PFHOrg.9-104/5, 111
$C_5F_9NO_2S$	$n-C_4F_9S(O)NCO$	F:	PFHOrg.SVol.3-6, 25
$C_5F_9NS_2$	$(CF_3)_3C_2NS_2$	F:	PFHOrg.SVol.2-43, 48
$C_5F_{10}H_2N_2$	$(C_2F_5)_2CNNH_2$	F:	PFHOrg.8-69, 92
$C_5F_{10}H_2OS$	$(C_2F_5)_2C(OH)SH$	F:	PFHOrg.SVol.2-64/5, 68
$C_5F_{10}H_3NO$	$(CF_3)_2C(OCH_2CF_3)NFH$	F:	PFHOrg.8-221
$C_5F_{10}H_4N_2O$	$C_3F_7C(OCH_3)(NFH)NF_2$	F:	PFHOrg.8-224
$C_5F_{10}H_6NP$	$(CF_3)_3P(F)N(CH_3)_2$	F:	PFHOrg.SVol.1-128
$C_5F_{10}N_2$	$CF_2N(CF_2)_3NCF_2$	F:	PFHOrg.9-131/2
–	$CF_2NCF_2CF(CF_3)NCF_2$	F:	PFHOrg.9-131/2, 142/3, 149
–	$CF_3NCFC(CF_3)NCF_3$	F:	PFHOrg.9-131/2, 143, 149

$C_5GaH_{13}O$	$Ga(C_2H_5)_2OCH_3$	Ga: Org.Comp.1-177
$C_5GaH_{13}O_3S$	$Ga(C_2H_5)_2OSO_2CH_3$	Ga: Org.Comp.1-207
$C_5GaH_{13}S$	$Ga(CH_3)_2SC_3H_7$	Ga: Org.Comp.1-229
–	$Ga(CH_3)_2SC_3H_7$-i	Ga: Org.Comp.1-229
$C_5GaH_{14}Na$	$Na[Ga(CH_3)_3C_2H_5]$	Ga: Org.Comp.1-316
$C_5GaH_{15}NOP$	$Ga(CH_3)_2OP(NCH_3)(CH_3)_2$	Ga: Org.Comp.1-205
$C_5GaH_{15}NP$	$Ga(CH_3)_2NP(CH_3)_3$	Ga: Org.Comp.1-273/5
$C_5GaH_{15}NPS$	$Ga(CH_3)_2SP(NCH_3)(CH_3)_2$	Ga: Org.Comp.1-230
$C_5GaH_{15}O$	$Ga(CH_3)_3 \cdot O(CH_3)_2$	Ga: Org.Comp.1-32/5
–	$Ga(CH_3)_3 \cdot O(CD_3)_2$	Ga: Org.Comp.1-33, 35
$C_5GaH_{15}OS$	$Ga(CH_3)_3 \cdot OS(CH_3)_2$	Ga: Org.Comp.1-33
$C_5GaH_{15}OSi$	$Ga(CH_3)_2OSi(CH_3)_3$	Ga: Org.Comp.1-206
$C_5GaH_{15}S$	$Ga(CH_3)_3 \cdot S(CH_3)_2$	Ga: Org.Comp.1-34, 38
$C_5GaH_{15}Se$	$Ga(CH_3)_3 \cdot Se(CH_3)_2$	Ga: Org.Comp.1-34, 38
$C_5GaH_{15}Te$	$Ga(CH_3)_3 \cdot Te(CH_3)_2$	Ga: Org.Comp.1-34, 38
$C_5GaH_{16}N$	$Ga(CH_3)_2H \cdot N(CH_3)_3$	Ga: Org.Comp.1-123
–	$Ga(CH_3)_2D \cdot N(CH_3)_3$	Ga: Org.Comp.1-123
–	$Ga(CH_3)_3 \cdot NH(CH_3)_2$	Ga: Org.Comp.1-41/2, 49
–	$Ga(CH_3)_3 \cdot NH_2C_2H_5$	Ga: Org.Comp.1-42
$C_5GaH_{16}P$	$Ga(CH_3)_3 \cdot PH(CH_3)_2$	Ga: Org.Comp.1-47
$C_5GaH_{17}N_2$	$Ga(CH_3)_3 \cdot NH_2CH_2CH_2NH_2$ ..	Ga: Org.Comp.1-44
$C_5GaH_{18}NSi$	$Ga(CH_3)_3 \cdot N(CH_3)_2SiH_3$	Ga: Org.Comp.1-43
$C_5GaH_{20}Mo_6N_3O_{24}$ ..	$C_5H_5NH(NH_4)_2[Ga(OH)_6Mo_6O_{18}] \cdot 3 H_2O$	
	$= 2 C_5H_5N \cdot 2 (NH_4)_2O \cdot Ga_2O_3$	
	$\cdot 12 MoO_3 \cdot 13 H_2O$	Mo: SVol.B4-241
$C_5GdH_3O_2S^{2+}$	$[Gd(SC_4H_3COO)]^{2+}$	Sc: MVol.D4-100
$C_5GdH_3O_3{}^{2+}$	$[Gd(OC_4H_3COO)]^{2+}$	Sc: MVol.D5-110
$C_5GdH_4NO_2{}^{2+}$	$[Gd(NC_5H_3(OH)O)]^{2+}$	Sc: MVol.D2-6
$C_5GdH_4O_4{}^+$	$[Gd(OOCC(CH_2)CH_2COO)]^+$	Sc: MVol.D5-206/7
$C_5GdH_5O_4{}^{2+}$	$[GdH(OOCC(CH_2)CH_2COO)]^{2+}$	Sc: MVol.D5-206/7
$C_5GdH_6NO_3{}^{2+}$	$[Gd(ONC(COCH_3)_2)]^{2+}$	Sc: MVol.D2-102
$C_5GdH_6O_4{}^+$	$[Gd(CH_3C_2H_3(COO)_2)]^+$	Sc: MVol.D5-174
$C_5GdH_6O_7{}^+$	$[Gd(OOC(CHOH)_3COO)]^+$	Sc: MVol.D5-339/40
$C_5GdH_7NO_4{}^+$	$[Gd(CH_3N(CH_2COO)_2)]^+$	Sc: MVol.D1-127
$C_5GdH_7N_2O_3{}^{2+}$	$[Gd(CH_3C(NO)C(NOH)COCH_3)]^{2+}$	Sc: MVol.D2-116/7
$C_5GdH_7N_3O_4S^+$	$Gd[H_2NC(S)NHN(CH_2COO)_2]^+$	Sc: MVol.D4-88/9
$C_5GdH_7O_2{}^{2+}$	$[Gd(CH_3COCHCOCH_3)]^{2+}$	Sc: MVol.D3-76/82
$C_5GdH_8NO_2{}^{2+}$	$Gd(HNC_4H_7COO)^{2+}$	Sc: MVol.D1-215
$C_5GdH_8N_3O_5S$	$Gd[H_2NC(S)NHN(CH_2COO)_2]OH \cdot 2 H_2O$ ..	Sc: MVol.D4-89/93
$C_5GdH_9N_2O_2S$	$Gd(HN(CH_2CH_2O)_2)(NCS)$	Sc: MVol.D2-2
$C_5GdH_9N_2O_3{}^{2+}$	$Gd(H_2NCHCH_3C(O)NHCH_2COO)^{2+}$	Sc: MVol.D4-282/3
–	$Gd(H_2NCH_2C(O)NHCHCH_3COO)^{2+}$	Sc: MVol.D4-282/3
$C_5GdH_9N_2O_4{}^{2+}$	$Gd(H_2NCH_2C(O)NHCH(CH_2OH)COO)^{2+}$	Sc: MVol.D4-282/3
$C_5GdH_9O_2{}^{2+}$	$[Gd((CH_3)_2CHCH_2COO)]^{2+}$	Sc: MVol.D5-86
$C_5GdH_9O_3{}^{2+}$	$[Gd(C_2H_5C(OH)(CH_3)COO)]^{2+}$	Sc: MVol.D5-270
–	$[Gd(n-C_3H_7CH(OH)COO)]^{2+}$	Sc: MVol.D5-273
$C_5GdH_9O_4{}^{2+}$	$[Gd(CH_3C(CH_2OH)_2COO)]^{2+}$	Sc: MVol.D5-268/9
–	$[Gd(H_3CCH(OH)C(OH)(CH_3)COO)]^{2+}$	Sc: MVol.D5-272
$C_5GdH_{10}NO_2{}^{2+}$	$Gd[(CH_3)_2CHCH(NH_2)COO]^{2+}$	Sc: MVol.D1-115
$C_5GdH_{10}NO_2S^{2+}$	$Gd[CH_3SC_2H_4CH(NH_2)COO]^{2+}$	Sc: MVol.D1-115

$C_5GdH_{20}I_3N_{10}O_5$....	$GdI_3 \cdot 5\ OC(NH_2)_2$	Sc:	MVol.D2-221
$C_5GdH_{20}I_{11}N_{10}O_5$...	$GdI_3 \cdot 5\ OC(NH_2)_2 \cdot 4\ I_2 \cdot 10\ H_2O$	Sc:	MVol.D2-221
C_5GdO_5	$Gd(CO)_5$	Sc:	MVol.D6-166
$C_5GdO_5{}^+$	$[Gd(C_5O_5)]^+$	Sc:	MVol.D3-250/1
$C_5HIMnN_2O_4$	$Mn[(OOC)(O)_2C_4IN_2H]$	Mn:	MVol.D4-336/7
$C_5HK_3MnN_5O$	$K_3[Mn(CN)_5OH]$	Mn:	MVol.D2-239
$C_5HK_3N_5Rh$	$K_3[Rh(CN)_5H]$	Rh:	SVol.B1-195
$-$	$K_3[Rh(CN)_5H] \cdot H_2O$	Rh:	SVol.B1-195, 196
$C_5HMnN_3O_6$	$Mn[(OOC)(O)_2(NO_2)C_4N_2H]$	Mn:	MVol.D4-336/7
$C_5HMnN_5O^{4-}$	$[Mn(CN)_5OH]^{4-}$	Mn:	MVol.D2-207
$C_5HMnN_5O^{5-}$	$[Mn(CN)_5OH]^{5-}$	Mn:	MVol.D2-198
$C_5HN_5Na_3Rh$	$Na_3[Rh(CN)_5H]$	Rh:	SVol.B1-194
$C_5HN_5ORh^{3-}$	$[Rh(CN)_5(OH)]^{3-}$	Rh:	SVol.B1-196
$C_5HN_5Rb_3Rh$	$Rb_3[Rh(CN)_5H]$	Rh:	SVol.B1-195
$C_5HN_5Rh^{3-}$	$[Rh(CN)_5H]^{3-}$	Rh:	SVol.B1-194
C_5HO_5Tc	$HTc(CO)_5$	Tc:	SVol.2-130
$C_5H_2K_3N_5O_3S_5U$ ...	$K_3[UO_2(NCS)_5(H_2O)] \cdot H_2O$	U:	SVol.C13-46
$C_5H_2MnN_2O_4$	$Mn[(OOC)(O)_2C_4HN_2H]$	Mn:	MVol.D4-336/7
$C_5H_2MnN_5O^{2-}$	$[Mn(CN)_5H_2O]^{2-}$	Mn:	MVol.D2-226
$C_5H_2MnN_5O^{3-}$	$[Mn(CN)_5H_2O]^{3-}$	Mn:	MVol.D2-203, 207
$C_5H_2N_5OOs^{2-}$	$[Os(CN)_5H_2O]^{2-}$	Os:	SVol.1-178
$C_5H_2N_5OOs^{3-}$	$[Os(CN)_5H_2O]^{3-}$	Os:	SVol.1-178
$C_5H_2N_5ORh^{2-}$	$[Rh(CN)_5(H_2O)]^{2-}$	Rh:	SVol.B1-196
$C_5H_2O_{10}U^{2-}$	$[UO_2(C_2O_4)(OOCCH_2COO)]^{2-}$	U:	SVol.D1-267
$C_5H_2O_{29}U_6$	$5\ UO_2CO_3 \cdot UO_2(OH)_2 \cdot 7\ H_2O$	U:	SVol.C13-4, 6
$-$	$(UO_2)_6(CO_3)_5(OH)_2 \cdot x\ H_2O$	U:	SVol.C13-4
$C_5H_3HoO_3{}^{2+}$	$[Ho(OC_4H_3COO)]^{2+}$	Sc:	MVol.D5-110
$C_5H_3IN_4Rh^{3-}$	$[Rh(CN)_4(CH_3)I]^{3-}$	Rh:	SVol.B1-187
$C_5H_3KO_{11}U_2$	$K[(UO_2)_2(OOC(CH(O))_3COO)] \cdot 3\ H_2O$	U:	SVol.C13-252/4
$C_5H_3LaO_2S^{2+}$	$[La(SC_4H_3COO)]^{2+}$	Sc:	MVol.D4-100
$C_5H_3LaO_3{}^{2+}$	$[La(OC_4H_3COO)]^{2+}$	Sc:	MVol.D5-110
$C_5H_3LuO_3{}^{2+}$	$[Lu(OC_4H_3COO)]^{2+}$	Sc:	MVol.D5-110
$C_5H_3MnN_2O_4{}^+$	$Mn[(OOC)(O)_2C_4HN_2H_2]^+$	Mn:	MVol.D4-336/7
$C_5H_3MnN_4O^+$	$Mn[6-(O)-1,3,7,9-C_5H_2N_4H]^+$	Mn:	MVol.D4-20
$C_5H_3MnN_6O$	$H_3[Mn(CN)_5NO]$	Mn:	MVol.D2-262/3
$C_5H_3N_3OS$	$SN_2C_5H_2(O)NH$	S:	S-N Comp.3-225/6
$-$	$SN_2C_5H_2(OH)N$	S:	S-N Comp.3-225/6
$C_5H_3N_3S$	$SN_2C_5H_3N$	S:	S-N Comp.3-224, 225
$C_5H_3N_4O_2Tc^{2-}$	$[TcO(OCH_3)(CN)_4]^{2-}$	Tc:	SVol.2-200
$C_5H_3N_8Rh_2$	$Rh_2(CN)_5 \cdot 3\ NH \cdot 8.5\ H_2O$	Rh:	SVol.B1-197
$C_5H_3NdO_2S^{2+}$	$[Nd(SC_4H_3COO)]^{2+}$	Sc:	MVol.D4-100
$C_5H_3NdO_3{}^{2+}$	$[Nd(OC_4H_3COO)]^{2+}$	Sc:	MVol.D5-110
$C_5H_3O_2SSm^{2+}$	$[Sm(SC_4H_3COO)]^{2+}$	Sc:	MVol.D4-100
$C_5H_3O_3Pr^{2+}$	$[Pr(OC_4H_3COO)]^{2+}$	Sc:	MVol.D5-110
$C_5H_3O_3Sm^{2+}$	$[Sm(OC_4H_3COO)]^{2+}$	Sc:	MVol.D5-110
$C_5H_3O_3Tb^{2+}$	$[Tb(OC_4H_3COO)]^{2+}$	Sc:	MVol.D5-110
$C_5H_3O_3Tm^{2+}$	$[Tm(OC_4H_3COO)]^{2+}$	Sc:	MVol.D5-110
$C_5H_3O_3Yb^{2+}$	$[Yb(OC_4H_3COO)]^{2+}$	Sc:	MVol.D5-110
$C_5H_3O_4SU^+$........	$[UO_2(OOCC_4H_3S)]^+$	U:	SVol.D1-252
$C_5H_4KNO_8Rh_2S$	$K[Rh_2(O_2CH)_4(NCS)] \cdot 2\ H_2O$	Rh:	SVol.B2-12

$C_5H_5NaO_9U$	$Na[UO_2(OOCCHOHCH(O)CHOHCOO)] \cdot 2\ H_2O$		
		U:	SVol.C13-252/4
$C_5H_5NaO_{12}U_2$	$Na(HCOO) \cdot 2\ UO(HCOO)_2 \cdot H_2O$	U:	SVol.C13-104
$C_5H_5NdO_4{}^{2+}$	$[NdH(OOCC(CH_2)CH_2COO)]^{2+}$	Sc:	MVol.D5-206/7
$C_5H_5NdO_7$	$Nd(OCH(CHOHCOO)_2) \cdot H_2O = Nd(OH)$		
	$(OOC(CHOH)_3COO)$	Sc:	MVol.D5-339/40
$C_5H_5OPrS^{2+}$	$Pr[(SCH_2)C_4H_3O]^{2+}$	Sc:	MVol.D4-51
$C_5H_5OSSm^{2+}$	$Sm[(SCH_2)C_4H_3O]^{2+}$	Sc:	MVol.D4-51
$C_5H_5O_4Sm^{2+}$	$[SmH(OOCC(CH_2)CH_2COO)]^{2+}$	Sc:	MVol.D5-206/7
$C_5H_5O_8Pr^{2-}$	$[Pr(OH)(HOCH(CH(O)COO)_2)]^{2-}$	Sc:	MVol.D5-339/40
$C_5H_5O_8SU^-$	$[UO_2(OOCCH_2COO)(HSCH_2COO)]^-$	U:	SVol.D1-267
$C_5H_5O_9U^-$	$[UO_2(HOCH_2COO)(OOCCH_2COO)]^-$	U:	SVol.D1-265
C_5H_5Sb	SbC_5H_5	Sb:	Org.Comp.2-154/6
$C_5H_5U^{3+}$	$[U(C_5H_5)]^{3+}$	U:	SVol.A5-96, 105
$C_5H_6HoNO_3{}^{2+}$	$[Ho(ONC(COCH_3)_2)]^{2+}$	Sc:	MVol.D2-102
$C_5H_6HoO_4{}^+$	$[Ho(CH_3C_2H_3(COO)_2)]^+$	Sc:	MVol.D5-174
$C_5H_6HoO_7{}^+$	$[Ho(OOC(CHOH)_3COO)]^+$	Sc:	MVol.D5-339/40
$C_5H_6KNO_9Rh_2$	$K[Rh_2(O_2CH)_4(CN)(H_2O)]$	Rh:	SVol.B2-14
$C_5H_6KNO_{11}SU$	$K[U(C_2O_4)_2(NCS)(H_2O)_3]$	U:	SVol.C13-230/2
$C_5H_6LaNO_3{}^{2+}$	$[La(ONC(COCH_3)_2)]^{2+}$	Sc:	MVol.D2-102
$C_5H_6LaNO_7P^-$	$La[(OOCCH_2)_2NCH_2PO_3]^-$	Sc:	MVol.D4-148
$C_5H_6LaO_4{}^+$	$[La(CH_2(CH_2COO)_2)]^+$	Sc:	MVol.D5-174/5
$-$	$[La(CH_3C_2H_3(COO)_2)]^+$	Sc:	MVol.D5-174
$C_5H_6LaO_7{}^+$	$[La(OOC(CHOH)_3COO)]^+$	Sc:	MVol.D5-339/40
$C_5H_6LuNO_3{}^{2+}$	$[Lu(ONC(COCH_3)_2)]^{2+}$	Sc:	MVol.D2-102
$C_5H_6LuO_4{}^+$	$[Lu(CH_3C_2H_3(COO)_2)]^+$	Sc:	MVol.D5-174
$C_5H_6LuO_7{}^+$	$[Lu(OOC(CHOH)_3COO)]^+$	Sc:	MVol.D5-339/40
$C_5H_6MnN_2{}^{2+}$	$Mn(H_2NC_5H_4N)^{2+}$	Mn:	MVol.D3-134
$C_5H_6MnN_2O^{2+}$	$Mn[NH_2C_5H_4NO]^{2+}$	Mn:	MVol.D3-164
$C_5H_6MnO_4$	$Mn[(CH_2)_3(COO)_2]$	Mn:	MVol.D2-137
$C_5H_6NNdO_3{}^{2+}$	$[Nd(ONC(COCH_3)_2)]^{2+}$	Sc:	MVol.D2-102
$C_5H_6NNdO_4{}^{2+}$	$[Nd(CH_2CHC(O)NHCH(OH)COO)]^{2+}$	Sc:	MVol.D2-206
$C_5H_6NO_3Pr^{2+}$	$[Pr(ONC(COCH_3)_2)]^{2+}$	Sc:	MVol.D2-102
$C_5H_6NO_3Sm^{2+}$	$[Sm(ONC(COCH_3)_2)]^{2+}$	Sc:	MVol.D2-102
$C_5H_6NO_3Tb^{2+}$	$[Tb(ONC(COCH_3)_2)]^{2+}$	Sc:	MVol.D2-102
$C_5H_6NO_3Tm^{2+}$	$[Tm(ONC(COCH_3)_2)]^{2+}$	Sc:	MVol.D2-102
$C_5H_6NO_3Y^{2+}$	$[Y(ONC(COCH_3)_2)]^{2+}$	Sc:	MVol.D2-102
$C_5H_6NO_3Yb^{2+}$	$[Yb(ONC(COCH_3)_2)]^{2+}$	Sc:	MVol.D2-102
$C_5H_6NO_4Pr^{2+}$	$[Pr(CH_2CHC(O)NHCH(OH)COO)]^{2+}$	Sc:	MVol.D2-206
$C_5H_6NO_4Sm^{2+}$	$[Sm(CH_2CHC(O)NHCH(OH)COO)]^{2+}$	Sc:	MVol.D2-206
$C_5H_6NO_4Y^{2+}$	$[Y(CH_2CHC(O)NHCH(OH)COO)]^{2+}$	Sc:	MVol.D2-206
$C_5H_6NO_6OsS_2{}^+$	$[OsO_2(S_2CN(CH_2COOH)_2)]^+$	Os:	SVol.1-290
$C_5H_6NO_6PU$	$C_5H_5NHUO_2PO_4$	U:	SVol.C14-95
$C_5H_6NO_7PSc^-$	$Sc[(OOCCH_2)_2NCH_2PO_3]^-$	Sc:	MVol.D4-148
$C_5H_6NO_8U^-$	$[UO_2(NH_2CH_2COO)(OOCCH_2COO)]^-$	U:	SVol.D1-264
$C_5H_6N_2O_2S$	$SN_2C_2HCH_2COOCH_3$	S:	S-N Comp.3-93/8, 124
$-$	$SN_2C_2HCOOC_2H_5$	S:	S-N Comp.3-93/8, 172
$-$	$SN_2C_2HOCH_2CHCH_2O$	S:	S-N Comp.3-93/8, 157
$-$	$SN_2C_2CH_2COOHCH_3$	S:	S-N Comp.3-93/8, 126
$C_5H_6N_2O_2S_2$	$SN_2C_2HSCH_2COOCH_3$	S:	S-N Comp.3-93/8, 162

$C_5H_9HoO_4^{2+}$	$[Ho(H_3CCH(OH)C(OH)(CH_3)COO)]^{2+}$	Sc:	MVol.D5-272
C_5H_9ISn	$(CH_2CH)_2(CH_3)SnI$	Sn:	Org.Comp.8-79, 81
$C_5H_9I_4NOTh$	$ThI_4 \cdot HN(CH_2)_4CO$	Th:	SVol.E-63/4
$C_5H_9LaN_2O_3^{2+}$	$La[OOCCH(NH_2)C_2H_4C(O)NH_2]^{2+}$	Sc:	MVol.D1-115
$C_5H_9LaO_2^{2+}$	$[La((CH_3)_2CHCH_2COO)]^{2+}$	Sc:	MVol.D5-86
$C_5H_9LaO_3^{2+}$	$[La(C_2H_5C(OH)(CH_3)COO)]^{2+}$	Sc:	MVol.D5-270
–	$[La(n-C_3H_7CH(OH)COO)]^{2+}$	Sc:	MVol.D5-273
$C_5H_9LaO_4^{2+}$	$[La(CH_3C(CH_2OH)_2COO)]^{2+}$	Sc:	MVol.D5-268/9
–	$[La(H_3CCH(OH)C(OH)(CH_3)COO)]^{2+}$	Sc:	MVol.D5-272
$C_5H_9LuO_2^{2+}$	$[Lu((CH_3)_2CHCH_2COO)]^{2+}$	Sc:	MVol.D5-86
$C_5H_9LuO_3^{2+}$	$[Lu(C_2H_5C(OH)(CH_3)COO)]^{2+}$	Sc:	MVol.D5-270
–	$[Lu(n-C_3H_7CH(OH)COO)]^{2+}$	Sc:	MVol.D5-273
$C_5H_9LuO_4^{2+}$	$[Lu(CH_3C(CH_2OH)_2COO)]^{2+}$	Sc:	MVol.D5-268/9
–	$[Lu(H_3CCH(OH)C(OH)(CH_3)COO)]^{2+}$	Sc:	MVol.D5-272
$C_5H_9MnNO_2^{2+}$	$Mn[2-(OOC)-C_4H_7NH_2]^{2+}$	Mn:	MVol.D4-304
$C_5H_9MnNO_2S$	$Mn[(CH_3)_2C(S)CH(NH_2)COO]$	Mn:	MVol.D4-292
$C_5H_9MnNO_5$	$Mn(OCHNH_2)(CH_3COO)_2$	Mn:	MVol.D5-92
$C_5H_9MnN_2O_3^+$	$Mn[H_2NCH_2C(O)N(CH_3)CH_2COO]^+$	Mn:	MVol.D4-341/4
–	$Mn[H_2NCH_2C(O)NHCH(CH_3)COO]^+$	Mn:	MVol.D4-341/4
–	$Mn[H_2NC(O)C_2H_4CH(NH_2)COO]^+$	Mn:	MVol.D4-285
–	$Mn[H_3CCH(NH_2)C(O)NHCH_2COO]^+$	Mn:	MVol.D4-341/4
$C_5H_9MnN_2O_7P$	$Mn[O_3POCH_2CH(NH_3)C(O)NHCH_2COO]$	Mn:	MVol.D4-346
$C_5H_9MnN_3^{2+}$	$Mn[NHCHNC(CH_2CH_2NH_2)CH]^{2+}$	Mn:	MVol.D3-312
$C_5H_9NO_5U^{2+}$	$[UO_2H(OOCC_4H_7N(OH))]^{2+}$	U:	SVol.D1-220
$C_5H_9NO_6SU$	$UO_2[(NH_2C_2H_4COO)(HSCH_2COO)]$	U:	SVol.D1-258
$C_5H_9NO_7U$	$UO_2[(NH_2C_2H_4COO)(HOCH_2COO)]$	U:	SVol.D1-257
$C_5H_9N_2NdO_3^{2+}$	$Nd(H_2NCHCH_3C(O)NHCH_2COO)^{2+}$	Sc:	MVol.D4-282/3
–	$Nd(H_2NCH_2C(O)NHCHCH_3COO)^{2+}$	Sc:	MVol.D4-282/3
–	$Nd[OOCCH(NH_2)C_2H_4C(O)NH_2]^{2+}$	Sc:	MVol.D1-115/6
$C_5H_9N_2NdO_4^{2+}$	$Nd(H_2NCH_2C(O)NHCH(CH_2OH)COO)^{2+}$	Sc:	MVol.D4-282/3
$C_5H_9N_2O_2Sb$	$(CH_3)_3Sb(NCO)_2$	Sb:	Org.Comp.4-102/3
$C_5H_9N_2O_3PS_2$	$SN_2C_2OP(S)(OCH_3)_2CH_3$	S:	S-N Comp.3-93/8, 160
$C_5H_9N_2O_3Pr^{2+}$	$Pr[OOCCH(NH_2)C_2H_4C(O)NH_2]^{2+}$	Sc:	MVol.D1-115/6
$C_5H_9N_2O_3Sc^{2+}$	$Sc[OOCCH(NH_2)C_2H_4C(O)NH_2]^{2+}$	Sc:	MVol.D1-115
$C_5H_9N_2O_3Y^{2+}$	$Y[OOCCH(NH_2)C_2H_4C(O)NH_2]^{2+}$	Sc:	MVol.D1-115
$C_5H_9N_2O_3Yb^{2+}$	$Yb(H_2NCHCH_3C(O)NHCH_2COO)^{2+}$	Sc:	MVol.D4-282/3
–	$Yb(H_2NCH_2C(O)NHCHCH_3COO)^{2+}$	Sc:	MVol.D4-282/3
$C_5H_9N_2O_4Yb^{2+}$	$Yb(H_2NCH_2C(O)NHCH(CH_2OH)COO)^{2+}$	Sc:	MVol.D4-282/3
$C_5H_9N_2O_5U^+$	$[UO_2(OOCCH(NH_2)CH_2CH_2CONH_2)]^+$	U:	SVol.D1-219
$C_5H_9N_2S_2Sb$	$(CH_3)_3Sb(NCS)_2$	Sb:	Org.Comp.4-104/5
$C_5H_9N_3O_2S$	$(O)SN_2C_2(OC_2H_5)(NHCH_3)$	S:	S-N Comp.3-250/7
$C_5H_9N_3O_2S_2$	$[C(O)NCH_3C(SCH_3)NS(O)NCH_3]$	S:	S-N Comp.4-23/7
$C_5H_9N_3S$	$SN_2C_2HNHC_3H_7-n$	S:	S-N Comp.3-93/8, 142
$C_5H_9N_5S$	$SN_2C_2NH_2NCHN(CH_3)_2$	S:	S-N Comp.3-93/8, 148
$C_5H_9NdO_2^{2+}$	$[Nd(CH_3(CH_2)_3COO)]^{2+}$	Sc:	MVol.D5-85
–	$[Nd((CH_3)_2CHCH_2COO)]^{2+}$	Sc:	MVol.D5-86
$C_5H_9NdO_3^{2+}$	$[Nd(C_2H_5C(OH)(CH_3)COO)]^{2+}$	Sc:	MVol.D5-270
–	$[Nd(n-C_3H_7CH(OH)COO)]^{2+}$	Sc:	MVol.D5-273
$C_5H_9NdO_4^{2+}$	$[Nd(CH_3C(CH_2OH)_2COO)]^{2+}$	Sc:	MVol.D5-268/9
–	$[Nd(H_3CCH(OH)C(OH)(CH_3)COO)]^{2+}$	Sc:	MVol.D5-272

$C_5H_{10}NO_2SSm^{2+}$	$Sm[CH_3SC_2H_4CH(NH_2)COO]^{2+}$	Sc:	MVol.D1-115
$C_5H_{10}NO_2SY^{2+}$	$Y[CH_3SC_2H_4CH(NH_2)COO]^{2+}$	Sc:	MVol.D1-115
$C_5H_{10}NO_2SYb^{2+}$	$Yb[CH_3SC_2H_4CH(NH_2)COO]^{2+}$	Sc:	MVol.D1-115
$C_5H_{10}NO_2S_2U^+$	$[UO_2(SCSN(C_2H_5)_2)]^+$	U:	SVol.E2-56
$C_5H_{10}NO_2Sm^{2+}$	$Sm[(CH_3)_2CHCH(NH_2)COO]^{2+}$	Sc:	MVol.D1-115
$C_5H_{10}NO_2Tb^{2+}$	$Tb[(CH_3)_2CHCH(NH_2)COO]^{2+}$	Sc:	MVol.D1-115
$C_5H_{10}NO_2U^{3+}$	$[U(OOCCH(NH_2)C_3H_7-i)]^{3+}$	U:	SVol.D1-217
$C_5H_{10}NO_2Y^{2+}$	$Y[(CH_3)_2CHCH(NH_2)COO]^{2+}$	Sc:	MVol.D1-115
$C_5H_{10}NO_2Yb^{2+}$	$Yb[(CH_3)_2CHCH(NH_2)COO]^{2+}$	Sc:	MVol.D1-115
$C_5H_{10}NO_4OsS_2^+$	$[OsO_2(S_2CN(C_2H_4OH)_2)]^+$	Os:	SVol.1-290
$C_5H_{10}NO_4PrS$	$Pr(SCH_2CH(NH_2)COO)(OC_2H_4OH)$	Sc:	MVol.D4-60
$C_5H_{10}NO_4SU^+$	$[UO_2(OOCCH(NH_2)CH_2CH_2SCH_3)]^+$	U:	SVol.D1-218
$C_5H_{10}NO_4U^+$	$[UO_2(OOCCH(NH_2)C_3H_7-i)]^+$	U:	SVol.D1-217/8
$C_5H_{10}N_2O_4Pt$	$C_2H_4(NH_2)_2Pt(O_2C)_2CH_2$		
	Cytostatic activity	Pt:	SVol.A1-332/4
$C_5H_{10}N_2O_6SU$	$UO_2(CH_3COO)_2((NH_2)_2CS)$	U:	SVol.A6-65
$C_5H_{10}N_2S$	$[SN(CH_2)_5N]$	S:	S-N Comp.4-145
$C_5H_{10}N_4O_3Rh^{3+}$	$[Rh((NH_2CH_2CONH)_2CO)]^{3+}$	Rh:	SVol.B2-321
$C_5H_{10}N_4S_2$	$(S_2N_3C(N(C_2H_5)_2))_n$	S:	S-N Comp.4-5
$C_5H_{10}N_6S_3$	$S_3N_5C(N(C_2H_5)_2)$	S:	S-N Comp.4-146/9
$C_5H_{10}O_2SSn$	$(CH_3)_2Sn(SCH(CH_3)COO)$	Sn:	Org.Comp.10-230
–	$(CH_3)_2Sn(SCH_2CH_2COO)$	Sn:	Org.Comp.10-234
$C_5H_{10}O_3Sn$	$(CH_3)_2SnO_2CCH(CH_3)O$	Sn:	Org.Comp.14-107/8
–	$(C_2H_5)_2SnOOC(O)$	Sn:	Org.Comp.14-182
$C_5H_{11}IO_2Sn$	$(CH_3)_3SnOOCCH_2I$	Sn:	Org.Comp.11-93
$C_5H_{11}I_2MnNO_2$	$MnI_2 \cdot (CH_3)_2CHCH(NH_2)COOH$	Mn:	MVol.D4-284
$C_5H_{11}MnNO_2^{2+}$	$Mn[(CH_3)_2CHCH(NH_3)COO]^{2+}$	Mn:	MVol.D4-283/4
$C_5H_{11}MnNO_3$	$[Mn(OH)((CH_3)_2CHCH(NH_2)COO)]$	Mn:	MVol.D4-283/4
$C_5H_{11}MnN_2O_2^+$	$Mn[H_2NC_3H_6CH(NH_2)COO]^+$	Mn:	MVol.D4-288/9
$C_5H_{11}NO_2S$	$[OS(O)NCH_3CHCH_3CH_2CH_2]$	S:	S-N Comp.4-86/95
$C_5H_{11}NO_2SSn$	$(CH_3)_2Sn(SCH_2CH(NH_2)COO)$	Sn:	Org.Comp.10-234
$C_5H_{11}NO_3Os$	$OsO_3(NCH_2C(CH_3)_3)$	Os:	SVol.1-240
$C_5H_{11}N_2S_4Sb$	$CH_3Sb(SC(S)NHCH_3)_2$	Sb:	Org.Comp.2-135
$C_5H_{11}N_3O_3S_2$	$[S(O)NCH_3NCCH_3CH_2N(SO_2CH_3)]$	S:	S-N Comp.4-14/7
$C_5H_{11}O_2Sb$	$(CH_3)_2SbC_2H_4COOH$	Sb:	Org.Comp.1-110
–	$CH_3Sb(OCHCH_3)_2$	Sb:	Org.Comp.2-141
$C_5H_{11}O_3Sb$	$(CH_3)_3Sb(OCH_2CO_2)$	Sb:	Org.Comp.4-143, 148
$C_5H_{11}O_3Sm$	$Sm(OC_3H_7)(OCH_2)_2$	Sc:	MVol.D3-31/2
$C_5H_{11}O_3U$	$U(OC_2H_5)(O_2C_3H_6)$	U:	SVol.C13-91/2
$C_5H_{11}O_4Sb$	$(CH_3)_3Sb(O_2CH)_2$	Sb:	Org.Comp.4-163/4
$C_5H_{11}Sb$	$(CH_3)_2SbCH_2CHCH_2$	Sb:	Org.Comp.1-112
–	$CH_3Sb(CH_2)_4$	Sb:	Org.Comp.1-147
$C_5H_{12}INS_2Sn$	$(CH_3)_2SnI(SC(S)N(CH_3)_2)$	Sn:	Org.Comp.10-215
$C_5H_{12}I_2SSn$	$CH_3SnI_2(SC_4H_9-t)$	Sn:	Org.Comp.10-254/5
$C_5H_{12}I_3Sb$	$(CH_3)_3Sb(CH_2I)(CHI_2)$	Sb:	Org.Comp.3-26
$C_5H_{12}MnN_2O_2^{2+}$	$Mn[H_2NC_3H_6CH(NH_3)COO]^{2+}$	Mn:	MVol.D4-288/9
$C_5H_{12}MnN_4O_2^{2+}$	$Mn(CH_2(CH_2C(O)NHNH_2)_2)^{2+}$	Mn:	MVol.D5-205
$C_5H_{12}MnN_6O_7S$	$Mn(OCHNH_2)_4(NO_3)(NCS)$	Mn:	MVol.D5-91/2
$C_5H_{12}MnN_9O$	$(NH_4)_3[Mn(CN)_5NO] \cdot H_2O$	Mn:	MVol.D2-267/8
$C_5H_{12}NSSb$	$(CH_3)_4SbNCS$	Sb:	Org.Comp.3-67

$C_5H_{12}NSSb$	$(CH_3)_4SbSCN$	Sb:	Org.Comp.3-67/8
$C_5H_{12}NSb$	$(CH_3)_4SbCN$	Sb:	Org.Comp.3-66/7
$C_5H_{12}N_2O_3S_2$	$[(O)SNCH_3C(CH_3)_2NSO_2CH_3]$	S:	S-N Comp.3-60
$C_5H_{12}N_2O_4Pt$	$(NH_3)_2Pt(O_2C)_2CHC_2H_5$		
	Cytostatic activity	Pt:	SVol.A1-332/4
$C_5H_{12}N_2O_6SU$	$UO_2SO_3 \cdot ((CH_3)_2N)_2CO$	U:	SVol.C10-139
$C_5H_{12}N_2O_9U$	$(NH_4)_2[UO_2(OOCCH(O)CHOHCH(O)COO)]$		
	$\cdot$ 2 H_2O	U:	SVol.C13-252/4
$C_5H_{12}N_2Sb_2$	$[(CH_3)_2Sb]_2CN_2$	Sb:	Org.Comp.1-176/7
$C_5H_{12}N_4O_{10}SU$	$(NH_4)_3[UO_2(C_2O_4)_2(NCS)]$	U:	SVol.C13-230/2
−	$(NH_4)_3[UO_2(C_2O_4)_2(NCS)] \cdot 2 H_2O$	U:	SVol.C13-230/2
−	$(NH_4)_3[UO_2(C_2O_4)_2(NCS)] \cdot 3 H_2O$	U:	SVol.C13-230/2
$C_5H_{12}N_6OS_3$	$N_3S_3(N(CH_3)_2)(NCH_3C(O)NCH_3)$	S:	S-N Comp.4-153/4
$C_5H_{12}N_6O_9U$	$(CN_3H_6)_2[U(CO_3)_3]$	U:	SVol.C13-7/8
$C_5H_{12}N_6O_{12}STh$	$Th(NO_3)_4 \cdot ((CH_3)_2N)_2CS$	Th:	SVol.C5-33/4
$C_5H_{12}N_6S_4$	$[C_5H_{10}NH_2]S_4N_5$	S:	S-N Comp.2-259, 261, 263
$C_5H_{12}N_8O_2S_5U$	$(NH_4)_3[UO_2(NCS)_5]$	U:	SVol.C13-43, 46
−	$(NH_4)_3[UO_2(NCS)_5] \cdot H_2O$	U:	SVol.C13-43
−	$(NH_4)_3[UO_2(NCS)_5] \cdot 2 H_2O$	U:	SVol.C13-43, 46
$C_5H_{12}Na_4O_{21}U_2$	$Na_4[U_2(C_2O_4)(CO_3)_3(OH)_4(H_2O)_4]$	U:	SVol.C13-228/9
$C_5H_{12}OSSn$	$(CH_3)_3SnSC(O)CH_3$	Sn:	Org.Comp.9-48
$C_5H_{12}OSn$	$(CH_3)_2(CH_2CH)SnOCH_3$	Sn:	Org.Comp.13-238
−	$(CH_3)_2SnC_3H_6O$	Sn:	Org.Comp.13-276
$C_5H_{12}O_2SSn$	$(CH_3)_3SnOOCCH_2SH$	Sn:	Org.Comp.11-97, 105
$C_5H_{12}O_2Sn$	$(CH_3)_3SnO_2C_2H_3(CH_3)$	Sn:	Org.Comp.14-47
−	$(CH_3)_3SnOOCCH_3$	Sn:	Org.Comp.11-82/8
−	$(CH_3)_3SnOOCCD_3$	Sn:	Org.Comp.11-86
$C_5H_{12}O_3Sn$	$(CH_3)_3SnOOCCH_2OH$	Sn:	Org.Comp.11-94
$C_5H_{12}O_6Os$	$OsO_2(OH)_2(O_2C_2H(CH_3)_3)$	Os:	SVol.1-190
$C_5H_{12}SSn$	$(CH_3)_3SnSCHCH_2$	Sn:	Org.Comp.9-35
$C_5H_{12}S_2Sn$	$(CH_3)_2Sn(SCH(CH_3)CH_2S)$	Sn:	Org.Comp.10-31
−	$(CH_3)_2Sn(SCH_2)_2CH_2$	Sn:	Org.Comp.10-33
−	$(CH_3)_3SnSC(S)CH_3$	Sn:	Org.Comp.9-49, 50/1
$C_5H_{13}ISSn$	$(CH_3)_2SnI(SC_3H_7-i)$	Sn:	Org.Comp.10-214/5
$C_5H_{13}IS_2Sn$	$CH_3SnI(SC_2H_5)_2$	Sn:	Org.Comp.10-251
$C_5H_{13}ISn$	$(CH_3)_2(C_3H_7)SnI$	Sn:	Org.Comp.8-49
−	$(CH_3)_2(i-C_3H_7)SnI$	Sn:	Org.Comp.8-49, 56
−	$(C_2H_5)_2(CH_3)SnI$	Sn:	Org.Comp.8-60
$C_5H_{13}NO_2Sn$	$(CH_3)_3SnON(O)CHCH_3$	Sn:	Org.Comp.11-131/7
−	$(CH_3)_3SnOOCCH_2NH_2$	Sn:	Org.Comp.11-95, 103/4
$C_5H_{13}NaO_8Rh_2$	$NaH_3[((CH_3CO)_2CH)Rh(O_2)_2Rh(OH)(H_2O)]$	Rh:	SVol.B2-83
$C_5H_{13}OSb$	$(CH_3)_2SbCH_2CH(OH)CH_3$	Sb:	Org.Comp.1-110
−	$(C_2H_5)_2SbOCH_3$	Sb:	Org.Comp.2-28
$C_5H_{13}O_2Sb$	$(CH_3)_4SbO_2CH$	Sb:	Org.Comp.3-119/20
−	$CH_3Sb(OC_2H_5)_2$	Sb:	Org.Comp.2-126/7
$C_5H_{13}O_3Sb$	$(CH_3)_4SbHCO_3$	Sb:	Org.Comp.3-75
$C_5H_{13}S_2Sb$	$CH_3Sb(SC_2H_5)_2$	Sb:	Org.Comp.2-132
$C_5H_{13}Sb$	$(C_2H_5)_2SbCH_3$	Sb:	Org.Comp.1-118
$C_5H_{14}ISb$	$(CH_3)_3C_2H_5SbI$	Sb:	Org.Comp.3-147

$C_5H_{14}N_2OSn$	$(CH_3)_3SnONC(CH_3)NH_2$	Sn:	Org.Comp.11-129
$C_5H_{14}N_2O_{11}U$	$C_2H_4(NH_3)_2[U(CO_3)_3(H_2O)_2] \cdot 2 H_2O$	U:	SVol.C13-7/8
$C_5H_{14}N_6O_8U$	$[C(NH_2)_3]_2[UO_2(CH_2(COO)_2)(OO)] \cdot H_2O$	U:	SVol.C13-238/40
$C_5H_{14}N_8O_3S_5U$	$(NH_4)_3[UO_2(NCS)_5(H_2O)] \cdot H_2O$	U:	SVol.C13-46
$C_5H_{14}OSSn$	$(CH_3)_3SnSCH_2CH_2OH$	Sn:	Org.Comp.9-32
$C_5H_{14}OSn$	$(CH_3)_3SnOC_2H_5$	Sn:	Org.Comp.11-52/4
$C_5H_{14}O_2SeSn$	$(CH_3)_3SnOSe(O)C_2H_5$	Sn:	Org.Comp.11-152
$C_5H_{14}O_2Sn$	$(CH_3)_3SnOOC_2H_5$	Sn:	Org.Comp.11-125
$C_5H_{14}SSn$	$(CH_3)_3SnSC_2H_5$	Sn:	Org.Comp.9-28/9
$C_5H_{14}S_3Sn$	$C_2H_5Sn(SCH_3)_3$	Sn:	Org.Comp.10-178
$C_5H_{14}Sb^+$	$(CH_3)_3(C_2H_5)Sb^+$	Sb:	Org.Comp.3-38
$C_5H_{14}Sb_2$	$[(CH_3)_2Sb]_2CH_2$	Sb:	Org.Comp.1-166
$C_5H_{15}IOSSn$	$(CH_3)_3SnI \cdot (CH_3)_2SO$	Sn:	Org.Comp.8-19
$C_5H_{15}I_3InSb$	$(CH_3)_4Sb[CH_3InI_3]$	Sb:	Org.Comp.3-92
$C_5H_{15}NO_2SSn$	$(CH_3)_3SnOS(O)N(CH_3)_2$	Sn:	Org.Comp.11-142
$C_5H_{15}NSSn$	$(CH_3)_3SnSCH_2CH_2NH_2$	Sn:	Org.Comp.9-32
$C_5H_{15}N_2O_2S_2^+$	$[(CH_3)_2NS(O)OS(CH_3)(N(CH_3)_2)]^+$	S:	S-N Comp.3-1
$C_5H_{15}N_2Sb$	$CH_3Sb[N(CH_3)_2]_2$	Sb:	Org.Comp.2-130
$C_5H_{15}N_3SSi_2$	$S(NSi(CH_3)_2)_2NCH_3$	S:	S-N Comp.3-7
$C_5H_{15}N_5NiS_4Si_2$	$[Ni((S_2N_2Si(CH_3)_2)_2NCH_3)]$	S:	S-N Comp.2-291, 295, 296
$C_5H_{15}N_7S_5Si$	$S_4N_5NS(CH_3)_2NSi(CH_3)_3$	S:	S-N Comp.2-268/9
$C_5H_{15}OPSSn$	$(CH_3)_3SnSP(O)(CH_3)_2$	Sn:	Org.Comp.9-62
$C_5H_{15}OSb$	$(CH_3)_4SbOCH_3$	Sb:	Org.Comp.3-100/1
–	$(CH_3)_4SbOCD_3$	Sb:	Org.Comp.3-101
$C_5H_{15}OSbSi$	$(CH_3)_2SbOSi(CH_3)_3$	Sb:	Org.Comp.2-34
$C_5H_{15}O_2PS_2Sn$	$(CH_3)_3SnSP(S)(OCH_3)_2$	Sn:	Org.Comp.9-63
$C_5H_{15}O_2PSn$	$(CH_3)_3SnOP(O)(CH_3)_2$	Sn:	Org.Comp.11-155/9
$C_5H_{15}O_2SSb$	$(CH_3)_4SbO_2SCH_3$	Sb:	Org.Comp.3-132/3
$C_5H_{15}O_2Sb$	$(CH_3)_3Sb(OCH_3)_2$	Sb:	Org.Comp.4-131
$C_5H_{15}O_3PSSn$	$(CH_3)_3SnSP(O)(OCH_3)_2$	Sn:	Org.Comp.9-62
$C_5H_{15}O_3PSn$	$(CH_3)_3SnOP(O)(CH_3)OCH_3$	Sn:	Org.Comp.11-156
$C_5H_{15}O_3SSb$	$(CH_3)_4SbO_3SCH_3$	Sb:	Org.Comp.3-135
$C_5H_{15}O_5U$	$U(OCH_3)_5$	U:	SVol.C13-63/7
$C_5H_{15}PS_2Sn$	$(CH_3)_3SnSP(S)(CH_3)_2$	Sn:	Org.Comp.9-63
$C_5H_{15}SSb$	$(CH_3)_4SbSCH_3$	Sb:	Org.Comp.3-139/40
$C_5H_{15}S_2Sb$	$(CH_3)_3Sb(SCH_3)_2$	Sb:	Org.Comp.4-205
$C_5H_{15}Sb$	$Sb(CH_3)_5$	Sb:	Org.Comp.3-1/4
$C_5H_{15}SbZn$	$Zn(CH_3)_2[Sb(CH_3)_3]$	Sb:	Org.Comp.1-15
$C_5H_{16}IN_4O_3Rh$	$Rh(NH_2C_2H_4NH_2)_2(O_2CO)I$	Rh:	SVol.B2-173
$C_5H_{16}N_6O_9U$	$[C(NH_2)_3]_2[UO_2(CH_2(COO)_2)(OO)(H_2O)]$	U:	SVol.C13-238/40
$C_5H_{16}N_{15}Rh_3$	$Rh_3(CN)_5 \cdot N_2H_4 \cdot 4 N_2H_3 \cdot 14 H_2O$	Rh:	SVol.B1-197
$C_5H_{16}O_3PSb$	$(CH_3)_4SbOP(O)(OH)CH_3$	Sb:	Org.Comp.3-138/9
$C_5H_{17}MnNO_{10}S$	$[Mn(2-(OOC)-C_4H_7NH_2)(H_2O)_4SO_4]$	Mn:	MVol.D4-305
$C_5H_{18}N_4O_4Rh^+$	$[Rh(NH_2C_2H_4NH_2)_2(OCO_2)(OH_2)]^+$	Rh:	SVol.B2-170
$C_5H_{18}N_5ORhS^{2+}$	$[Rh(NH_2C_2H_4NH_2)_2(OH_2)(NCS)]^{2+}$	Rh:	SVol.B2-184
–	$[Rh(NH_2C_2H_4NH_2)_2(OH_2)(SCN)]^{2+}$	Rh:	SVol.B2-184
$C_5H_{20}I_3LaN_{10}O_5$	$LaI_3 \cdot 5 OC(NH_2)_2$	Sc:	MVol.D2-221
$C_5H_{20}I_3N_{10}NdO_5$	$NdI_3 \cdot 5 OC(NH_2)_2$	Sc:	MVol.D2-221
$C_5H_{20}I_3N_{10}O_5Pr$	$PrI_3 \cdot 5 OC(NH_2)_2$	Sc:	MVol.D2-221

$C_5H_{20}I_3N_{10}O_5Sm$ ...	$SmI_3 \cdot 5\ OC(NH_2)_2$..................	Sc:	MVol.D2-221
$C_5H_{20}I_{11}LaN_{10}O_5$ ..	$LaI_3 \cdot 5\ OC(NH_2)_2 \cdot 4\ I_2 \cdot 10\ H_2O$.......	Sc:	MVol.D2-221
$C_5H_{20}I_{11}N_{10}NdO_5$ ...	$NdI_3 \cdot 5\ OC(NH_2)_2 \cdot 4\ I_2 \cdot 10\ H_2O$.......	Sc:	MVol.D2-221
$C_5H_{20}I_{11}N_{10}O_5Pr$ ...	$PrI_3 \cdot 5\ OC(NH_2)_2 \cdot 4\ I_2 \cdot 10\ H_2O$	Sc:	MVol.D2-221
$C_5H_{20}I_{11}N_{10}O_5Sm$...	$SmI_3 \cdot 5\ OC(NH_2)_2 \cdot 4\ I_2 \cdot 10\ H_2O$	Sc:	MVol.D2-221
$C_5H_{20}MnNO_8{}^+$	$[Mn(H_2O)_6(2-(OOC)-C_4H_7NH)]^+$	Mn:	MVol.D4-304
$C_5H_{20}N_{10}O_{13}S_2Th$..	$Th(SO_4)_2 \cdot 5\ (NH_2)_2CO \cdot 3\ H_2O$	Th:	SVol.E-38, 42
$C_5H_{20}N_{10}O_{17}S_3Sc_2$..	$Sc_2(SO_4)_3 \cdot 5\ CO(NH_2)_2$	Sc:	MVol.C8-70
$C_5H_{20}N_{12}O_{13}U$	$[UO_2(CO(NH_2)_2)_5](NO_3)_2$.	U:	SVol.A6-36
$C_5H_{20}N_{14}O_{17}Th$.....	$Th(NO_3)_4 \cdot 5\ (NH_2)_2CO \cdot 3\ H_2O$	Th:	SVol.E-38, 41
$C_5H_{20}N_{16}NiO_{23}Th$...	$NiTh(NO_3)_6 \cdot 5\ (NH_2)_2CO \cdot 3\ H_2O$	Th:	SVol.E-38, 41
$C_5H_{20}N_{17}Rh_3$	$Rh_3(CN)_5 \cdot 2\ N_2H_4 \cdot 4\ N_2H_3 \cdot 6\ H_2O$.....	Rh:	SVol.B1-197
$C_5H_{21}N_8O_8Rh$	$[Rh(NH_2C_2H_4NH_2)_2(NH_2CH_3)(NO_2)](NO_3)_2$..	Rh:	SVol.B2-175
$C_5H_{22}MnNO_8{}^+$	$[Mn(H_2O)_6((CH_3)_2CHCH(NH_2)COO)]^+$	Mn:	MVol.D4-283/4
$C_5H_{24}N_4O_{21}U_2$	$(NH_4)_4[(UO_2)_2(C_2O_4)(CO_3)_3(H_2O)_4]$	U:	SVol.C13-228/9
$C_5H_{24}N_5O_2Rh^{2+}$	$[Rh(NH_3)_5(OCOC_4H_9-t)]^{2+}$	Rh:	SVol.B2-134
$C_5H_{24}N_{14}Os_2S_5$.....	$[Os_2N(NH_3)_8(NCS)_2](NCS)_3$	Os:	SVol.1-169
$C_5H_{27}N_5ORh^{3+}$	$[Rh(NH_2CH_3)_5(H_2O)]^{3+}$	Rh:	SVol.B2-156
$C_5H_{28}N_6O_{21}U_2$	$(NH_4)_6[(UO_2)_2(CO_3)_5(H_2O)_2] \cdot H_2O$	U:	SVol.C13-20/2
$C_5H_{32}MnN_{15}O_{40}SiW_{11}$			
	$(CH_6N_3)_5H_2SiMnW_{11}O_{40} \cdot 4\ H_2O$	Mn:	MVol.C8-333
C_5HoO_5	$Ho(CO)_5$.	Sc:	MVol.D6-166
$C_5HoO_5{}^+$.	$[Ho(C_5O_5)]^+$	Sc:	MVol.D3-250/1
$C_5IK_3N_5Rh$	$K_3[Rh(CN)_5I]$	Rh:	SVol.B1-198/9
$C_5IN_5Rh^{3-}$	$[Rh(CN)_5I]^{3-}$	Rh:	SVol.B1-198/9
C_5IO_5Tc.	$Tc(CO)_5I$	Tc:	SVol.1-316/7
		Tc:	SVol.2-129
$-$	$^{99m}Tc(CO)_5I$.	Tc:	SVol.1-258/60
$C_5K_2MnN_6O$	$K_2[Mn(CN)_5NO]$.	Mn:	MVol.D2-274, 276/7
$-$	$K_2[Mn(CN)_5NO] \cdot 4\ H_2O$	Mn:	MVol.D2-277
$-$	$K_2[Mn(CN)_5NO] \cdot 6\ H_2O$	Mn:	MVol.D2-274, 277
$C_5K_2N_6OOs$........	$K_2[Os(NO)(CN)_5] \cdot 2\ H_2O$	Os:	SVol.1-229/30
$C_5K_3MnN_5$.	$K_3Mn(CN)_5$	Mn:	MVol.D2-216
$C_5K_3MnN_6O$	$K_3[Mn(CN)_5NO]$.	Mn:	MVol.D2-263/4
$-$	$K_3[Mn(CN)_5NO] \cdot 2\ H_2O$	Mn:	MVol.D2-265/7
$-$	$K_3[Mn(CN)_5NO] \cdot 2\ D_2O$	Mn:	MVol.D2-265/7
$-$	$K_3[Mn(CN)_5{}^{15}NO] \cdot 2\ H_2O$.	Mn:	MVol.D2-265/7
$C_5K_3NO_{10}SU$	$K_3[UO_2(C_2O_4)_2(NCS)]$	U:	SVol.C13-230/2
$-$	$K_3[UO_2(C_2O_4)_2(NCS)] \cdot 3\ H_2O$	U:	SVol.C13-230/2
$C_5K_3N_5O_2S_5U$	$K_3[UO_2(NCS)_5]$	U:	SVol.C13-43, 46
$-$	$K_3[UO_2(NCS)_5] \cdot H_2O$	U:	SVol.C13-43
$-$	$K_3[UO_2(NCS)_5] \cdot 2\ H_2O$	U:	SVol.C13-43, 46
$C_5K_3N_5O_{10}U_2$	$K_3[U_2O_5(NCO)_5] \cdot H_2O$	U:	SVol.C13-33/4
$C_5K_3N_6O_2Rh$........	$K_3[Rh(CN)_5NO_2]$	Rh:	SVol.B1-197
$C_5K_3N_8Rh$	$K_3[Rh(CN)_5N_3] \cdot H_2O$	Rh:	SVol.B1-197
$C_5K_6O_{15}U$	$K_6[U(CO_3)_5] \cdot 6\ H_2O$	U:	SVol.C13-8
$C_5K_8O_{21}U_2$	$K_8[(UO_2)_2O_2(CO_3)_5]$	U:	SVol.C13-24/5
$-$	$K_8[(UO_2)_2O_2(CO_3)_5] \cdot 3\ H_2O$.	U:	SVol.C13-24/5
$C_5LaO_5{}^+$	$[La(C_5O_5)]^+$.	Sc:	MVol.D3-250/1
$C_5Li_6O_{19}U_2$	$Li_6[(UO_2)_2(CO_3)_5] \cdot x\ H_2O$	U:	SVol.C13-20/2

$C_5LuO_5^+$	$[Lu(C_5O_5)]^+$	Sc:	MVol.D3-250/1
$C_5MnN_5^{3-}$	$[Mn(CN)_5]^{3-}$	Mn:	MVol.D2-207
$C_5MnN_5S_5^{3-}$	$[Mn(CNS)_5]^{3-}$	Mn:	MVol.D2-292
$C_5MnN_5Se_5^{3-}$	$[Mn(NCSe)_5]^{3-}$	Mn:	MVol.D2-293
$C_5MnN_6Na_2O$	$Na_2[Mn(CN)_5NO]$	Mn:	MVol.D2-276
$C_5MnN_6O^{2-}$	$[Mn(CN)_5NO]^{2-}$	Mn:	MVol.D2-271/6
$C_5MnN_6O^{3-}$	$[Mn(CN)_5NO]^{3-}$	Mn:	MVol.D2-253/62
$C_5MnN_6O^{4-}$	$[Mn(CN)_5NO]^{4-}$	Mn:	MVol.D2-260, 280
$C_5MnN_6O^{5-}$	$[Mn(CN)_5NO]^{5-}$	Mn:	MVol.D2-259
C_5MnN_6OZn	$Zn[Mn(CN)_5NO]$	Mn:	MVol.D2-274, 278/9
–	$Zn[Mn(CN)_5NO] \cdot H_2O$	Mn:	MVol.D2-274, 278/9
–	$Zn[Mn(CN)_5NO] \cdot 3.3\ H_2O$	Mn:	MVol.D2-274, 278/9
$C_5Mo_3U_3$	$U_3Mo_3C_5 = UMoC_{1.67}$	U:	SVol.C12-191
$C_5NNa_3O_{10}SU$	$Na_3[UO_2(C_2O_4)_2(NCS)]$	U:	SVol.C13-230/2
–	$Na_3[UO_2(C_2O_4)_2(NCS)] \cdot 5\ H_2O$	U:	SVol.C13-230/2
$C_5NO_{10}SU^{3-}$	$[UO_2(C_2O_4)_2(NCS)]^{3-}$	U:	SVol.C13-231
$C_5N_5OS_5Tc^{2-}$	$[TcO(NCS)_5]^{2-}$	Tc:	SVol.2-190, 201
$C_5N_5OTc^{2-}$	$[TcO(CN)_5]^{2-}$	Tc:	SVol.2-200
$C_5N_5O_2S_5U^{3-}$	$[UO_2(NCS)_5]^{3-}$	U:	SVol.C13-43/8
		U:	SVol.D1-180
		U:	SVol.D2-376
$C_5N_5O_2U^{3-}$	$[UO_2(CN)_5]^{3-}$	U:	SVol.D1-178
$C_5N_5O_2U^{4-}$	$[UO_2(CN)_5]^{4-}$	U:	SVol.C13-28
$C_5N_5Os^{4-}$	$[Os(CN)_5]^{4-}$	Os:	SVol.1-177
$C_5N_5Rh^{4-}$	$[Rh(CN)_5]^{4-}$	Rh:	SVol.B1-188
$C_5N_6OOs^{2-}$	$[Os(NO)(CN)_5]^{2-}$	Os:	SVol.1-176, 179, 229/30
$C_5N_6OOsS^{4-}$	$[Os(NOS)(CN)_5]^{4-}$	Os:	SVol.1-179
$C_5N_6O_2Os^{4-}$	$[Os(NO_2)(CN)_5]^{4-}$	Os:	SVol.1-179
$C_5N_8Rh^{3-}$	$[Rh(CN)_5N_3]^{3-}$	Rh:	SVol.B1-196
$C_5Na_6O_{15}U$	$Na_6[U(CO_3)_5] \cdot 11\ H_2O$	U:	SVol.C13-7/8
–	$Na_6[U(CO_3)_5] \cdot 12\ H_2O$	U:	SVol.C13-8
$C_5Na_6O_{19}U_2$	$Na_6[(UO_2)_2(CO_3)_5]$	U:	SVol.C13-20/2
$C_5Na_6O_{28}U_5$	$Na_6[(UO_2)_2(CO_3)_5(UO_3)_3]$	U:	SVol.C13-20/2
C_5NdO_5	$Nd(CO)_5$	Sc:	MVol.D6-166
$C_5NdO_5^+$	$[Nd(C_5O_5)]^+$	Sc:	MVol.D3-250/1
C_5O_5Pr	$Pr(CO)_5$	Sc:	MVol.D6-166
$C_5O_5Pr^+$	$[Pr(C_5O_5)]^+$	Sc:	MVol.D3-250/1
$C_5O_5Sm^+$	$[Sm(C_5O_5)]^+$	Sc:	MVol.D3-250/1
$C_5O_5Tb^+$	$[Tb(C_5O_5)]^+$	Sc:	MVol.D3-250/1
C_5O_5Tc	$Tc(CO)_5$ (radical)	Tc:	SVol.1-316
–	$^{99m}Tc(CO)_5$ (radical)	Tc:	SVol.1-258/61
$C_5O_5Tc^-$	$Tc(CO)_5^-$	Tc:	SVol.2-130
$C_5O_5Tm^+$	$[Tm(C_5O_5)]^+$	Sc:	MVol.D3-250/1
C_5O_5U	$U(CO)_5$	U:	SVol.E2-171/2
$C_5O_5Y^+$	$[Y(C_5O_5)]^+$	Sc:	MVol.D3-250/1
C_5O_5Yb	$Yb(CO)_5$	Sc:	MVol.D6-166
$C_5O_5Yb^+$	$[Yb(C_5O_5)]^+$	Sc:	MVol.D3-250/1
$C_5O_{15}U^{6-}$	$[U(CO_3)_5]^{6-}$	U:	SVol.C13-9
		U:	SVol.D1-169/72
$C_5O_{19}U_2^{6-}$	$[(UO_2)_2(CO_3)_5]^{6-}$	U:	SVol.C13-10

$C_5O_{19}U_2^{6-}$ .... $[(UO_2)_2(CO_3)_5]^{6-}$ .... U: SVol.D1-177

C_5U .... UC_5 .... U: SVol.A6-180

                                             U: SVol.C12-54/5

$C_{5.5}Cl_3H_{5.5}LaN_{1.1}$ ... $LaCl_3 \cdot 1.1\ C_5H_5N$ .... Sc: MVol.D1-34

$C_{5.5}H_{10}N_3O_{9.5}Th$ .... $Th(HCOO)_4 \cdot 1.5\ (NH_2)_2CO$ .... Th: SVol.E-39, 42

$C_6CaF_{13}I$ .... $C_6F_{13}CaI$ .... F: PFHOrg.SVol.1-27

$C_6CaKO_6RhS_6$ ..... $KCa[Rh(S_2C_2O_2)_3] \cdot 4\ H_2O$ .... Rh: SVol.B3-24/5

$C_6CaMn_2N_6$ .... $CaMn[Mn(CN)_6]$ .... Mn: MVol.D2-215

$C_6Ca_2MnN_6$ .... $Ca_2[Mn(CN)_6] \cdot n\ H_2O$ .... Mn: MVol.D2-216

$C_6Ca_3H_4Mg_3O_{26}U_2$.. $(CaMg)_3[UO_2(CO_3)_3]_2(OH)_4 \cdot 18\ H_2O$ ..... U: SVol.C13-22/4

$C_6Ca_3K_2O_{22}U_2$ ..... $K_2Ca_3[UO_2(CO_3)_3]_2 \cdot 10\ H_2O$ .... U: SVol.C13-15

– .... $K_2Ca_3[UO_2(CO_3)_3]_2 \cdot x\ H_2O$ (x = 9 to 10) U: SVol.C13-15, 16, 18

$C_6CdCl_2H_4N_2S$ ..... $SN_2C_6H_4 \cdot CdCl_2$ .... S: S-N Comp.3-217

$C_6CdFeH_8N_2O_4$ ..... $[Cd(NH_2CH_2CH_2NH_2)Fe(CO)_4]_n$ ...... Fe: Org.Comp.C7-94

C_6CdMnN_6 .... $Cd[Mn(CN)_6]$ .... Mn: MVol.D2-251/2

$C_6CeCl_3H_5O$ .... $Ce(OC_6H_5)Cl_3$ .... Sc: MVol.D3-37

$C_6CeCl_3H_5O_2$ ...... $CeCl_3(OC_6H_4OH)$ .... Sc: MVol.D3-43/4

$C_6CeCl_3H_5O_3$ ...... $CeCl_3(OC_6H_3(OH)_2)$ .... Sc: MVol.D3-44

$C_6CeCl_3H_6O_6$ ...... $Ce(CH_2ClCOO)_3$ .... Sc: MVol.D5-54/6, 61/3

– .... $Ce(CH_2ClCOO)_3 \cdot H_2O$ .... Sc: MVol.D5-61/3

– .... $Ce(CH_2ClCOO)_3 \cdot 3\ H_2O$ .... Sc: MVol.D5-61/3

$C_6CeCl_3H_{11}N$ ...... $CeCl_3 \cdot (CH_3)_2NCH_2CCCH_3 \cdot H_2O$ .... Sc: MVol.D1-18

$C_6CeCl_3H_{12}O_{1.5}$ ..... $CeCl_3 \cdot 1.5\ C_4H_8O$ .... Sc: MVol.D3-272

$C_6CeCl_3H_{15}N_3O_6$ .... $Ce(NH_2CH_2COOH)_3Cl_3 \cdot 3\ H_2O$ .... Sc: MVol.D1-104

$C_6CeCl_3H_{15}O_4P$ .... $CeCl_3 \cdot (C_2H_5O)_3PO$ .... Sc: MVol.D4-187

$C_6CeCl_3H_{18}O_3$ ..... $CeCl_3 \cdot 3\ C_2H_5OH$ .... Sc: MVol.D3-18

$C_6CeCl_3H_{24}N_{12}O_6$... $CeCl_3 \cdot 6\ OC(NH_2)_2$ .... Sc: MVol.D2-220

$C_6CeCl_3H_{24}N_{12}S_6$... $CeCl_3 \cdot 6\ (NH_2)_2CS$ .... Sc: MVol.D4-77

$C_6CeCl_4H_{12}N$ ...... $CeCl_3 \cdot (CH_3)_2NCH_2(CH)_2CH_2Cl \cdot H_2O$ ... Sc: MVol.D1-18

$C_6CeCl_4H_{18}O_3S_3$ .... $CeCl_4 \cdot 3\ (CH_3)_2SO$ .... Sc: MVol.D4-15/6

$C_6CeCl_6H_3O_6$ ...... $Ce(CHCl_2COO)_3$ .... Sc: MVol.D5-54/6, 63

$C_6CeCl_9O_6$ .... $Ce(CCl_3COO)_3$ .... Sc: MVol.D5-63/4

– .... $Ce(CCl_3COO)_3 \cdot 3\ H_2O$ .... Sc: MVol.D5-64

$C_6CeF_9O_6$ .... $Ce(CF_3COO)_3 \cdot 3\ H_2O$ .... Sc: MVol.D5-57/9

$C_6CeH_2N_3O_7^{2+}$ ..... $[Ce(OC_6H_2(NO_2)_3)]^{2+}$ .... Sc: MVol.D3-39

$C_6CeH_3K_3N_3O_9$ ..... $K_3[Ce(HNC(O)COO)_3] \cdot 3\ H_2O$ .... Sc: MVol.D2-257

$C_6CeH_3N_2O_5^{2+}$ ..... $[Ce(OC_6H_3(NO_2)_2)]^{2+}$ .... Sc: MVol.D3-38/9

$C_6CeH_3O_6$ .... $Ce(OOCCHC(COO)CH_2COO)$ .... Sc: MVol.D5-217/8

– .... $Ce(OOCCHC(COO)CH_2COO) \cdot 2\ H_2O$ ..... Sc: MVol.D5-217/8

– .... $Ce(OOCCHC(COO)CH_2COO) \cdot 3\ H_2O$ ..... Sc: MVol.D5-217/8

– .... $Ce(OOCCHC(COO)CH_2COO) \cdot 5\ H_2O$ ..... Sc: MVol.D5-217/8

$C_6CeH_4KO_8$ .... $K[Ce(CH_2(COO)_2)_2] \cdot 2\ H_2O$ .... Sc: MVol.D5-161/2

$C_6CeH_4LiO_8$ .... $Li[Ce(CH_2(COO)_2)_2] \cdot 3\ H_2O$ .... Sc: MVol.D5-161/2

$C_6CeH_4NO_2^{2+}$ ..... $Ce(NC_5H_4COO)^{2+}$ .... Sc: MVol.D1-216/8

$C_6CeH_4NO_3^{2+}$ ..... $[Ce(OC_6H_4NO_2)]^{2+}$ .... Sc: MVol.D3-38

– .... $[Ce(OOCC_5H_4NO)]^{2+}$ .... Sc: MVol.D2-158

$C_6CeH_4NaO_8$ .... $Na[Ce(CH_2(COO)_2)_2] \cdot 2\ H_2O$ .... Sc: MVol.D5-161/2

$C_6CeH_4O_3^+$ .... $[Ce(O_2C_6H_3OH)]^+$ .... Sc: MVol.D3-44/5

$C_6CeH_4O_8^-$ .... $[Ce(CH_2(COO)_2)_2]^-$ .... Sc: MVol.D5-150, 153

$C_6CeH_5N_2O^{2+}$	$[Ce(ONCHC_5H_4N)]^{2+}$	Sc:	MVol.D2-100
$C_6CeH_5O_4^{2+}$	$[Ce(HOCH_2C_5H_2O_3)]^{2+}$	Sc:	MVol.D3-291/3
$C_6CeH_5O_6$	$Ce(CH(CH_2COO)_2COO)$	Sc:	MVol.D5-216
—	$Ce(CH(CH_2COO)_2COO) \cdot 4 H_2O$	Sc:	MVol.D5-216
$C_6CeH_5O_6S$	$Ce[OOCCH_2SCH(COO)CH_2COO]$	Sc:	MVol.D4-66/7
$C_6CeH_5O_7$	$Ce(HOC(CH_2COO)_2COO)$	Sc:	MVol.D5-344/51
—	$Ce(HOC(CH_2COO)_2COO) \cdot 3.5 H_2O$	Sc:	MVol.D5-352/4
$C_6CeH_5O_7^+$	$[Ce(HOC(CH_2COO)_2COO)]^+$	Sc:	MVol.D5-350
$C_6CeH_5O_8$	$[Ce(CH_2(COO)_2)_2H]$.	Sc:	MVol.D5-152, 154
—	$HCe(CH_2(COO)_2)_2 \cdot 2 H_2O$	Sc:	MVol.D5-159/60
—	$HCe(CH_2(COO)_2)_2 \cdot 3 H_2O$	Sc:	MVol.D5-159/60
$C_6CeH_6I_3O_6$	$Ce(CH_2ICOO)_3 \cdot 3 H_2O$	Sc:	MVol.D5-66
$C_6CeH_6NO_6$	$Ce[N(CH_2COO)_3]$.	Sc:	MVol.D1-137/42
—	$Ce[N(CH_2COO)_3] \cdot n H_2O$	Sc:	MVol.D1-143
$C_6CeH_6O_7^+$	$[Ce(HOC(CH_2COO)_2COOH)]^+$	Sc:	MVol.D5-344/51
$C_6CeH_6O_7^{2+}$	$[Ce(HOC(CH_2COO)_2COOH)]^{2+}$	Sc:	MVol.D5-350
$C_6CeH_7N_3O^{2+}$	$Ce[4-H_2NNHC(O)C_5H_4N]^{2+}$	Sc:	MVol.D2-248
$C_6CeH_7O_6^{2+}$	$[Ce((HOCH_2CHOH)(HO)C_4HO_3)]^{2+}$	Sc:	MVol.D3-286
$C_6CeH_7O_7^{2+}$	$[Ce(HOC(CH_2COOH)_2COO)]^{2+}$	Sc:	MVol.D5-344/51
$C_6CeH_7O_8$	$Ce(C_6H_7O_8) \cdot 6 H_2O = [Ce(C_6H_7O_8)$ $(H_2O)_3] \cdot 3 H_2O$	Sc:	MVol.D5-340/1
$C_6CeH_8NO_8$	$NH_4[Ce(CH_2(COO)_2)_2] \cdot 3 H_2O$	Sc:	MVol.D5-161/2
$C_6CeH_8N_3O_2^{2+}$	$[Ce(N_2C_3H_3CH_2CH(NH_2)COO)]^{2+}$	Sc:	MVol.D1-111
$C_6CeH_8O_6^+$	$[Ce(C_2H_4(OCH_2COO)_2)]^+$	Sc:	MVol.D5-193/5
$C_6CeH_9NO_5^+$	$[Ce(HOC_2H_4N(CH_2COO)_2)]^+$	Sc:	MVol.D1-128/9
$C_6CeH_9N_2O_8$	$Ce[HOC_2H_4N(CH_2COO)_2](NO_3)$	Sc:	MVol.D1-130
$C_6CeH_9O_3^{2+}$	$[Ce((CH_2)_4C(OH)COO)]^{2+}$	Sc:	MVol.D5-278/9
$C_6CeH_9O_6$	$Ce(CH_3COO)_3$	Sc:	MVol.D5-23/8, 31/3, 43
—	$Ce(CH_3COO)_3 \cdot 0.7 H_2O$	Sc:	MVol.D5-35/7, 39
—	$Ce(CH_3COO)_3 \cdot H_2O$	Sc:	MVol.D5-34/5, 43
—	$Ce(CH_3COO)_3 \cdot 1.5 H_2O$	Sc:	MVol.D5-34/6, 39, 43/4
—	$Ce(CH_3COO)_3 \cdot 2 H_2O$.	Sc:	MVol.D5-34/5
$C_6CeH_9O_6S_3$	$Ce(OOCCH_2SH)_3 \cdot 3 H_2O$	Sc:	MVol.D4-54/5
$C_6CeH_9O_7$	$Ce[HOCH_2(CHOH)_2(CH(O))_2COO]$	Sc:	MVol.D5-275/8
$C_6CeH_9O_9$	$Ce(HOCH_2COO)_3$	Sc:	MVol.D5-222/30, 234, 238/9
$C_6CeH_9O_{12}$	$Ce((HO)_2CHCOO)_3$	Sc:	MVol.D5-249/50
$C_6CeH_{10}N_2O_4^+$	$Ce[C_2H_4(NHCH_2COO)_2]^+$	Sc:	MVol.D1-148/9
$C_6CeH_{10}N_2O_4S_2^-$	$[Ce(SCH_2CH(NH_2)COO)_2]^-$	Sc:	MVol.D4-59/60
$C_6CeH_{10}N_3O_2S_2$	$Ce(HN(CH_2CH_2O)_2H)(NCS)_2$	Sc:	MVol.D2-2
$C_6CeH_{10}N_4O_4^+$	$[Ce(CH_3C(NO)CHNOH)_2]^+$	Sc:	MVol.D2-116/7
$C_6CeH_{10}O_4^+$	$[Ce(C_2H_5COO)_2]^+$	Sc:	MVol.D5-71/4
$C_6CeH_{10}O_6^+$	$[Ce(CH_3OCH_2COO)_2]^+$	Sc:	MVol.D5-69/70
—	$[Ce(HOC_2H_4COO)_2]^+$	Sc:	MVol.D5-260/1
—	$[Ce(H_3CCHOHCOO)_2]^+$	Sc:	MVol.D5-250/5
$C_6CeH_{10}O_7^+$	$[Ce(OCH_2(CHOH)_4COO)]^+$	Sc:	MVol.D5-275/8
$C_6CeH_{11}O_2S^{2+}$	$[Ce(SCH_2COOC_4H_9)]^{2+}$	Sc:	MVol.D4-58
$C_6CeH_{11}O_3^{2+}$	$[Ce(n-C_3H_7C(OH)(CH_3)COO)]^{2+}$	Sc:	MVol.D5-273
$C_6CeH_{11}O_4^{2+}$	$[Ce(H_3C(C(OH)CH_3)_2COO)]^{2+}$	Sc:	MVol.D5-272
$C_6CeH_{11}O_7^{2+}$	$[Ce(HOCH_2(CHOH)_4COO)]^{2+}$	Sc:	MVol.D5-275/8

$C_6CeH_{12}NO_2{}^{2+}$	$Ce[(CH_3)_2CHCH_2CH(NH_2)COO]^{2+}$	Sc:	MVol.D1-116
$C_6CeH_{12}NO_4{}^{2+}$	$[Ce(N(C_2H_4OH)_2CH_2COO)]^{2+}$	Sc:	MVol.D1-106/7
$C_6CeH_{12}NO_4{}^{3+}$	$[Ce(N(C_2H_4OH)_2CH_2COO)]^{3+}$	Sc:	MVol.D1-107
$C_6CeH_{12}N_2O_{12}P_4{}^{5-}$	$Ce(C_2H_4(N(CH_2PO_3)_2)_2)^{5-}$	Sc:	MVol.D4-154/5
$C_6CeH_{12}N_6O_6{}^{2-}$	$[Ce(CH_2NO)_6]^{2-}$	Sc:	MVol.D2-97
$C_6CeH_{13}N_2O_2{}^{2+}$	$Ce[NH_2(CH_2)_4CH(NH_2)COO]^{2+}$	Sc:	MVol.D1-116
$C_6CeH_{13}N_2O_{12}P_4{}^{4-}$	$Ce[H(C_2H_4(N(CH_2PO_3)_2)_2)]^{4-}$	Sc:	MVol.D4-154/5
$C_6CeH_{13}N_4O_2{}^{2+}$	$Ce[NHC(NH_2)NH(CH_2)_3CH(NH_2)COO]^{2+}$	Sc:	MVol.D1-116
$C_6CeH_{13}O_{11}$	$[Ce(C_6H_7O_8)(H_2O)_3] \cdot 3 H_2O$		
	$= Ce(C_6H_7O_8) \cdot 6 H_2O$	Sc:	MVol.D5-340/1
$C_6CeH_{14}N_2O_{12}P_4{}^{3-}$	$Ce[H_2(C_2H_4(N(CH_2PO_3)_2)_2)]^{3-}$	Sc:	MVol.D4-154/5
$C_6CeH_{14}S_2{}^+$	$Ce(SC_3H_7-i)_2{}^+$	Sc:	MVol.D4-48
$C_6CeH_{15}N_2O_{12}P_4{}^{2-}$	$Ce[H_3(C_2H_4(N(CH_2PO_3)_2)_2)]^{2-}$	Sc:	MVol.D4-154/5
$C_6CeH_{15}O_{12}S_3$	$Ce(C_2H_5SO_4)_3 \cdot 9 H_2O$	Sc:	MVol.C8-362/414
$C_6CeH_{16}N_2O_{12}P_4{}^-$	$Ce[H_4(C_2H_4(N(CH_2PO_3)_2)_2)]^-$	Sc:	MVol.D4-154/5
$C_6CeH_{17}N_2O_{12}P_4$	$Ce[H_5(C_2H_4(N(CH_2PO_3)_2)_2)]$	Sc:	MVol.D4-154/5
$C_6CeH_{21}N_6O_6$	$Ce(N_2H_4)_3(CH_3COO)_3 \cdot 2 H_2O$	Sc:	MVol.D1-13
$C_6CeH_{21}O_9P_3$	$Ce(HP(O)(OH)OC_2H_5)_3$	Sc:	MVol.D4-157/8
$C_6CeO_{12}{}^{3-}$	$[Ce(C_2O_4)_3]^{3-}$	Sc:	MVol.D5-120/4
$C_6Ce_2H_4O_{10}$	$[Ce_2(C_2H_4(COO)_2)(CO_3)_2]$	Sc:	MVol.D5-173
–	$[Ce_2(C_2H_4(COO)_2)(CO_3)_2] \cdot 2 H_2O$	Sc:	MVol.D5-173
$C_6Ce_2H_6N_6O_{12}$	$Ce_2(ONHC(O)C(O)NHO)_3 \cdot 4 H_2O$	Sc:	MVol.D2-260
$C_6Ce_2H_6O_7{}^{4+}$	$[Ce_2(HOC(CH_2COO)_2COOH)]^{4+}$	Sc:	MVol.D5-349
$C_6Ce_2H_{16}N_8O_{12}$	$Ce_2(N_2H_4)_4(C_2O_4)_3 \cdot 2.8 H_2O$	Sc:	MVol.D1-13
$C_6Ce_2H_{18}O_{15}P_6$	$Ce_2[((O)_2PCH_2)_2O]_3$	Sc:	MVol.D4-159
$C_6Ce_2H_{24}N_{12}O_{18}S_3$	$Ce_2(SO_4)_3 \cdot 6 CO(NH_2)_2$	Sc:	MVol.C8-115
		Sc:	MVol.D2-222/3
$C_6Ce_2O_{12}$	$Ce_2(C_2O_4)_3$	Sc:	MVol.D5-132/6
–	$Ce_2(C_2O_4)_3 \cdot 9 H_2O$	Sc:	MVol.D5-128
–	$Ce_2(C_2O_4)_3 \cdot 10 H_2O$	Sc:	MVol.D5-124/6, 130/5
–	$Ce_2(C_2O_4)_3 \cdot 10.5 H_2O$	Sc:	MVol.D5-128
–	$Ce_2(C_2O_4)_3 \cdot n H_2O$	Sc:	MVol.D5-136/7
$C_6CICsGaH_{15}$	$Cs[Ga(C_2H_5)_3CI]$	Ga:	Org.Comp.1-327
C_6CICuH_2S	$4-(SC_4H_2CI-1)CCCu$	Cu:	Org.Comp.3-22
C_6CICuH_4	$CIC_6H_4Cu \cdot x C_5H_5N \cdot y MgI_2$	Cu:	Org.Comp.1-398/9
–	$CIC_6H_4Cu \cdot x MgBr_2$	Cu:	Org.Comp.1-217, 272
–	$CIC_6H_4Cu \cdot x MgI_2$	Cu:	Org.Comp.1-381
$C_6CICuH_6O_2$	$C_2H_2 \cdot CuCl \cdot 2 CH_2CO$	Cu:	Org.Comp.4-52
$C_6CICuH_7NOS_2$	$[(S)(O)C_3H_2SN(CH_2CHCH_2)]CuCl$	Cu:	Org.Comp.4-13
C_6CICuH_{10}	$(c-C_6H_{10}Cu)Cl$	Cu:	Org.Comp.4-20
$C_6CICuH_{10}MgO$	$CIMgO(CH_2)_4CHCHCu \cdot x LiMgBrCl_2$	Cu:	Org.Comp.1-300
$C_6CICuH_{10}O_4$	$(c-C_6H_{10}Cu)(ClO_4)$	Cu:	Org.Comp.4-20
$C_6CICuH_{12}MgO$	$CIMgO(CH_2)_6Cu \cdot x LiMgBrCl_2$	Cu:	Org.Comp.1-301
–	$CIMgO(CH_2)_6Cu \cdot x MgCII$	Cu:	Org.Comp.1-220, 284
$C_6CICuH_{12}O$	$[CH_3CH(OH)CH_2C(CH_3)CH_2]CuCl$	Cu:	Org.Comp.4-13
$C_6CICuH_{14}Mg$	$(n-C_3H_7)_2CuMgCl$	Cu:	Org.Comp.2-213, 217/22
–	$(i-C_3H_7)_2CuMgCl$	Cu:	Org.Comp.2-213, 217/25
$C_6CICuH_{14}O_7$	$[(HOCH_2CHCH_2)_2Cu(H_2O)](ClO_4)$	Cu:	Org.Comp.4-37
$C_6CICuH_{17}OPS$	$[(CH_3)_2S(O)CH_2CuP(CH_3)_3]Cl$	Cu:	Org.Comp.1-467
$C_6CIDyH_4O_3{}^{2+}$	$[Dy(CICH_2C_5H_2O_3)]^{2+}$	Sc:	MVol.D3-294

$C_6ClF_{12}NO$	$(CF_3)_2CFNClC(O)C_2F_5$	F:	PFHOrg.8-162/3, 187, 203
$C_6ClF_{13}N_2$	$CF_2(CF_2CFNF_2)_2CFCl$	F:	PFHOrg.8-166, 191
–	$(CF_3)_2CFNNCCl(CF_3)_2$	F:	PFHOrg.8-74/5, 98
$C_6ClF_{13}OS$	$C_6F_{13}S(O)Cl$	F:	PFHOrg.SVol.3-11, 35
$C_6ClF_{13}O_2S$	$C_6F_{13}SO_2Cl$	F:	PFHOrg.SVol.3-158/9, 168, 175
$C_6ClF_{13}O_3S$	$CF_2ClCF_2OSO_2C_4F_9$	F:	PFHOrg.SVol.3-57, 58, 74, 109
–	$Cl(CF_2CF_2)_2OCF_2CF_2SO_2F$	F:	PFHOrg.SVol.3-155/6, 163
$C_6ClF_{14}N$	$CF_2ClCF(CF_3)N(CF_3)C_2F_5$	F:	PFHOrg.9-165, 174
–	$CF_3CFClCF_2N(CF_3)C_2F_5$	F:	PFHOrg.9-165, 174
$C_6ClF_{14}P$	$(i-C_3F_7)_2PCl$	F:	PFHOrg.SVol.1-105, 116
$C_6ClF_{15}N_2O_4S$	$(CF_3)_2NOCF_2CF(SO_2Cl)ON(CF_3)_2$	F:	PFHOrg.SVol.3-159, 168
$C_6ClF_{18}N_2O_2P$	$Cl(CF_3)_2P[ON(CF_3)_2]_2$	F:	PFHOrg.SVol.1-89, 93/5
C_6ClGaH_{14}	$Ga(C_3H_7)_2Cl$	Ga:	Org.Comp.1-128
$C_6ClGaH_{15}Na$	$Na[Ga(C_2H_5)_3Cl]$	Ga:	Org.Comp.1-327
$C_6ClGaH_{16}O$	$Ga(CH_3)_2Cl \cdot O(C_2H_5)_2$	Ga:	Org.Comp.1-129, 135
$C_6ClGaH_{17}P$	$Ga(CH_3)_2Cl \cdot PH(C_2H_5)_2$	Ga:	Org.Comp.1-130
$C_6ClGaH_{18}N_2$	$Ga(C_2H_5)_2Cl \cdot NH_2CH_2CH_2NH_2$	Ga:	Org.Comp.1-131
$C_6ClGdH_4O_3{}^{2+}$	$[Gd(ClCH_2C_5H_2O_3)]^{2+}$	Sc:	MVol.D3-294
$C_6ClGdH_{10}N_2O_4$	$Gd[C_2H_4(NHCH_2COO)_2]Cl \cdot 4 H_2O$	Sc:	MVol.D1-149
$C_6ClGeH_{15}O_3S$	$(C_2H_5)_3GeOSO_2Cl$	S:	SVol.3-321
$C_6ClH_4I_2Sb$	$ClC_6H_4SbI_2$	Sb:	Org.Comp.2-116/7
$C_6ClH_4LaO_3{}^{2+}$	$[La(ClCH_2C_5H_2O_3)]^{2+}$	Sc:	MVol.D3-294
$C_6ClH_4NdO_3{}^{2+}$	$[Nd(ClCH_2C_5H_2O_3)]^{2+}$	Sc:	MVol.D3-294
C_6ClH_4OSb	ClC_6H_4SbO	Sb:	Org.Comp.2-121
$C_6ClH_4O_3Pr^{2+}$	$[Pr(ClCH_2C_5H_2O_3)]^{2+}$	Sc:	MVol.D3-294
$C_6ClH_4O_3Sm^{2+}$	$[Sm(ClCH_2C_5H_2O_3)]^{2+}$	Sc:	MVol.D3-294
$C_6ClH_4O_3Y^{2+}$	$[Y(ClCH_2C_5H_2O_3)]^{2+}$	Sc:	MVol.D3-294
$C_6ClH_4O_3Yb^{2+}$	$[Yb(ClCH_2C_5H_2O_3)]^{2+}$	Sc:	MVol.D3-294
$C_6ClH_5I_2Sn$	$C_6H_5SnI_2Cl$	Sn:	Org.Comp.8-144
$C_6ClH_5Mo_3N_2O_{10}$ . . .	$ClC_6H_4N_2[HMo_3O_{10}]$	Mo:	SVol.B4-158
$C_6ClH_8N_2O_4Rh^{2-}$ . . .	$[Rh(((O_2CCH_2)N)_2C_2H_4)Cl]^{2-}$	Rh:	SVol.B2-75
$C_6ClH_8N_3OS$	$SN_2C_2ClNC_4H_8O$	S:	S-N Comp.3-93/8, 145/6
$C_6ClH_9N_2OS$	$SN_2C_2ClOC_4H_9-n$	S:	S-N Comp.3-93/8, 157/9
–	$SN_2C_2ClOC_4H_9-i$	S:	S-N Comp.3-93/8, 157/9
$C_6ClH_9N_2O_2S$	$SN_2C_2ClO(CH_2)_2OC_2H_5$	S:	S-N Comp.3-93/8, 157/8
$C_6ClH_9N_3O_5RhS$	$Rh(OOCCH_2CH_2C(NNHCSNH_2)COO)(H_2O)Cl$		
		Rh:	SVol.B3-43
$C_6ClH_9O_6U$	$UCl(CH_3COO)_3$	U:	SVol.C13-112, 114
$C_6ClH_{10}HoN_2O_4$	$Ho[C_2H_4(NHCH_2COO)_2]Cl \cdot 4 H_2O$	Sc:	MVol.D1-149
$C_6ClH_{10}LaN_2O_4$	$La[C_2H_4(NHCH_2COO)_2]Cl \cdot 3 H_2O$	Sc:	MVol.D1-149
–	$La[C_2H_4(NHCH_2COO)_2]Cl \cdot 5 H_2O$	Sc:	MVol.D1-149
$C_6ClH_{10}LuN_2O_4$	$Lu[C_2H_4(NHCH_2COO)_2]Cl \cdot n H_2O$	Sc:	MVol.D1-149/50
$C_6ClH_{10}N_2NdO_4$	$Nd[C_2H_4(NHCH_2COO)_2]Cl \cdot 2 H_2O$	Sc:	MVol.D1-149
$C_6ClH_{10}N_2O_4Pr$	$Pr[C_2H_4(NHCH_2COO)_2]Cl \cdot 2 H_2O$	Sc:	MVol.D1-149
$C_6ClH_{10}N_2O_4Sm$	$Sm[C_2H_4(NHCH_2COO)_2]Cl \cdot 4 H_2O$	Sc:	MVol.D1-149
$C_6ClH_{10}N_2O_4Tb$	$Tb[C_2H_4(NHCH_2COO)_2]Cl \cdot 4 H_2O$	Sc:	MVol.D1-149
$C_6ClH_{10}N_2O_4Tm$	$Tm[C_2H_4(NHCH_2COO)_2]Cl \cdot 4 H_2O$	Sc:	MVol.D1-149

$C_6Cl_2H_5Sb$ . . . . $C_6H_5SbCl_2$ . . . .   Sb: Org.Comp.2-95/8
$C_6Cl_2H_6MnN_2O_2$ . . . . $Mn(ONC_3H_3)_2Cl_2$ . . . .   Mn: MVol.D4-213/5
— . . . . $Mn(ONC_5H_4CONH_2)Cl_2$ . . . .   Mn: MVol.D5-112
$C_6Cl_2H_6MnN_6O_6$ $[MnCl_2(NH(CONH)_2CO)_2]$ · 0.5 H_2O . . . . . .   Mn: MVol.D4-73
$C_6Cl_2H_6NOSb$ . . . . . . $(NH_2)(HO)C_6H_3SbCl_2$ · HCl . . . .   Sb: Org.Comp.2-101
$C_6Cl_2H_6NO_2SSb$ . . . . $NH_2SO_2C_6H_4SbCl_2$ . . . .   Sb: Org.Comp.2-105
$C_6Cl_2H_6NSb$ . . . . $NH_2C_6H_4SbCl_2$ · HCl . . . .   Sb: Org.Comp.2-100
$C_6Cl_2H_7MnN$ . . . . $Mn(3-CH_3C_5H_4N)Cl_2$ . . . .   Mn: MVol.D3-121
— . . . . $Mn(4-CH_3C_5H_4N)Cl_2$ . . . .   Mn: MVol.D3-121
$C_6Cl_2H_7MnNO$ . . . . . $Mn(2-CH_3C_5H_4NO)Cl_2$ . . . .   Mn: MVol.D3-157
— . . . . $Mn(2-CH_3C_5H_4NO)Cl_2$ · H_2O . . . .   Mn: MVol.D3-157
— . . . . $Mn(3-CH_3C_5H_4NO)Cl_2$ . . . .   Mn: MVol.D3-157
— . . . . $Mn(3-CH_3C_5H_4NO)Cl_2$ · H_2O . . . .   Mn: MVol.D3-157
— . . . . $Mn(4-CH_3C_5H_4NO)Cl_2$ . . . .   Mn: MVol.D3-157
$C_6Cl_2H_7MnNO_2$ . . . . $Mn(CH_3OC_5H_4NO)Cl_2$ . . . .   Mn: MVol.D3-162/3
— . . . . $Mn(CH_3OC_5H_4NO)Cl_2$ · H_2O . . . .   Mn: MVol.D3-163
$C_6Cl_2H_7MnN_3O$ . . . . $Mn(NC_5H_4C(O)NHNH_2)Cl_2$ . . . .   Mn: MVol.D5-188/9, 192
— . . . . $Mn(NC_5H_4C(O)NDND_2)Cl_2$ . . . .   Mn: MVol.D5-188
$C_6Cl_2H_7MnN_3O_2$ . . . . $Mn(O_2NC_6H_4NHNH_2)Cl_2$ . . . .   Mn: MVol.D3-73/4
$C_6Cl_2H_7MnN_5$ . . . . . . $Mn[(NH_2)(CH_3)C_5H_2N_4]Cl_2$ . . . .   Mn: MVol.D4-18
— . . . . $Mn[(NH_2)(CH_3)C_5H_2N_4]Cl_2$ · 2 H_2O . . . . . .   Mn: MVol.D4-18
$C_6Cl_2H_8MnN_2$ . . . . . . $Mn(C_6H_4(NH_2)_2)Cl_2$ . . . .   Mn: MVol.D3-44
— . . . . $Mn(H_2NCH_2C_5H_4N)Cl_2$ . . . .   Mn: MVol.D3-135/6
$C_6Cl_2H_8MnN_2O$ . . . . . $Mn(CH_2CHNC(CH_2OH)NCHCH)Cl_2$ . . . .   Mn: MVol.D3-311
$C_6Cl_2H_8MnN_2O_8$ . . . . $Mn(C_6H_4(NH_2)_2)(ClO_4)_2$ · C_2H_5OH . . . . . .   Mn: MVol.D3-46
$C_6Cl_2H_8MnN_4$ . . . . . . $Mn(NHCHNCHCH)_2Cl_2$ . . . .   Mn: MVol.D3-293
— . . . . $Mn(NHN(CH)_3)_2Cl_2$ . . . .   Mn: MVol.D3-274/6
$C_6Cl_2H_8N_2Pt$ . . . . $(C_6H_4(NH_2)_2)PtCl_2$
                                 Cytostatic activity . . . .   Pt: SVol.A1-332
$C_6Cl_2H_8N_2S$ . . . . $SN_2C_2Cl(CH_2)_4Cl$ . . . .   S: S-N Comp.3-93/8, 136/7
$C_6Cl_2H_8O_6U$ . . . . $UO_2(CH_2ClCH_2COO)_2$ . . . .   U: SVol.C13-165
                                                                                   U: SVol.D1-201
— . . . . $UO_2(CH_3CHClCOO)_2$ . . . .   U: SVol.C13-165
$C_6Cl_2H_9Sb$ . . . . $(CH_2CH)_3SbCl_2$ . . . .   Sb: Org.Comp.4-26/7
$C_6Cl_2H_{10}MnN_2O_2$ . . . $Mn(OCC_2H_3NH_2)_2Cl_2$ . . . .   Mn: MVol.D5-109
$C_6Cl_2H_{10}MnN_6O_2$ . . . $Mn(NCCH_2C(O)NHNH_2)_2Cl_2$ · H_2O . . . .   Mn: MVol.D5-171
$C_6Cl_2H_{10}O_4Sn$ . . . . . . $(CH_3)_2Sn(OOCCH_2Cl)_2$ . . . .   Sn: Org.Comp.14-87, 90/1
$C_6Cl_2H_{10}Pd_2$ . . . . $[(C_3H_5)PdCl]_2$
                                 Cytostatic activity . . . .   Pt: SVol.A1-336
$C_6Cl_2H_{11}N_4O_4Rh$ . . . $H[Rh(CH_3CNOCHNOH)_2Cl_2]$ . . . .   Rh: SVol.B2-219
$C_6Cl_2H_{12}HgMnN_2O_2S_2$
                                 $HgMn(SCN)_2Cl_2$ · 2 C_2H_5OH . . . .   Mn: MVol.C7-239
$C_6Cl_2H_{12}KMnNO_2$ . . . $MnCl_2$ · $K[i-C_4H_9CH(NH_2)COO]$ . . . .   Mn: MVol.D4-286
$C_6Cl_2H_{12}MnN_2O_4S_2$ $[Mn(HSCH_2CH(NH_2)COO)_2Cl_2]$ . . . .   Mn: MVol.D4-291
$C_6Cl_2H_{12}MnN_4O_4$ . . . $Mn(CH_2(CONH_2)_2)_2Cl_2$ . . . .   Mn: MVol.D5-124
— . . . . $[Mn(CH_3C(O)NHC(O)NH_2)_2Cl_2]_n$ . . . .   Mn: MVol.D5-154
— . . . . $MnCl_2$ · 2 $CH_2(CONH_2)_2$ · 2 H_2O . . . .   Mn: MVol.D5-125
$C_6Cl_2H_{12}N_2O_5S_3$ . . . . $OS(N(SO_2CH_3)CHCH_2Cl)_2$ . . . .   S: S-N Comp.3-266/7
$C_6Cl_2H_{12}N_6S_2$ . . . . . . $((CH_3)_2N)_2C_2N_4S_2Cl_2$ . . . .   S: S-N Comp.4-133/4

$C_6Cl_2H_{12}N_6S_2$ . . . . . . [(SClNC(N(CH_3)_2)N)_2] . . . . S: S–N Comp.4-145
$C_6Cl_2H_{13.5}N_{1.5}O_3Rh$ $Rh(O_2C(CH_2)_3NH_2)Cl \cdot HCl$
                          $\cdot \, 0.5 \, NH_2(CH_2)_3CO_2H \cdot 2 \, H_2O$ . . . .   Rh: SVol.B2-64
$C_6Cl_2H_{14}MnN_2O_2$ . . . $Mn(OCHN(CH_3)_2)_2Cl_2$ . . . .   Mn: MVol.D5-94
–   . . . . $Mn(OCHN(CH_3)_2)_2Cl_2 \cdot 2 \, H_2O$. . . . .   Mn: MVol.D5-95
$C_6Cl_2H_{14}MnN_2O_4$ . . . $Mn[OOCCH(CH_3)NH_3]_2Cl_2$ . . . .   Mn: MVol.D4-267/8
$C_6Cl_2H_{14}MnN_4$ . . . . . $Mn((C(NH)C(NH))(NHC_2H_5)_2)Cl_2 \cdot 4 \, H_2O$ . .   Mn: MVol.D5-160
$C_6Cl_2H_{14}MnN_4O_2$ . . . $Mn(C_2H_4(CH_2C(O)NHNH_2)_2)Cl_2 \cdot H_2O$ . . . .   Mn: MVol.D5-206
–   . . . . $Mn(C_2H_4(CH_2C(O)NHNH_2)_2)Cl_2 \cdot 2 \, H_2O$ . . .   Mn: MVol.D5-206
$C_6Cl_2H_{14}N_2O_{12}Rh_2$. $[H_3N(CH_2)_2NH_3][Rh_2(HCO_3)_4Cl_2]$ . . . .   Rh: SVol.B1-186
$C_6Cl_2H_{14}N_2Pt$ . . . . . . $(C_3H_5NH_2)_2PtCl_2$
                          Cytostatic activity . . . .   Pt: SVol.A1-332/4

–   . . . . $(C_6H_{10}(NH_2)_2)PtCl_2$
                          Cytostatic activity . . . .   Pt: SVol.A1-332

$C_6Cl_2H_{14}N_3O_6Rh$ . . . $Rh(O_2CCH_2NH_2)Cl_2(NH_2CH_2CO_2H)_2$ . . . . . .   Rh: SVol.B2-66
$C_6Cl_2H_{15}LaO_9$ . . . . . . $[La(HOCH_2(CHOH)_4COO)(H_2O)_2]Cl_2$ . . . . . .   Sc: MVol.D5-278
$C_6Cl_2H_{15}MnN$ . . . . . . $Mn((C_2H_5)_3N)Cl_2$ . . . .   Mn: MVol.D3-25
$C_6Cl_2H_{15}MnNO_3$ . . . . $Mn[N(C_2H_4OH)_3]Cl_2$ . . . .   Mn: MVol.D4-232
$C_6Cl_2H_{15}N_9OOs$ . . . . $[Os(N_2)_2(C_6H_6N_2O)(NH_3)_3]Cl_2$ . . . .   Os: SVol.1-249
$C_6Cl_2H_{15}O_3U$ . . . . . . $U(OC_2H_5)_3Cl_2$ . . . .   U: SVol.C13-72/3
$C_6Cl_2H_{15}Sb$ . . . . $(C_2H_5)_3SbCl_2$. . . . .   Sb: Org.Comp.4-20/1
$C_6Cl_2H_{16}MnN_8O_4$ . . . $Mn(CH_2(C(O)NHNH_2)_2)_2Cl_2$ . . . .   Mn: MVol.D5-200
–   . . . . $Mn(CH_2(C(O)NHNH_2)_2)_2Cl_2 \cdot 2 \, H_2O$ . . . . .   Mn: MVol.D5-200/3
$C_6Cl_2H_{16}N_6O_8Rh_2$ . . . $[C(NH_2)_3]_2[Rh_2(O_2CH)_4Cl_2]$ . . . .   Rh: SVol.B2-12
$C_6Cl_2H_{16}N_6O_{12}Rh_2$ . . $[H_2NC(NH_2)_2]_2[Rh_2(HCO_3)_4Cl_2] \cdot 2 \, H_2O$ . . .   Rh: SVol.B1-186
$C_6Cl_2H_{17}N_7O_2Os$ . . . . $[Os(C_6H_5NO_2)(N_2)(NH_3)_4]Cl_2$ . . . .   Os: SVol.1-249
$C_6Cl_2H_{18}IRhTe_3$ . . . . $Rh[Te(CH_3)_2]_3Cl_2I$. . . . .   Rh: SVol.B3-64
$C_6Cl_2H_{18}MnN_2O_2$ . . . $Mn((CH_3)_3NO)_2Cl_2$ . . . .   Mn: MVol.D3-24
$C_6Cl_2H_{18}MnN_4O_8$ . . . $Mn(H_2N(CH_2)_2NHCH_2)_2(ClO_4)_2$. . . . .   Mn: MVol.D3-57
–   . . . . $Mn(H_2N(CH_2)_2NHCH_2)_2(ClO_4)_2 \cdot n \, H_2O$ . . .   Mn: MVol.D3-57
$C_6Cl_2H_{18}MnN_6O_3$ . . . $Mn(OCCH_3NHNH_2)_3Cl_2$. . . . .   Mn: MVol.D5-169
$C_6Cl_2H_{18}N_2Pt$ . . . . . . $(i\text{-}C_3H_7NH_2)_2PtCl_2$
                          Cytostatic activity . . . .   Pt: SVol.A1-332

$C_6Cl_2H_{18}N_4Rh^+$ . . . . $[Rh((NH_2C_2H_4NH)_2C_2H_4)Cl_2]^+$ . . . .   Rh: SVol.B2-196
$C_6Cl_2H_{18}N_8OOs$ . . . . $[Os(C_6H_6N_2O)(N_2)(NH_3)_4]Cl_2$ . . . .   Os: SVol.1-249
$C_6Cl_2H_{18}N_8O_8Rh_2$ . . $[(NHC(NH_2)NHNH_2)H]_2[Rh_2(O_2CH)_4Cl_2]$. . . .   Rh: SVol.B2-12
$C_6Cl_2H_{18}N_{11}O_3Rh$ . . $[Rh(N_5C_2H_6CH_3)_2Cl_2](NO_3) \cdot 4 \, H_2O$ . . . . . .   Rh: SVol.B2-223
$C_6Cl_2H_{18}O_5S_3U$ . . . . . $UO_2Cl_2((CH_3)_2SO)_3$. . . . .   U: SVol.A5-217, 219
$C_6Cl_2H_{18}PdSb_2$ . . . . . $PdCl_2[Sb(CH_3)_3]_2$ . . . .   Sb: Org.Comp.1-15
$C_6Cl_2H_{18}PtSb_2$ . . . . . $PtCl_2[Sb(CH_3)_3]_2$.   Sb: Org.Comp.1-15
$C_6Cl_2H_{19}NSb_2$ . . . . . . $[(CH_3)_3SbCl]_2NH$.   Sb: Org.Comp.1-14
$C_6Cl_2H_{20}N_2O_2Pt$ . . . . $(i\text{-}C_3H_7NH_2)_2Pt(OH)_2Cl_2$
                          Cytostatic activity . . . .   Pt: SVol.A1-332/4

$C_6Cl_2H_{20}N_5O_3Rh$ . . . $[Rh(NH_2CH_2CH(CH_3)NH_2)_2Cl_2]NO_3$ . . . .   Rh: SVol.B2-188
$C_6Cl_2H_{22}IN_6ORhS$. . . $[Rh((NHC_2H_4NH_2)_2SO)((NH_2)_2C_2H_4)]Cl_2I$ . . .   Rh: SVol.B2-186
$C_6Cl_2H_{22}IN_6O_2RhS$. . $[Rh((NHC_2H_4NH_2)_2SO_2)((NH_2)_2C_2H_4)]Cl_2I$ . .   Rh: SVol.B2-186
$C_6Cl_2H_{24}MnN_{12}O_{14}$ $Mn(OC(NH_2)_2)_6(ClO_4)_2$ . . . .   Mn: MVol.D5-139/40
$C_6Cl_2H_{24}N_5O_8Rh$ . . . $[Rh(NH_2CH_2CH(CH_3)NH_2)_2(NH_3)H](ClO_4)_2$. .   Rh: SVol.B2-187
$C_6Cl_2H_{24}NdO_6^+$ . . . . $[Nd(CH_3OH)_6Cl_2]^+$ . . . .   Sc: MVol.D4-317

$C_6Cl_2H_{30}MnN_6$	$Mn(CH_3NH_2)_6Cl_2$	Mn:	MVol.D3-21/2
$C_6Cl_2O_{12}U_2$	$U_2Cl_2(C_2O_4)_3$	U:	SVol.C13-219/20
–	$U_2Cl_2(C_2O_4)_3 \cdot 2 H_2O$	U:	SVol.C13-219/20
–	$U_2Cl_2(C_2O_4)_3 \cdot 4 H_2O$	U:	SVol.C13-219
–	$U_2Cl_2(C_2O_4)_3 \cdot 12 H_2O$	U:	SVol.C13-219/20
$C_6Cl_3CsH_6O_8U$	$Cs[UO_2(ClCH_2COO)_3]$	U:	SVol.C13-150
$C_6Cl_3DyH_{15}N_3O_6$	$Dy(NH_2CH_2COOH)_3Cl_3 \cdot 3 H_2O$	Sc:	MVol.D1-104
$C_6Cl_3DyH_{20}N_4$	$Dy(NH_2C_3H_6NH_2)_2Cl_3$	Sc:	MVol.D1-26
$C_6Cl_3DyH_{24}N_6O_{12}$	$Dy(C_2H_8N_2)_3(ClO_4)_3$	Sc:	MVol.D1-19/21
$C_6Cl_3ErH_6O_6$	$Er(CH_2ClCOO)_3 \cdot 5 H_2O$	Sc:	MVol.D5-61/3
$C_6Cl_3ErH_{18}N_4O_{12}$	$Er[(NH_2C_2H_4)_3N](ClO_4)_3$	Sc:	MVol.D1-34
$C_6Cl_3ErH_{24}N_6$	$Er(C_2H_8N_2)_3Cl_3$	Sc:	MVol.D1-22/3
$C_6Cl_3ErH_{24}N_6O_{12}$	$Er(C_2H_8N_2)_3(ClO_4)_3$	Sc:	MVol.D1-19/21
$C_6Cl_3EuH_6O_6$	$Eu(CH_2ClCOO)_3$	Sc:	MVol.D5-55/6
–	$Eu(CH_2ClCOO)_3 \cdot 2 H_2O$	Sc:	MVol.D5-61/3
$C_6Cl_3EuH_{20}N_4$	$Eu(NH_2C_3H_6NH_2)_2Cl_3$	Sc:	MVol.D1-26
$C_6Cl_3EuH_{24}N_6O_{12}$	$Eu(C_2H_8N_2)_3(ClO_4)_3$	Sc:	MVol.D1-19/21
$C_6Cl_3F_3HNS$	$CF_3(SH)Cl_3C_5N$	F:	PFHOrg.SVol.2-64, 67, 81
$C_6Cl_3F_4HS_3$	$SC_4H(SCF_2Cl)_2Cl$	F:	PFHOrg.SVol.2-167
$C_6Cl_3F_4H_7N_2S_2$	$i\text{-}C_3H_7N(SCF_2Cl)C(Cl)NSCF_2Cl$	F:	PFHOrg.SVol.2-144
$C_6Cl_3F_4P$	$4\text{-}ClC_6F_4PCl_2$	F:	PFHOrg.SVol.1-132/3
$C_6Cl_3F_5Ge$	$C_6F_5GeCl_3$	F:	PFHOrg.SVol.1-52/63
$C_6Cl_3F_5Si$	$2,5\text{-}Cl_2C_6F_3SiF_2Cl$	F:	PFHOrg.SVol.1-42/6
–	$4\text{-}ClC_6F_4SiFCl_2$	F:	PFHOrg.SVol.1-42/5
–	$C_6F_5SiCl_3$	F:	PFHOrg.SVol.1-43, 45, 47
$C_6Cl_3F_6H_3O_2$	$Cl_3C(CF_2)_3C(O)OCH_3$	F:	PFHOrg.9-154
$C_6Cl_3F_6NO_2$	$NC(CF_2)_3COOCCl_3$	F:	PFHOrg.9-54/5, 70
$C_6Cl_3F_6P$	$C_6F_5PFCl_3$	F:	PFHOrg.SVol.1-133/6
$C_6Cl_3F_7Si_2$	$C_6F_4(SiF_2Cl)SiFCl_2$	F:	PFHOrg.SVol.1-42/3
$C_6Cl_3F_8HO_2$	$CCl_3(CF_2)_4C(O)OH$	F:	PFHOrg.9-150
$C_6Cl_3F_{11}O_3S$	$CF_2ClCCl_2OSO_2C_4F_9$	F:	PFHOrg.SVol.3-57, 58, 74
$C_6Cl_3F_{12}HN_2$	$(CF_3)_2NN(CF_3)C_2Cl_3HCF_3$	F:	PFHOrg.8-105
$C_6Cl_3GdH_6O_6$	$Gd(CH_2ClCOO)_3$	Sc:	MVol.D5-54/6
–	$Gd(CH_2ClCOO)_3 \cdot 2 H_2O$	Sc:	MVol.D5-61/3
$C_6Cl_3GdH_{18}N_4O_{12}$	$Gd[(NH_2C_2H_4)_3N](ClO_4)_3$	Sc:	MVol.D1-34
$C_6Cl_3GdH_{18}O_3$	$GdCl_3 \cdot 3 C_2H_5OH$	Sc:	MVol.D3-19
$C_6Cl_3GdH_{24}N_6$	$Gd(C_2H_8N_2)_3Cl_3$	Sc:	MVol.D1-22/3
$C_6Cl_3GdH_{24}N_6O_{12}$	$Gd(C_2H_8N_2)_3(ClO_4)_3$	Sc:	MVol.D1-19/21
$C_6Cl_3H_4HgSb$	$ClHgC_6H_4SbCl_2$	Sb:	Org.Comp.2-105
$C_6Cl_3H_4Sb$	$ClC_6H_4SbCl_2$	Sb:	Org.Comp.2-103
$C_6Cl_3H_5NO_4U^-$	$[UO_2Cl_3(C_6H_5NO_2)]^-$	U:	SVol.D2-376
$C_6Cl_3H_5NSb$	$NH_2C_6H_3ClSbCl_2 \cdot HCl$	Sb:	Org.Comp.2-100
$C_6Cl_3H_5O_6RhS^{3-}$	$[Rh(O_2CCH_2SCH(CO_2)CH_2CO_2)Cl_3]^{3-}$	Rh:	SVol.B3-34
$C_6Cl_3H_6HoO_6$	$Ho(CH_2ClCOO)_3 \cdot 4 H_2O$	Sc:	MVol.D5-61/3
$C_6Cl_3H_6LaO_6$	$La(CH_2ClCOO)_3$	Sc:	MVol.D5-54/6, 61/3
–	$La(CH_2ClCOO)_3 \cdot H_2O$	Sc:	MVol.D5-61/3
–	$La(CH_2ClCOO)_3 \cdot 2 H_2O$	Sc:	MVol.D5-61/3
–	$La(CH_2ClCOO)_3 \cdot 3 H_2O$	Sc:	MVol.D5-61/3
$C_6Cl_3H_6LuO_6$	$Lu(CH_2ClCOO)_3 \cdot 3 H_2O$	Sc:	MVol.D5-61/3
$C_6Cl_3H_6NaO_8U$	$Na[UO_2(ClCH_2COO)_3] \cdot H_2O$	U:	SVol.C13-150

$C_6Cl_3H_6NdO_6$	Nd(CH$_2$ClCOO)$_3$	Sc:	MVol.D5-55/6, 61/3
—	Nd(CH$_2$ClCOO)$_3$ · 2 H$_2$O	Sc:	MVol.D5-61/3
—	Nd(CH$_2$ClCOO)$_3$ · 3 H$_2$O	Sc:	MVol.D5-61/3
$C_6Cl_3H_6O_6Pr$	Pr(CH$_2$ClCOO)$_3$	Sc:	MVol.D5-54/6, 61/3
—	Pr(CH$_2$ClCOO)$_3$ · H$_2$O	Sc:	MVol.D5-61/3
—	Pr(CH$_2$ClCOO)$_3$ · 2 H$_2$O	Sc:	MVol.D5-61/3
—	Pr(CH$_2$ClCOO)$_3$ · 3 H$_2$O	Sc:	MVol.D5-61/3
$C_6Cl_3H_6O_6Sc$	Sc(CH$_2$ClCOO)$_3$ · x H$_2$O	Sc:	MVol.D5-61
$C_6Cl_3H_6O_6Sm$	Sm(CH$_2$ClCOO)$_3$	Sc:	MVol.D5-54/6, 61/3
—	Sm(CH$_2$ClCOO)$_3$ · 2 H$_2$O	Sc:	MVol.D5-61/3
—	Sm(CH$_2$ClCOO)$_3$ · 3 H$_2$O	Sc:	MVol.D5-61/3
$C_6Cl_3H_6O_6Tm$	Tm(CH$_2$ClCOO)$_3$	Sc:	MVol.D5-54
—	Tm(CH$_2$ClCOO)$_3$ · 3 H$_2$O	Sc:	MVol.D5-61/3
$C_6Cl_3H_6O_6Y$	Y(CH$_2$ClCOO)$_3$	Sc:	MVol.D5-55/6, 61/3
$C_6Cl_3H_6O_6Yb$	Yb(CH$_2$ClCOO)$_3$	Sc:	MVol.D5-54
—	Yb(CH$_2$ClCOO)$_3$ · 3 H$_2$O	Sc:	MVol.D5-61/3
$C_6Cl_3H_6O_8RbU$	Rb[UO$_2$(ClCH$_2$COO)$_3$]	U:	SVol.C13-150
$C_6Cl_3H_6O_8U^-$	[UO$_2$(CH$_2$ClCOO)$_3$]$^-$	U:	SVol.D1-201
$C_6Cl_3H_6Sb$	Sb(CHCHCl)$_3$	Sb:	Org.Comp.1-39/40
$C_6Cl_3H_6Ti$	C$_6$H$_6$ · TiCl$_3$ · 2 C$_4$H$_8$O	Ti:	Org.Comp.4-76
$C_6Cl_3H_7NOSb$	(NH$_2$)(HO)C$_6$H$_3$SbCl$_2$ · HCl	Sb:	Org.Comp.2-101
$C_6Cl_3H_7NSb$	NH$_2$C$_6$H$_4$SbCl$_2$ · HCl	Sb:	Org.Comp.2-100
$C_6Cl_3H_8MnN$	(CH$_3$C$_5$H$_4$NH)MnCl$_3$		
	Spectra	Mn:	MVol.C10-40
$C_6Cl_3H_8N_4Rh$	[Rh(NHCHNCHCH)$_2$Cl$_2$]Cl	Rh:	SVol.B2-241
$C_6Cl_3H_8N_4Sc$	ScCl$_3$ · 2 C$_3$H$_4$N$_2$ · C$_2$H$_5$OH	Sc:	MVol.D1-82/3
$C_6Cl_3H_9LaN_3O_2$	La(N$_2$C$_3$H$_3$CH$_2$CH(NH$_2$)COOH)Cl$_3$ · H$_2$O	Sc:	MVol.D1-113
$C_6Cl_3H_9N_3NdO_2$	Nd(N$_2$C$_3$H$_3$CH$_2$CH(NH$_2$)COOH)Cl$_3$ · H$_2$O	Sc:	MVol.D1-113
$C_6Cl_3H_9N_3ORh$	Rh(NC$_5$H$_2$(OCH$_3$)(NH$_2$)$_2$)Cl$_3$ · 2 H$_2$O	Rh:	SVol.B2-266
$C_6Cl_3H_9N_3Rh$	RhCl$_3$(CH$_3$CN)$_3$	Rh:	SVol.B2-119/20
—	RhCl$_3$(CD$_3$CN)$_3$	Rh:	SVol.B2-120
$C_6Cl_3H_9N_3Y$	YCl$_3$ · 3 CH$_3$CN	Sc:	MVol.D1-102
$C_6Cl_3H_{11}NPr$	PrCl$_3$ · (CH$_3$)$_2$NCH$_2$CCCH$_3$ · 3 H$_2$O	Sc:	MVol.D1-18
$C_6Cl_3H_{12}LaO_{1.5}$	LaCl$_3$ · 1.5 C$_4$H$_8$O	Sc:	MVol.D3-272
$C_6Cl_3H_{12}N_4NdO_4$	NdCl$_3$ · 2 H$_2$NC(O)NHC(O)CH$_3$ · 6 H$_2$O	Sc:	MVol.D2-241
$C_6Cl_3H_{12}N_4S_2Sm$	SmCl$_3$ · 2 CH$_2$CHNHC(S)NH$_2$	Sc:	MVol.D4-85/6
$C_6Cl_3H_{12}O_2Sb$	(CH$_3$)$_4$SbO$_2$CCCl$_3$	Sb:	Org.Comp.3-126
$C_6Cl_3H_{12}O_2Sc$	ScCl$_3$ · 2 C$_3$H$_5$OH	Sc:	MVol.D3-29
$C_6Cl_3H_{13}MnN_2$	[Mn(HN(C$_2$H$_4$)$_3$N)Cl$_3$]	Mn:	MVol.D4-51
$C_6Cl_3H_{13}OSSn$	(CH$_3$)$_3$SnOCH(CCl$_3$)SCH$_3$	Sn:	Org.Comp.9-26
		Sn:	Org.Comp.11-59, 61
$C_6Cl_3H_{14}N_2P_2Rh$	[Rh(PClNC$_3$H$_7$-i)$_2$Cl]$_n$	Rh:	SVol.B3-93
$C_6Cl_3H_{14}N_2Rh$	Rh(NH$_2$C$_3$H$_5$)$_2$Cl$_3$	Rh:	SVol.B2-160
$C_6Cl_3H_{14}O_2U$	U(OC$_3$H$_7$-i)$_2$Cl$_3$	U:	SVol.C13-72/3
$C_6Cl_3H_{15}LaN_3O_6$	La(NH$_2$CH$_2$COOH)$_3$Cl$_3$ · 3 H$_2$O	Sc:	MVol.D1-104
$C_6Cl_3H_{15}NSc$	ScCl$_3$ · (C$_2$H$_5$)$_3$N	Sc:	MVol.D1-16/7
$C_6Cl_3H_{15}N_3NdO_6$	Nd(NH$_2$CH$_2$COOH)$_3$Cl$_3$ · 3 H$_2$O	Sc:	MVol.D1-104
$C_6Cl_3H_{15}N_3O_6Pr$	Pr(NH$_2$CH$_2$COOH)$_3$Cl$_3$ · 3 H$_2$O	Sc:	MVol.D1-104
$C_6Cl_3H_{15}N_3O_6Rh$	RhCl$_3$(NH$_2$CH$_2$CO$_2$H)$_3$	Rh:	SVol.B2-67
$C_6Cl_3H_{15}N_3O_6Sm$	Sm(NH$_2$CH$_2$COOH)$_3$Cl$_3$ · 3 H$_2$O	Sc:	MVol.D1-104

$C_6Cl_3H_{15}N_3O_6Y$	$Y(NH_2CH_2COOH)_3Cl_3 \cdot 3 H_2O$	Sc:	MVol.D1-104
$C_6Cl_3H_{15}N_3Rh$	$Rh(NHCH_2CH_2)_3Cl_3$	Rh:	SVol.B2-159
$C_6Cl_3H_{15}N_3RhS_3$	$Rh(CH_3CSNH_2)_3Cl_3$	Rh:	SVol.B3-52
$C_6Cl_3H_{15}N_9NdO_6$	$NdCl_3 \cdot 3 (H_2NCO)_2NH \cdot 3 H_2O$	Sc:	MVol.D2-241
$C_6Cl_3H_{15}N_9O_6Pr$	$PrCl_3 \cdot 3 (H_2NCO)_2NH \cdot 3 H_2O$	Sc:	MVol.D2-241
$C_6Cl_3H_{15}NdO_4P$	$NdCl_3 \cdot (C_2H_5O)_3PO$	Sc:	MVol.D4-187
$C_6Cl_3H_{15}O_{1.5}Yb$	$YbCl_3 \cdot 1.5 t-C_4H_9OH$	Sc:	MVol.D3-28
$C_6Cl_3H_{15}O_4PPr$	$PrCl_3 \cdot (C_2H_5O)_3PO$	Sc:	MVol.D4-187
$C_6Cl_3H_{15}O_4PSm$	$SmCl_3 \cdot (C_2H_5O)_3PO$	Sc:	MVol.D4-187
$C_6Cl_3H_{15}Rh_2S_3$	$[Rh_2(SC_2H_5)_3Cl_3]_n$	Rh:	SVol.B3-3/5
$C_6Cl_3H_{16}LaO_2$	$LaCl_3 \cdot 2 i-C_3H_7OH \cdot H_2O$	Sc:	MVol.D3-21
$C_6Cl_3H_{16}N_2Rh$	$Rh(NH_2C(CH_3)_2C(CH_3)_2NH_2)Cl_3$	Rh:	SVol.B2-189
$C_6Cl_3H_{16}N_{10}Rh$	$[Rh((N_5C_2H_6)_2C_2H_4)Cl_2]Cl \cdot 2 H_2O$	Rh:	SVol.B2-224
$C_6Cl_3H_{16}NdO_2$	$NdCl_3 \cdot 2 n-C_3H_7OH$	Sc:	MVol.D3-20
–	$NdCl_3 \cdot 2 i-C_3H_7OH \cdot H_2O$	Sc:	MVol.D3-21
$C_6Cl_3H_{16}O_2Pr$	$PrCl_3 \cdot 2 n-C_3H_7OH$	Sc:	MVol.D3-20
–	$PrCl_3 \cdot 2 i-C_3H_7OH \cdot H_2O$	Sc:	MVol.D3-21
$C_6Cl_3H_{16}O_2Sc$	$ScCl_3 \cdot 2 n-C_3H_7OH$	Sc:	MVol.D3-20
$C_6Cl_3H_{16}O_2Sm$	$SmCl_3 \cdot 2 n-C_3H_7OH \cdot H_2O$	Sc:	MVol.D3-20
–	$SmCl_3 \cdot 2 i-C_3H_7OH \cdot H_2O$	Sc:	MVol.D3-21
$C_6Cl_3H_{16}SbSi_2$	$[(CH_3)_2SiClCH_2]_2SbCl$	Sb:	Org.Comp.2-9
$C_6Cl_3H_{17}N_5O_2Os$ . . .	$[OsCl \cdot (C_6H_5NO_2)(NH_3)_4]Cl_2$	Os:	SVol.1-249
$C_6Cl_3H_{18}LaO_3$	$LaCl_3 \cdot 3 C_2H_5OH$	Sc:	MVol.D3-18/9
$C_6Cl_3H_{18}N_2Rh$	$Rh(NH_2C_3H_7-n)_2Cl_3$	Rh:	SVol.B2-160
$C_6Cl_3H_{18}N_2Sc$	$ScCl_3 \cdot 2 (i-C_3H_7)NH_2$	Sc:	MVol.D1-16/7
$C_6Cl_3H_{18}N_4O_4Rh$ . . .	$[Rh((NH_2C_2H_4NH)_2C_2H_4)Cl_2]ClO_4$	Rh:	SVol.B2-196/7
$C_6Cl_3H_{18}N_4O_{12}Pr$ . . .	$Pr[(NH_2C_2H_4)_3N](ClO_4)_3$	Sc:	MVol.D1-34
$C_6Cl_3H_{18}N_4Rh$	$[Rh(N(C_2H_4NH_2)_3)Cl_2]Cl$	Rh:	SVol.B2-199
–	$[Rh((NH_2C_2H_4NH)_2C_2H_4)Cl_2]Cl$	Rh:	SVol.B2-196
–	$[Rh((NH_2C_2H_4NH)_2C_2H_4)Cl_2]Cl \cdot H_2O$	Rh:	SVol.B2-196
–	$[Rh((NH_2C_2H_4NH)_2C_2H_4)Cl_2]Cl \cdot 2 H_2O$ . . .	Rh:	SVol.B2-196
$C_6Cl_3H_{18}NdO_3$	$NdCl_3 \cdot 3 C_2H_5OH$	Sc:	MVol.D3-18/9
$C_6Cl_3H_{18}ORhS_3$	$RhCl_3[(CH_3)_2SO][(CH_3)_2S]_2$	Rh:	SVol.B1-147
$C_6Cl_3H_{18}O_2RhS_3$	$RhCl_3[(CH_3)_2SO]_2[(CH_3)_2S]$	Rh:	SVol.B1-147
$C_6Cl_3H_{18}O_3OsS_3$	$Os(SO(CH_3)_2)_3Cl_3$	Os:	SVol.1-167
$C_6Cl_3H_{18}O_3Pr$	$PrCl_3 \cdot 3 C_2H_5OH$	Sc:	MVol.D3-18/9
$C_6Cl_3H_{18}O_3RhS_3$	$RhCl_3[(CH_3)_2SO]_3$	Rh:	SVol.B1-144/5
$C_6Cl_3H_{18}O_3S_3Th^+$ . .	$ThCl_3^+ \cdot 3 (CH_3)_2SO$	Th:	SVol.D2-105
$C_6Cl_3H_{18}O_3S_3Y$	$YCl_3 \cdot 3 (CH_3)_2SO$	Sc:	MVol.D4-10/2
$C_6Cl_3H_{18}O_3Sc$	$ScCl_3 \cdot 3 C_2H_5OH$	Sc:	MVol.D3-18
$C_6Cl_3H_{18}O_3Sm$	$SmCl_3 \cdot 3 C_2H_5OH$	Sc:	MVol.D3-18/9
$C_6Cl_3H_{18}O_3Yb$	$YbCl_3 \cdot 3 C_2H_5OH$	Sc:	MVol.D3-19
$C_6Cl_3H_{18}RhS_3$	$Rh[S(CH_3)_2]_3Cl_3$	Rh:	SVol.B3-3/4
$C_6Cl_3H_{18}RhSe_3$	$Rh(Se(CH_3)_2)_3Cl_3$	Rh:	SVol.B3-59
$C_6Cl_3H_{18}RhTe_3$	$Rh[Te(CH_3)_2]_3Cl_3$	Rh:	SVol.B3-64
$C_6Cl_3H_{19}N_7O_{12}Rh$ . . .	$[Rh(NH_3)_5(NC_5H_4CN)](ClO_4)_3$	Rh:	SVol.B2-140
$C_6Cl_3H_{20}N_4O_4Rh$ . . .	$[Rh(NH_2(CH_2)_3NH_2)_2Cl_2]ClO_4$	Rh:	SVol.B2-193
–	$[Rh(NH_2C_2H_4NHCH_3)_2Cl_2]ClO_4$	Rh:	SVol.B2-191/2
$C_6Cl_3H_{20}N_4Rh$	$[Rh(NH_2CH_2CH(CH_3)NH_2)_2Cl_2]Cl$	Rh:	SVol.B2-188
$C_6Cl_3H_{20}N_4Yb$	$Yb(NH_2C_3H_6NH_2)_2Cl_3$	Sc:	MVol.D1-26

$C_6Cl_3H_{21}MnN_3$	$Mn((CH_3)_2NH)_3Cl_3$	Mn: MVol.D3-23
–	$Mn(C_2H_5NH_2)_3Cl_3$	Mn: MVol.D3-23
$C_6Cl_3H_{21}N_3O_3Rh$	$RhCl_3(NH_2C_2H_4OH)_3$	Rh: SVol.B2-319
$C_6Cl_3H_{21}N_3Sc$	$ScCl_3 \cdot 3\ (CH_3)_2NH$	Sc: MVol.D1-16/7
–	$ScCl_3 \cdot 3\ C_2H_5NH_2$	Sc: MVol.D1-16/7
$C_6Cl_3H_{21}N_7OOs$	$[Os(C_6H_6N_2O)(NH_3)_5]Cl_3$	Os: SVol.1-248
$C_6Cl_3H_{21}N_{15}Rh$	$[Rh((NH_2CNH)_2NH)_3]Cl_3$	Rh: SVol.B2-220/1
$C_6Cl_3H_{22}N_6O_{12}Rh$	$[Rh(NH_3)_5(NC_5H_4CH_3)](ClO_4)_3$	Rh: SVol.B2-140
$C_6Cl_3H_{24}HoN_6O_{12}$	$Ho(C_2H_8N_2)_3(ClO_4)_3$	Sc: MVol.D1-19/21
$C_6Cl_3H_{24}HoN_{12}O_6$	$HoCl_3 \cdot 6\ OC(NH_2)_2$	Sc: MVol.D2-220
$C_6Cl_3H_{24}LaN_6O_{12}$	$La(C_2H_8N_2)_3(ClO_4)_3$	Sc: MVol.D1-19/21
$C_6Cl_3H_{24}LaN_{12}O_6$	$LaCl_3 \cdot 6\ OC(NH_2)_2$	Sc: MVol.D2-220
$C_6Cl_3H_{24}LuN_6O_{12}$	$Lu(C_2H_8N_2)_3(ClO_4)_3$	Sc: MVol.D1-19/21
$C_6Cl_3H_{24}MnN_6O_{12}$	$Mn(C_2H_8N_2)_3(ClO_4)_3$	Mn: MVol.D3-42
$C_6Cl_3H_{24}MnN_{12}O_{18}$	$Mn(OC(NH_2)_2)_6(ClO_4)_3$	Mn: MVol.D5-140/1
$C_6Cl_3H_{24}N_6NdO_{12}$	$Nd(C_2H_8N_2)_3(ClO_4)_3$	Sc: MVol.D1-19/21
$C_6Cl_3H_{24}N_6O_{12}Pr$	$Pr(C_2H_8N_2)_3(ClO_4)_3$	Sc: MVol.D1-19/21
$C_6Cl_3H_{24}N_6O_{12}Rh$	$[Rh(NH_2C_2H_4NH_2)_3](ClO_4)_3 \cdot H_2O$	Rh: SVol.B2-162/5
$C_6Cl_3H_{24}N_6O_{12}Sm$	$Sm(C_2H_8N_2)_3(ClO_4)_3$	Sc: MVol.D1-19/21
$C_6Cl_3H_{24}N_6O_{12}Tb$	$Tb(C_2H_8N_2)_3(ClO_4)_3$	Sc: MVol.D1-19/21
$C_6Cl_3H_{24}N_6O_{12}Y$	$Y(C_2H_8N_2)_3(ClO_4)_3$	Sc: MVol.D1-19/21
$C_6Cl_3H_{24}N_6O_{12}Yb$	$Yb(C_2H_8N_2)_3(ClO_4)_3$	Sc: MVol.D1-19/21
$C_6Cl_3H_{24}N_6Os$	$[Os(C_2H_4(NH_2)_2)_3]Cl_3 \cdot 3\ H_2O$	Os: SVol.1-250/1
$C_6Cl_3H_{24}N_6Rh$	$[Rh(NH_2C_2H_4NH_2)_3]Cl_3$	Rh: SVol.B2-164/6
–	$[Rh(NH_2C_2H_4NH_2)_3]Cl_3 \cdot H_2O$	Rh: SVol.B2-163/5
–	$[Rh(NH_2C_2H_4NH_2)_3]Cl_3 \cdot 2\ H_2O$	Rh: SVol.B2-163
–	$[Rh(NH_2C_2H_4NH_2)_3]Cl_3 \cdot 2.5\ H_2O$	Rh: SVol.B2-166
–	$[Rh(NH_2C_2H_4NH_2)_3]Cl_3 \cdot 3\ H_2O$	Rh: SVol.B2-162/4
–	$[Rh(NH_2C_2H_4NH_2)_3]Cl_3 \cdot x\ H_2O$	Rh: SVol.B2-164
$C_6Cl_3H_{24}N_6Sc$	$[Sc(C_2H_8N_2)_3]Cl_3$	Sc: MVol.D1-23
$C_6Cl_3H_{24}N_{12}NdO_6$	$NdCl_3 \cdot 6\ OC(NH_2)_2$	Sc: MVol.D2-220
$C_6Cl_3H_{24}N_{12}O_6Pr$	$PrCl_3 \cdot 6\ OC(NH_2)_2$	Sc: MVol.D2-220
$C_6Cl_3H_{24}N_{12}O_6Sm$	$SmCl_3 \cdot 6\ OC(NH_2)_2$	Sc: MVol.D2-220
$C_6Cl_3H_{24}N_{12}O_{18}Pr$	$Pr(ClO_4)_3 \cdot 6\ OC(NH_2)_2$	Sc: MVol.D2-220
$C_6Cl_3H_{24}N_{12}O_{18}Sc$	$Sc(ClO_4)_3 \cdot 6\ OC(NH_2)_2$	Sc: MVol.D2-219
$C_6Cl_3H_{24}N_{12}O_{18}Sm$	$Sm(ClO_4)_3 \cdot 6\ OC(NH_2)_2$	Sc: MVol.D2-220
$C_6Cl_3H_{24}N_{12}OsS_6$	$[Os(SC(NH_2)_2)_6]Cl_3 \cdot H_2O$	Os: SVol.1-288/9
$C_6Cl_3H_{24}N_{12}RhS_6$	$[Rh(SC(NH_2)_2)_6]Cl_3$	Rh: SVol.B3-9
$C_6Cl_3H_{25}N_{12}OOsS_6$	$[Os(SC(NH_2)_2)_6](OH)Cl_3$	Os: SVol.1-288
C_6Cl_4CuH	$Cl_4C_6HCu \cdot x\ LiHal$	Cu: Org.Comp.1-61, 112, 117
–	$Cl_4C_6HCu \cdot x\ MgBrCl$	Cu: Org.Comp.1-289
–	$Cl_4C_6HCu \cdot x\ MgHal_2$	Cu: Org.Comp.1-217, 290
$C_6Cl_4FNO_2$	$NO_2C_6Cl_4F$	F: PFHOrg.8-13/4, 27
$C_6Cl_4F_4Si$	$2,4-Cl_2C_6F_3SiFCl_2$	F: PFHOrg.SVol.1-42/3
–	$4-ClC_6F_4SiCl_3$	F: PFHOrg.SVol.1-43, 45
$C_6Cl_4F_5P$	$C_6F_5PCl_4$	F: PFHOrg.SVol.1-133/5
$C_6Cl_4F_6Si_2$	$C_6F_4(SiFCl_2)_2$	F: PFHOrg.SVol.1-42/3
$C_6Cl_4H_3N_2Sb$	$[C_6H_3Cl(N_2)SbCl_2]Cl$	Sb: Org.Comp.2-104
$C_6Cl_4H_5NO_3U$	$UCl_4 \cdot HOOCC_5H_4NO$	U: SVol.D2-61/2
$C_6Cl_4H_6NSb$	$NH_2C_6H_3ClSbCl_2 \cdot HCl$	Sb: Org.Comp.2-100

C$_6$Cl$_4$H$_6$Ti	TiCl$_4$ · C$_6$H$_6$	Ti:	Org.Comp.4-78
C$_6$Cl$_4$H$_7$NOTh	ThCl$_4$ · CH$_3$C$_5$H$_4$NO	Th:	SVol.E-69
C$_6$Cl$_4$H$_7$NTh	ThCl$_4$ · CH$_3$C$_5$H$_4$N	Th:	SVol.E-10, 12
C$_6$Cl$_4$H$_8$O$_4$Sn	(CH$_3$)$_2$Sn(OOCCHCl$_2$)$_2$	Sn:	Org.Comp.14-87
C$_6$Cl$_4$H$_{12}$NPr	PrCl$_3$ · (CH$_3$)$_2$NCH$_2$(CH)$_2$CH$_2$Cl · H$_2$O	Sc:	MVol.D1-18
C$_6$Cl$_4$H$_{12}$N$_6$S$_2$	[ClS$_2$N$_4$C$_2$(N(CH$_3$)$_2$)$_2$]Cl$_3$	S:	S-N Comp.4-143/4
C$_6$Cl$_4$H$_{12}$O$_2$U	[UCl$_4$((CH$_3$)$_2$CO)$_2$].	U:	SVol.D2-50
C$_6$Cl$_4$H$_{12}$O$_3$U^{2-}	[UO$_2$Cl$_4$(C$_6$H$_{11}$OH)]$^{2-}$	U:	SVol.D2-388
C$_6$Cl$_4$H$_{14}$N$_2$O$_2$Th	ThCl$_4$ · 2 HCON(CH$_3$)$_2$	Th:	SVol.E-45
C$_6$Cl$_4$H$_{14}$O$_3$U^{2-}	[U(OH)$_2$(Cl$_4$C$_6$H$_{11}$OH)]$^{2-}$	U:	SVol.D2-388
C$_6$Cl$_4$H$_{18}$O$_2$P$_2$Th	ThCl$_4$ · 2 (CH$_3$)$_3$PO	Th:	SVol.E-77, 78
C$_6$Cl$_4$H$_{18}$O$_3$S$_3$Th	ThCl$_4$ · 3 (CH$_3$)$_2$SO	Th:	SVol.D2-105
		Th:	SVol.E-100
C$_6$Cl$_4$H$_{18}$O$_3$S$_3$U	UCl$_4$ · 3 (CH$_3$)$_2$SO	U:	SVol.A5-106, 196
C$_6$Cl$_4$H$_{18}$O$_6$OsP$_2$	Os[P(OCH$_3$)$_3$]$_2$Cl$_4$	Os:	SVol.1-333
C$_6$Cl$_4$H$_{19}$N$_4$Rh	[Rh((N(C$_2$H$_4$NH$_2$)$_3$)H)Cl$_3$]Cl	Rh:	SVol.B2-199
C$_6$Cl$_4$H$_{20}$MnN$_2$	(C$_3$H$_7$NH$_3$)$_2$MnCl$_4$		
	Spectra	Mn:	MVol.C10-38
C$_6$Cl$_4$H$_{24}$N$_{12}$O$_6$Th	ThCl$_4$ · 6 (NH$_2$)$_2$CO · 2 H$_2$O	Th:	SVol.E-37, 39/40
C$_6$Cl$_4$H$_{25}$N$_6$Rh	[Rh(NH$_2$C$_2$H$_4$NH$_2$)$_2$(NH$_2$C$_2$H$_4$NH$_3$)Cl]Cl$_3$ · 2 H$_2$O		
		Rh:	SVol.B2-175
C$_6$Cl$_{4.5}$H$_{36.5}$N$_{22}$O$_{12}$Rh$_4$			
	Rh$_4$Cl$_{4.5}$(O$_2$CN$_2$H$_3$)$_6$(N$_2$H$_3$)$_{1.5}$(N$_2$H$_4$)$_{3.5}$	Rh:	SVol.B2-62
C$_6$Cl$_5$Cu	C$_6$Cl$_5$Cu · x C$_5$H$_5$N · y LiI	Cu:	Org.Comp.1-405
–	C$_6$Cl$_5$Cu · x LiBr	Cu:	Org.Comp.1-114, 116/7
–	C$_6$Cl$_5$Cu · x LiCl	Cu:	Org.Comp.1-61, 91,
			109/14
–	C$_6$Cl$_5$Cu · x LiHal	Cu:	Org.Comp.1-92/3, 118/21
–	C$_6$Cl$_5$Cu · x LiI	Cu:	Org.Comp.1-61, 91, 112,
			117
–	C$_6$Cl$_5$Cu · x MgClI	Cu:	Org.Comp.1-217, 291/3
–	C$_6$Cl$_5$Cu · x MgCl$_2$	Cu:	Org.Comp.1-217, 262,
			291/2
C$_6$Cl$_5$F$_3$Si	2,5-Cl$_2$C$_6$F$_3$SiCl$_3$	F:	PFHOrg.SVol.1-43, 45
C$_6$Cl$_5$F$_5$O$_4$S	CCl$_3$C(O)OC(CCl$_2$)CF(CF$_3$)SO$_2$F	F:	PFHOrg.SVol.3-157, 167
C$_6$Cl$_5$F$_5$Si$_2$	C$_6$F$_4$(SiFCl$_2$)SiCl$_3$	F:	PFHOrg.SVol.1-42/3
C$_6$Cl$_5$H$_6$Sb	(ClCHCH)$_3$SbCl$_2$	Sb:	Org.Comp.4-28/30
C$_6$Cl$_5$H$_{15}$O$_4$PSc^{2-}	[ScCl$_5$((C$_2$H$_5$O)$_3$PO)]$^{2-}$	Sc:	MVol.D4-187
C$_6$Cl$_5$H$_{15}$SbTi	(C$_2$H$_5$)$_3$SbCl$_2$ · TiCl$_3$	Sb:	Org.Comp.1-20
C$_6$Cl$_5$H$_{20}$N$_4$Rh	[Rh(H$_2$((NH$_2$C$_2$H$_4$NH)$_2$C$_2$H$_4$))Cl$_4$]Cl	Rh:	SVol.B2-197
C$_6$Cl$_5$H$_{42}$N$_{20}$O$_6$Rh$_4$	[Rh$_4$Cl$_5$(O$_2$CCH$_2$NH$_2$)$_3$(N$_2$H$_3$)$_4$(N$_2$H$_4$)$_{4.5}$]	Rh:	SVol.B2-62
C$_6$Cl$_6$Dy$_2$H$_{10}$O$_4$	2 DyCl$_3$ · (COOC$_2$H$_5$)$_2$	Sc:	MVol.D5-360/1
C$_6$Cl$_6$ErH$_3$O$_6$	Er(CHCl$_2$COO)$_3$ · 3 H$_2$O	Sc:	MVol.D5-63
C$_6$Cl$_6$F$_4$Si$_2$	C$_6$F$_2$Cl$_2$(SiFCl$_2$)$_2$	F:	PFHOrg.SVol.1-43, 46
–	C$_6$F$_3$Cl(SiFCl$_2$)SiCl$_3$	F:	PFHOrg.SVol.1-43, 46
–	C$_6$F$_4$(SiCl$_3$)$_2$	F:	PFHOrg.SVol.1-43
C$_6$Cl$_6$F$_5$HO$_3$S	C$_6$F$_5$Cl$_6$SO$_3$H	F:	PFHOrg.SVol.3-51
C$_6$Cl$_6$F$_8$N$_2$O$_2$P$_2$	[Cl$_3$PNC(O)CF$_2$CF$_2$]$_2$	F:	PFHOrg.9-40, 42, 45
C$_6$Cl$_6$FeH$_6$OSi$_2$	C$_5$H$_5$Fe(CO)(SiCl$_3$)$_2$H	Fe:	Org.Comp.B11-71/3
–	C$_5$H$_5$Fe(CO)(SiCl$_3$)$_2$D	Fe:	Org.Comp.B11-72

$C_6Cl_6GdH_3O_6$	$Gd(CHCl_2COO)_3$	Sc:	MVol.D5-54
$C_6Cl_6Gd_2H_{10}O_4$	$2\ GdCl_3 \cdot (COOC_2H_5)_2$.	Sc:	MVol.D5-360/1
$C_6Cl_6H_3HoO_6$	$Ho(CHCl_2COO)_3$	Sc:	MVol.D5-63
–	$Ho(CHCl_2COO)_3 \cdot 3\ H_2O$	Sc:	MVol.D5-63
$C_6Cl_6H_3LaO_6$.	$La(CHCl_2COO)_3$	Sc:	MVol.D5-54/6, 63
$C_6Cl_6H_3NdO_6$	$Nd(CHCl_2COO)_3$	Sc:	MVol.D5-55/6, 63
–	$Nd(CHCl_2COO)_3 \cdot 3\ H_2O$	Sc:	MVol.D5-63
$C_6Cl_6H_3O_6Pr$	$Pr(CHCl_2COO)_3$.	Sc:	MVol.D5-54/6, 63
–	$Pr(CHCl_2COO)_3 \cdot 3\ H_2O$	Sc:	MVol.D5-63
$C_6Cl_6H_3O_6Sc$.	$Sc(CHCl_2COO)_3$	Sc:	MVol.D5-63
–	$Sc(CHCl_2COO)_3 \cdot H_2O$.	Sc:	MVol.D5-63
$C_6Cl_6H_3O_6Sm$	$Sm(CHCl_2COO)_3$	Sc:	MVol.D5-54/6, 63
–	$Sm(CHCl_2COO)_3 \cdot 3\ H_2O$.	Sc:	MVol.D5-63
$C_6Cl_6H_3O_6Tm$	$Tm(CHCl_2COO)_3$	Sc:	MVol.D5-54
$C_6Cl_6H_3O_6Y$	$Y(CHCl_2COO)_3$	Sc:	MVol.D5-54
–	$Y(CHCl_2COO)_3 \cdot x\ H_2O$	Sc:	MVol.D5-63
$C_6Cl_6H_6O_4Sn$	$(CH_3)_2Sn(OOCCCl_3)_2$	Sn:	Org.Comp.14-88
$C_6Cl_6H_8MnO_6$	$Mn(CCl_3COO)_2 \cdot 2\ CH_3OH$	Mn:	MVol.D2-63
$C_6Cl_6H_{10}La_2O_4$	$2\ LaCl_3 \cdot (COOC_2H_5)_2$	Sc:	MVol.D5-360/1
$C_6Cl_6H_{10}O_4Sm_2$	$2\ SmCl_3 \cdot (COOC_2H_5)_2$	Sc:	MVol.D5-360/1
$C_6Cl_6H_{24}I_3N_6Os$	$[Os(C_2H_4(NH_2)_2)_3](Cl_2I)_3$	Os:	SVol.1-250/1
$C_6Cl_6H_{24}N_3Rh$.	$[(CH_3)_2NH_2]_3[RhCl_6]$	Rh:	SVol.B1-104
$C_6Cl_7F_4HO_3S$	$C_6F_4Cl_7SO_3H$	F:	PFHOrg.SVol.3-51
$C_6Cl_7H_{18}O_3RhS_3Sn_2$	$Rh[SnCl_2 \cdot OS(CH_3)_2]_2[OS(CH_3)_2]Cl_3$	Rh:	SVol.B3-245
$C_6Cl_7H_{24}N_{12}OsS_6Sn_2$			
	$[Os(SC(NH_2)_2)_6](SnCl_3)_2Cl$	Os:	SVol.1-289, 344
$C_6Cl_8F_3HO_3S$	$C_6F_3Cl_8SO_3H$	F:	PFHOrg.SVol.3-51
$C_6Cl_8H_{23}N_{15}Rh_2$	$((NH_2CNH)_2NH \cdot H_2)[Rh((NH_2C(NH))_2NH)Cl_4]_2$		
		Rh:	SVol.B2-221
$C_6Cl_8H_{26}N_6Rh_2$	$(NH_3C_2H_4NH_3)[Rh(NH_2C_2H_4NH_2)Cl_4]_2$	Rh:	SVol.B2-184
$C_6Cl_9CsO_8U$	$Cs[UO_2(CCl_3COO)_3]$	U:	SVol.C13-151
$C_6Cl_9DyO_6$	$Dy(CCl_3COO)_3 \cdot 3\ H_2O$	Sc:	MVol.D5-64
$C_6Cl_9ErO_6$	$Er(CCl_3COO)_3 \cdot 3\ H_2O$.	Sc:	MVol.D5-64
$C_6Cl_9EuO_6$	$Eu(CCl_3COO)_3$.	Sc:	MVol.D5-55/6, 64
–	$Eu(CCl_3COO)_3 \cdot 2\ H_2O$	Sc:	MVol.D5-64
–	$Eu(CCl_3COO)_3 \cdot 3\ H_2O$	Sc:	MVol.D5-64
$C_6Cl_9F_2HO_3S$	$C_6F_2Cl_9SO_3H$	F:	PFHOrg.SVol.3-51
$C_6Cl_9GdO_6$	$Gd(CCl_3COO)_3 \cdot 3\ H_2O$	Sc:	MVol.D5-64
$C_6Cl_9H_4NO_8U$	$NH_4[UO_2(CCl_3COO)_3]$	U:	SVol.C13-151/3
		U:	SVol.D2-249
$C_6Cl_9HoO_6$	$Ho(CCl_3COO)_3 \cdot 3\ H_2O$	Sc:	MVol.D5-64
$C_6Cl_9KO_8U$	$K[UO_2(CCl_3COO)_3]$	U:	SVol.C13-151/3
$C_6Cl_9LaO_6$	$La(CCl_3COO)_3$.	Sc:	MVol.D5-54/6, 63/5
–	$La(CCl_3COO)_3 \cdot 3\ H_2O$	Sc:	MVol.D5-64
–	$La(CCl_3COO)_3 \cdot 5\ H_2O$	Sc:	MVol.D5-64
$C_6Cl_9LiO_8U$	$Li[UO_2(CCl_3COO)_3]$	U:	SVol.C13-151
$C_6Cl_9LuO_6$	$Lu(CCl_3COO)_3 \cdot 3\ H_2O$	Sc:	MVol.D5-64
$C_6Cl_9NaO_8U$	$Na[UO_2(CCl_3COO)_3]$	U:	SVol.C13-151/3
$C_6Cl_9NdO_6$	$Nd(CCl_3COO)_3$	Sc:	MVol.D5-55/6, 63/5
–	$Nd(CCl_3COO)_3 \cdot 3\ H_2O$	Sc:	MVol.D5-64

$C_6Cl_9O_6Pr$	$Pr(CCl_3COO)_3$	Sc:	MVol.D5-55/6, 63/5
−	$Pr(CCl_3COO)_3 \cdot 3\ H_2O$	Sc:	MVol.D5-64
$C_6Cl_9O_6Sc$	$Sc(CCl_3COO)_3 \cdot x\ H_2O$	Sc:	MVol.D5-64
$C_6Cl_9O_6Sm$	$Sm(CCl_3COO)_3$	Sc:	MVol.D5-63/4
−	$Sm(CCl_3COO)_3 \cdot 3\ H_2O$	Sc:	MVol.D5-64
$C_6Cl_9O_6Tb$	$Tb(CCl_3COO)_3 \cdot 3\ H_2O$	Sc:	MVol.D5-64
$C_6Cl_9O_6Tm$	$Tm(CCl_3COO)_3 \cdot 3\ H_2O$	Sc:	MVol.D5-64
$C_6Cl_9O_6Y$	$Y(CCl_3COO)_3 \cdot 3\ H_2O$	Sc:	MVol.D5-64
$C_6Cl_9O_6Yb$	$Yb(CCl_3COO)_3 \cdot 3\ H_2O$	Sc:	MVol.D5-64
$C_6Cl_9O_8RbU$	$Rb[UO_2(CCl_3COO)_3]$	U:	SVol.C13-151
$C_6Cl_9O_8TlU$	$Tl[UO_2(CCl_3COO)_3]$	U:	SVol.C13-151
$C_6Cl_{10}FHO_3S$	$C_6FCl_{10}SO_3H$	F:	PFHOrg.SVol.3-51
$C_6Cl_{12}H_6MnN_6O_3Sb_2$	$[Mn(O(NCH)_2)_3][SbCl_6]_2$	Mn:	MVol.D4-217/8
$C_6Cl_{12}H_6Ti_3$	$3\ TiCl_4 \cdot C_6H_6$	Ti:	Org.Comp.4-78
$C_6Cl_{12}H_{16}NRhSn_3$ . . .	$[NH(C_2H_5)_3][Rh(SnCl_3)_3Cl_3]$	Rh:	SVol.B3-243
$C_6CoF_{21}NOP_3$	$[Co(NO)(FP(CF_3)_2)_3]$	F:	PFHOrg.SVol.1-130
$C_6CoFr_3N_6$	$Fr_3Co(CN)_6$	Fr:	MVol.-117
$C_6CoH_{18}MnN_{12}$	$[Co(NH_3)_6][Mn(CN)_6]$	Mn:	MVol.D2-247
$C_6CoK_3N_6$	$K_3[Co(CN)_6]$ solid solutions		
	$K_3[Co(CN)_6]-K_3[Mn(CN)_6]$	Mn:	MVol.D2-222
$C_6CrH_{18}MnN_{12}$	$[Cr(NH_3)_6][Mn(CN)_6]$	Mn:	MVol.D2-247
$C_6CrH_{30}MnN_6O_4$	$Mn(CH_3NH_2)_6CrO_4$	Mn:	MVol.D3-22/3
$C_6Cr_2H_4NO_7Th$	$ThCr_2O_7 \cdot 0.5\ C_{12}H_8N_2$	Th:	SVol.E-17, 21
C_6Cr_{23}	$Cr_{23}C_6$ solid solutions		
	$(Cr,Mn)_{23}C_6$	Mn:	MVol.C7-167/9
−	$Cr_{23}C_6$ systems		
	$Cr_{23}C_6-UC$	U:	SVol.C12-190/1
	$Cr_{23}C_6-UFe_2$	U:	SVol.C12-214
C_6CsF_5	CsC_6F_5	F:	PFHOrg.SVol.1-18
C_6CsF_5S	$CsSC_6F_5$	F:	PFHOrg.SVol.2-263
C_6CsF_{13}	$(CF_3)_2(C_3F_7)CCs$	F:	PFHOrg.SVol.1-17
$C_6CsH_8MnO_{10}$	$Cs[Mn(CH_2(COC)_2)_2(H_2O)_2]$	Mn:	MVol.D2-133, 135
$C_6CsH_9O_8Os$	$Cs[OsO_2(CH_3COO)_3]$	Os:	SVol.1-184
$C_6CsH_9O_8U$	$CsUO_2(CH_3COO)_3$	U:	SVol.A5-125, 220
		U:	SVol.A6-51, 64
		U:	SVol.C13-127, 133
$C_6CsH_{18}O_6U$	$Cs[U(OCH_3)_6]$	U:	SVol.C13-74
$C_6Cs_2H_4N_2O_{10}S_2U$ . .	$Cs_2[U(C_2O_4)_2(NCS)_2(H_2O)_2]$	U:	SVol.C13-230/2
$C_6Cs_2H_6O_{11}U$	$Cs_2[UO_2(CH_2(COO)_2)_2(H_2O)] \cdot 4\ H_2O$	U:	SVol.C13-238/40
$C_6Cs_2LiMnN_6$	$Cs_2Li[Mn(CN)_6]$	Mn:	MVol.D2-244/5
$C_6Cs_2LiN_6Rh$	$Cs_2Li[Rh(CN)_6]$	Rh:	SVol.B1-188/9, 191
$C_6Cs_2Mn_2O_{12}$	$Cs_2[Mn_2(C_2O_4)_3] \cdot 3\ H_2O$	Mn:	MVol.D2-106/7
$C_6Cs_2N_2O_8S_2U$	$Cs_2[U(C_2O_4)_2(NCS)_2]$	U:	SVol.C13-230/2
$C_6Cs_2O_{16}U_2$	$Cs_2[(UO_2)_2(C_2O_4)_3]$	U:	SVol.C13-204/6
$C_6Cs_3MnN_6$	$Cs_3[Mn(CN)_6]$	Mn:	MVol.D2-244
$C_6Cs_3N_6Rh$	$Cs_3[Rh(CN)_6]$	Rh:	SVol.B1-191
$C_6Cs_4FeN_6$	$Cs_4[Fe(CN)_6]$ systems		
	$Cs_4[Fe(CN)_6]-UO_2(NO_3)_2-H_2O$	U:	SVol.C13-32

C_6CuH_5	$C_6H_5Cu \cdot x\,S(C_3H_7\text{-}i)_2 \cdot y\,LiI$	Cu: Org.Comp.1-360, 376/7
C_6CuH_5ILi	$Li(C_6H_5CuI)$	Cu: Org.Comp.1-428
C_6CuH_5O	C_5H_5CuCO	Cu: Org.Comp.4-72
$C_6CuH_5O_2$	$CH_2CHCO_2CH_2CCCu$	Cu: Org.Comp.3-23
–	$CH_2CHCO_2CH_2CCCu \cdot x\,P(C_6H_5)_3$	Cu: Org.Comp.3-161
C_6CuH_6	C_6H_6Cu	Cu: Org.Comp.4-100
$C_6CuH_6IO_2$	$C_2H_2 \cdot CuI \cdot 2\,CH_2CO$	Cu: Org.Comp.4-52
C_6CuH_6K	$[(CH_3CC)_2Cu]K$	Cu: Org.Comp.3-182
C_6CuH_6N	$NC_5H_4CH_2Cu \cdot x\,LiBr$	Cu: Org.Comp.1-61, 85
–	$NC_5H_4CH_2Cu \cdot x\,LiCl$	Cu: Org.Comp.1-61/2, 85
–	$NC_5H_4CH_2Cu \cdot x\,LiI$	Cu: Org.Comp.1-61, 118
C_6CuH_7	$CH_2C(CH_3)CH_2CCCu$	Cu: Org.Comp.3-23
C_6CuH_7O	$CH_3CHCHCH(OH)CCCu$	Cu: Org.Comp.3-23
–	$C_2H_5OCHCHCCCu$	Cu: Org.Comp.3-23, 88
$C_6CuH_7O_2$	$CH_2CHCHC(CO_2CH_3)Cu \cdot x\,LiI$	Cu: Org.Comp.1-62
$C_6CuH_8^+$	$(c\text{-}1,3\text{-}C_6H_8Cu)^+$	Cu: Org.Comp.4-87
–	$(c\text{-}1,4\text{-}C_6H_8Cu)^+$	Cu: Org.Comp.4-88
C_6CuH_9	$CH_3(CHCH)_2CH_2Cu \cdot x\,LiI$	Cu: Org.Comp.1-62
–	$CH_3CC(CH_2)_3Cu \cdot x\,MgBrHal$	Cu: Org.Comp.1-224
–	$n\text{-}C_4H_9CCCu$	Cu: Org.Comp.3-5/7, 10/1, 23, 44/5, 77/129, 140, 144
–	$s\text{-}C_4H_9CCCu$	Cu: Org.Comp.3-41
–	$t\text{-}C_4H_9CCCu$	Cu: Org.Comp.3-2, 5, 7/11, 24, 45, 86, 88, 151
–	$n\text{-}C_4H_9CCCu \cdot x\,AsCl_3$	Cu: Org.Comp.3-169
–	$n\text{-}C_4H_9CCCu \cdot x\,LiBr$	Cu: Org.Comp.3-152
–	$t\text{-}C_4H_9CCCu \cdot x\,LiCl$	Cu: Org.Comp.1-62, 85
–	$n\text{-}C_4H_9CCCu \cdot x\,LiI$	Cu: Org.Comp.3-153, 155
–	$t\text{-}C_4H_9CCCu \cdot x\,LiI$	Cu: Org.Comp.3-153
–	$n\text{-}C_4H_9CCCu \cdot x\,MgBr_2$	Cu: Org.Comp.3-154/5
–	$n\text{-}C_4H_9CCCu \cdot x\,NaCl$	Cu: Org.Comp.3-154
–	$n\text{-}C_4H_9CCCu \cdot x\,NaI$	Cu: Org.Comp.3-154
–	$n\text{-}C_4H_9CCCu \cdot x\,PCl_3$	Cu: Org.Comp.3-161
–	$C_5H_8CHCu \cdot x\,LiSC_6H_5$	Cu: Org.Comp.1-184
–	$C_6H_9Cu \cdot x\,LiSC_6H_5$	Cu: Org.Comp.1-193
$C_6CuH_9Li_2$	$(CH_2CH)_3CuLi_2$	Cu: Org.Comp.2-240
$C_6CuH_9N_2$	$i\text{-}C_4H_9NCCuCN$	Cu: Org.Comp.3-217
–	$t\text{-}C_4H_9NCCuCN$	Cu: Org.Comp.3-217/8
C_6CuH_9O	$(CH_3)_2C(OCH_3)CCCu$	Cu: Org.Comp.3-5, 10, 24, 151, 157
–	$C_2H_5C(CH_3)(OH)CCCu$	Cu: Org.Comp.3-24
C_6CuH_9OS	$CH_2CHCCH_3C(S(O)CH_3)Cu \cdot x\,MgBrHal$ . .	Cu: Org.Comp.1-218
$C_6CuH_9O_2$	$CH_3CHC(CO_2C_2H_5)Cu \cdot x\,LiCN$	Cu: Org.Comp.1-125, 138/9
–	$CH_3CHC(C_3H_5O_2)Cu \cdot x\,LiSC_6H_5$	Cu: Org.Comp.1-166, 192/3
$C_6CuH_9O_2S$	$CH_2CHCCH_3C(SO_2CH_3)Cu \cdot x\,MgBrHal$ . . .	Cu: Org.Comp.1-218
$C_6CuH_9S_2$	$C_3H_6S_2CCHCH_2Cu \cdot x\,P(OCH_3)_3 \cdot y\,LiI$ . .	Cu: Org.Comp.1-430, 449
C_6CuH_{10}	$(CH_3CHCH)_2Cu$	Cu: Org.Comp.2-2
$C_6CuH_{10}^+$	$(c\text{-}C_6H_{10}Cu)^+$	Cu: Org.Comp.4-4
–	$(C_2H_5CHCHCHCH_2Cu)^+$	Cu: Org.Comp.4-86

$C_6EuH_6NO_6$	$Eu[N(CH_2COO)_3] \cdot 4\ H_2O$	Sc:	MVol.D1-144
$C_6EuH_6O_6^+$	$[EuH(CH(CH_2COO)_2COO)]^+$	Sc:	MVol.D5-216
−	$[Eu(H_3CCOCOO)_2]^+$	Sc:	MVol.D5-269/70
$C_6EuH_6O_8^+$	$[Eu(CH_2(COO)_2H)_2]^+$	Sc:	MVol.D5-152
$C_6EuH_7N_3O^{2+}$	$Eu[4-H_2NNHC(O)C_5H_4N]^{2+}$	Sc:	MVol.D2-248
$C_6EuH_7O_6^{2+}$	$[Eu((HOCH_2CHOH)(HO)C_4HO_3)]^{2+}$	Sc:	MVol.D3-286
$C_6EuH_7O_7^{2+}$	$[Eu(HOC(CH_2COOH)_2COO)]^{2+}$	Sc:	MVol.D5-344/51
$C_6EuH_7O_8$	$Eu(C_6H_7O_8) \cdot 6\ H_2O\ =\ [Eu(C_6H_7O_8)$		
	$(H_2O)_3] \cdot 3\ H_2O$	Sc:	MVol.D5-340/1
$C_6EuH_8N_3O_2^{2+}$	$[Eu(N_2C_3H_3CH_2CH(NH_2)COO)]^{2+}$	Sc:	MVol.D1-111
$C_6EuH_9NO_5^+$	$[Eu(HOC_2H_4N(CH_2COO)_2)]^+$	Sc:	MVol.D1-128/9
$C_6EuH_9N_2O_8$	$Eu[HOC_2H_4N(CH_2COO)_2](NO_3) \cdot H_2O$	Sc:	MVol.D1-130
$C_6EuH_9O_2S^{2+}$	$[Eu(CH_3C(S)CHCOOC_2H_5)]^{2+}$	Sc:	MVol.D4-100
$C_6EuH_9O_3^{2+}$	$[Eu((CH_2)_4C(OH)COO)]^{2+}$	Sc:	MVol.D5-278/9
−	$[Eu(CH_3COCHCOOC_2H_5)]^{2+}$	Sc:	MVol.D5-363
$C_6EuH_9O_6$	$Eu(CH_3COO)_3$	Sc:	MVol.D5-24, 32/3
−	$Eu(CD_3COO)_3$	Sc:	MVol.D5-32/3
−	$Eu(CH_3COO)_3 \cdot 2\ H_2O$	Sc:	MVol.D5-34/5
−	$Eu(CH_3COO)_3 \cdot 3.5\ H_2O$	Sc:	MVol.D5-43/4
−	$Eu(CH_3COO)_3 \cdot 4\ H_2O$	Sc:	MVol.D5-36
$C_6EuH_9O_6S_3$	$Eu(OOCCH_2SH)_3 \cdot 3\ H_2O$	Sc:	MVol.D4-54/5
$C_6EuH_9O_7$	$Eu[HOCH_2(CHOH)_2(CH(O))_2COO]$	Sc:	MVol.D5-275/8
$C_6EuH_9O_7^{2+}$	$[Eu(OCH(CHOH)_4COO)]^{2+}$	Sc:	MVol.D5-281
$C_6EuH_9O_9$	$Eu(HOCH_2COO)_3$	Sc:	MVol.D5-222/9, 233/6,
			238/9
$C_6EuH_9O_{12}$	$Eu((HO)_2CHCOO)_3$	Sc:	MVol.D5-249/50
$C_6EuH_{10}N_2O_4^+$	$Eu[C_2H_4(NHCH_2COO)_2]^+$	Sc:	MVol.D1-148/9
$C_6EuH_{10}N_2O_4S_2^-$ ...	$[Eu(SCH_2CH(NH_2)COO)_2]^-$	Sc:	MVol.D4-59/60
$C_6EuH_{10}N_3O_2S_2$	$Eu(HN(CH_2CH_2O)_2H)(NCS)_2$	Sc:	MVol.D2-2
$C_6EuH_{10}N_3O_4^{2+}$	$Eu(H_2N(CH_2C(O)NH)_2CH_2COO)^{2+}$	Sc:	MVol.D4-282/3
$C_6EuH_{10}N_4O_4^+$	$[Eu(CH_3C(NO)CHNOH)_2]^+$	Sc:	MVol.D2-116/7
$C_6EuH_{10}O_4^+$	$[Eu(C_2H_5COO)_2]^+$	Sc:	MVol.D5-71/4
$C_6EuH_{10}O_6^+$	$[Eu(CH_3OCH_2COO)_2]^+$	Sc:	MVol.D5-69/70
−	$[Eu(H_3CCHOHCOO)_2]^+$	Sc:	MVol.D5-250/5
$C_6EuH_{10}O_7^+$	$[Eu(OCH_2(CHOH)_4COO)]^+$	Sc:	MVol.D5-275/8
$C_6EuH_{10}O_8^+$	$[Eu(HOCH_2CHOHCOO)_2]^+$	Sc:	MVol.D5-260/1
$C_6EuH_{11}O_2^{2+}$	$[Eu(C_2H_5C(CH_3)_2COO)]^{2+}$	Sc:	MVol.D5-85
$C_6EuH_{11}O_3^{2+}$	$[Eu(i-C_3H_7C(OH)(CH_3)COO)]^{2+}$	Sc:	MVol.D5-271
$C_6EuH_{11}O_4^{2+}$	$[Eu(H_3C(C(OH)CH_3)_2COO)]^{2+}$	Sc:	MVol.D5-272
$C_6EuH_{11}O_7^{2+}$	$[Eu(HOCH_2(CHOH)_4COO)]^{2+}$	Sc:	MVol.D5-275/8
$C_6EuH_{11}O_9P^+$	$Eu[O_3POCH_2(CHOH)_4CHO]^+$	Sc:	MVol.D4-284
$C_6EuH_{12}NO_4^{2+}$	$[Eu(N(C_2H_4OH)_2CH_2COO)]^{2+}$	Sc:	MVol.D1-106/7
$C_6EuH_{12}N_2O_{12}P_4^{5-}$	$Eu(C_2H_4(N(CH_2PO_3)_2)_2)^{5-}$	Sc:	MVol.D4-154/5
$C_6EuH_{13}O_{11}$	$[Eu(C_6H_7O_8)(H_2O)_3] \cdot 3\ H_2O$		
	$=\ Eu(C_6H_7O_8) \cdot 6\ H_2O$	Sc:	MVol.D5-340/1
$C_6EuH_{15}O_{12}S_3$	$Eu(C_2H_5SO_4)_3 \cdot 9\ H_2O$	Sc:	MVol.C8-363, 369, 400/5,
			412/4
$C_6EuH_{17}N_2O_{12}P_4$ ...	$Eu[H_5(C_2H_4(N(CH_2PO_3)_2)_2)]$	Sc:	MVol.D4-154/5
−	$Eu[H_5(C_2H_4(N(CH_2PO_3)_2)_2)] \cdot 6\ H_2O$	Sc:	MVol.D4-155
$C_6EuH_{18}N_3O_{11}P_2$ ...	$Eu(NO_3)_3 \cdot 2\ ((CH_3)_3PO)$	Sc:	MVol.D4-117/8

$C_6F_3H_3O_6S_3$	4-(HS)-1,3-(HOSO$_2$)$_2$C$_6$F$_3$ · 2 H$_2$O	F:	PFHOrg.SVol.3-52, 54, 67
$C_6F_3H_3O_9S_3$	1,2,4-(HOSO$_2$)$_3$C$_6$F$_3$ · 2.5 H$_2$O	F:	PFHOrg.SVol.3-52, 54, 66
−	1,3,5-(HOSO$_2$)$_3$C$_6$F$_3$ · 4.5 H$_2$O	F:	PFHOrg.SVol.3-53, 66
$C_6F_3H_4N$	CF$_3$C$_5$H$_4$N	F:	PFHOrg.9-84
$C_6F_3H_4NO_2$	OC$_4$H$_3$C(O)NHCF$_3$	F:	PFHOrg.9-151
$C_6F_3H_4NO_3S$	2-CF$_3$SO$_2$OC$_5$H$_4$N	F:	PFHOrg.SVol.3-177
$C_6F_3H_4NS$	3-(CF$_3$S)C$_5$H$_4$N	F:	PFHOrg.SVol.2-161
$C_6F_3H_4N_5S$	HN$_4$C$_5$H$_2$(NHSCF$_3$)	F:	PFHOrg.SVol.2-168
$C_6F_3H_5N_2OS_2$	NSC$_3$H(SCF$_3$)NHC(O)CH$_3$	F:	PFHOrg.SVol.2-165
$C_6F_3H_5S_2$	SC$_4$H$_2$(CH$_3$)(SCF$_3$)	F:	PFHOrg.SVol.2-165
$C_6F_3H_6NO_2$	[ONC(CF$_3$)CHC(OC$_2$H$_5$)]	F:	PFHOrg.8-1
$C_6F_3H_6NO_3$	[ONC(CF$_3$)CH$_2$CH(OC(O)CH$_3$)]	F:	PFHOrg.8-1
$C_6F_3H_6NO_3Rh$	Rh((CH$_3$CO)$_2$CN(CF$_3$)O)	Rh:	SVol.B2-85
$C_6F_3H_6NS$	(CH$_3$)NC$_4$H$_3$(SCF$_3$)	F:	PFHOrg.SVol.2-163
$C_6F_3H_6NdO_6$	Nd(CH$_2$FCOO)$_3$	Sc:	MVol.D5-55/6
−	Nd(CH$_2$FCOO)$_3$ · 3 H$_2$O	Sc:	MVol.D5-56/7
$C_6F_3H_6O_6Pr$	Pr(CH$_2$FCOO)$_3$	Sc:	MVol.D5-55/6
−	Pr(CH$_2$FCOO)$_3$ · 3 H$_2$O	Sc:	MVol.D5-56/7
$C_6F_3H_6O_6Sm$	Sm(CH$_2$FCOO)$_3$ · 3 H$_2$O	Sc:	MVol.D5-56/7
$C_6F_3H_7N_2S$	(CF$_3$S)(CH$_3$)C$_3$HN$_2$(CH$_3$)	F:	PFHOrg.SVol.2-164
$C_6F_3H_7O_2S$	CH$_3$C(O)CH(SCF$_3$)C(O)CH$_3$	F:	PFHOrg.SVol.2-130
$C_6F_3H_8NO_2$	[ONC(CF$_3$)CH$_2$CH(OC$_2$H$_5$)]	F:	PFHOrg.8-1
$C_6F_3H_9OS$	CF$_3$C(O)SC$_4$H$_9$-n	F:	PFHOrg.9-121
$C_6F_3H_{10}NO$	(CH$_3$)$_3$CC(O)NHCF$_3$	F:	PFHOrg.9-151
$C_6F_3H_{10}NO_2$	CF$_3$NHC(O)O(CH$_2$)$_3$CH$_3$	F:	PFHOrg.9-116
−	C$_2$H$_5$OCF$_2$NCFOC$_2$H$_5$	F:	PFHOrg.9-150
$C_6F_3H_{11}N_2$	CF$_3$C(NH)N(C$_2$H$_5$)$_2$	F:	PFHOrg.9-79
−	CF$_3$C(NH)NHC(CH$_3$)$_3$	F:	PFHOrg.9-79
−	CF$_3$C(NH)NH(C$_4$H$_9$-n)	F:	PFHOrg.9-79
$C_6F_3H_{11}N_2O$	CF$_3$NHC(O)N(C$_2$H$_5$)$_2$	F:	PFHOrg.9-117
$C_6F_3H_{12}OSn$	(CH$_3$)$_3$SnOC(CH$_3$)CF$_3$ (radical)	Sn:	Org.Comp.11-164
$C_6F_3H_{12}O_2Sb$	(CH$_3$)$_4$SbO$_2$CCF$_3$	Sb:	Org.Comp.3-126
$C_6F_3H_{12}S_2$	CF$_3$SS(C$_4$H$_9$-t)CH$_3$ (radical)	F:	PFHOrg.SVol.2-201
−	CF$_3$SS(C$_4$H$_9$-t)CD$_3$ (radical)	F:	PFHOrg.SVol.2-201
$C_6F_3H_{13}NOPS$	CF$_3$NP(OC$_2$H$_5$)(SC$_2$H$_5$)CH$_3$	F:	PFHOrg.8-116
$C_6F_3H_{24}N_6Rh$	[Rh(C$_2$H$_8$N$_2$)$_3$]F$_3$	Rh:	SVol.B2-164
$C_6F_3Li_3$	1,3,5-Li$_3$C$_6$F$_3$	F:	PFHOrg.SVol.1-2
$C_6F_3MnO_8S$	[Mn(CO)$_5$](CF$_3$SO$_3$)	F:	PFHOrg.SVol.3-61, 92/3
$C_6F_3N_3O_3$	C$_6$F$_3$(NO$_2$)(N$_2$)O	F:	PFHOrg.8-13, 29, 44
$C_6F_4HN^+$	[C$_6$F$_4$NH]$^+$	F:	PFHOrg.8-81
$C_6F_4HNO_3$	NO$_2$C$_6$F$_4$OH	F:	PFHOrg.8-14, 27/8
$C_6F_4H_2N^+$	[C$_6$F$_4$NH$_2$]$^+$	F:	PFHOrg.8-81
$C_6F_4H_2N_2^+$	[C$_6$F$_4$N$_2$H$_2$]$^+$	F:	PFHOrg.8-81
$C_6F_4H_2O_3S_2$	4-HSC$_6$F$_4$SO$_3$H · 2.5 H$_2$O	F:	PFHOrg.SVol.3-51, 53, 64
$C_6F_4H_2O_4$	C$_2$F$_2$(CFCOOH)$_2$	F:	PFHOrg.9-78
$C_6F_4H_2O_6S_2$	1,2-(HOSO$_2$)$_2$C$_6$F$_4$ · H$_2$O	F:	PFHOrg.SVol.3-52/3, 65
−	1,3-(HOSO$_2$)$_2$C$_6$F$_4$ · 2.5 H$_2$O	F:	PFHOrg.SVol.3-52/3, 65
−	1,4-(HOSO$_2$)$_2$C$_6$F$_4$	F:	PFHOrg.SVol.3-52/4, 65
−	1,4-(HOSO$_2$)$_2$C$_6$F$_4$ · 2 H$_2$O	F:	PFHOrg.SVol.3-52/3, 64
$C_6F_4H_3IN_2$	IC$_6$F$_4$NHNH$_2$	F:	PFHOrg.8-61/2, 88, 123/4

$C_6F_5H_4NS$	$NH_4SC_6F_5$	F:	PFHOrg.SVol.2-252, 254
$C_6F_5H_5IN$	$CF_3IF_2 \cdot NC_5H_5$	F:	PFHOrg.SVol.3-266
$C_6F_5H_6NO_3$	$NO_2CF_2C(CF_3)(OH)CH_2CHCH_2$	F:	PFHOrg.8-39
$C_6F_5H_{10}ISn$	$(C_2H_5)_2(C_2F_5)SnI$	Sn:	Org.Comp.8-59, 61, 63
$C_6F_5H_{10}O_2P$	$C_2F_5P(OC_2H_5)_2$	F:	PFHOrg.SVol.1-122
$C_6F_5H_{27}MnN_6$	$MnF_2 \cdot 3 C_2H_8N_2 \cdot 3 HF$	Mn:	MVol.D3-41/2
$C_6F_5IN_2O_6$	$C_6F_5I(NO_3)_2$	F:	PFHOrg.SVol.3-255, 258, 262/5
C_6F_5IO	C_6F_5IO	F:	PFHOrg.SVol.3-253, 257, 263/5
$C_6F_5IO_2$	$C_6F_5IO_2$	F:	PFHOrg.SVol.3-253, 257
C_6F_5Li	C_6F_5Li	F:	PFHOrg.SVol.1-2, 10/4
$C_6F_5LiO_2S$	$LiOS(O)C_6F_5 \cdot H_2O$	F:	PFHOrg.SVol.3-4, 22, 40
C_6F_5LiS	$LiSC_6F_5$	F:	PFHOrg.SVol.2-76, 252, 254, 263/4
$C_6F_5Li_2N$	$C_6F_5NLi_2$	F:	PFHOrg.9-40/1, 47/8
C_6F_5NOS	C_6F_5NSO	F:	PFHOrg.9-6/7, 24, 33
$C_6F_5NOS_2$	C_6F_5SNSO	F:	PFHOrg.SVol.2-102, 108, 110, 114
$C_6F_5NO_2$	$C_6F_5NO_2$	F:	PFHOrg.8-13/4, 19, 27, 44/6
$C_6F_5NO_3$	$OC(CFCF)_2CFNO_2$	F:	PFHOrg.8-13, 29, 43/4
$C_6F_5NS_8$	$C_6F_5SNS_7$	F:	PFHOrg.SVol.2-102, 108, 110
$C_6F_5N_3$	$C_6F_5N_3$	F:	PFHOrg.8-60/1, 86, 103, 116/7
$C_6F_5NaO_2S$	$NaOS(O)C_6F_5$	F:	PFHOrg.SVol.3-4, 22
C_6F_5NaS	$NaSC_6F_5$	F:	PFHOrg.SVol.2-263
$C_6F_5O_2S$	$C_6F_5SO_2$ (radical)	F:	PFHOrg.SVol.3-121
C_6F_5STl	$TlSC_6F_5$	F:	PFHOrg.SVol.2-76, 252, 265
$C_6F_6FeH_4NO_3P$	$[C_4H_4Fe(CO)_2NO]PF_6$	Fe:	Org.Comp.B6-95, 100/1
$C_6F_6FeH_8OP_2$	$CH_2CHC(CH_3)CH_2Fe(PF_3)_2CO$	Fe:	Org.Comp.B6-27, 29, 41
–	$CH_2CHCHCH(CH_3)Fe(PF_3)_2CO$	Fe:	Org.Comp.B6-27/9, 40/1
$C_6F_6GeH_9N$	$(CH_3)_3GeNC(CF_3)_2$	F:	PFHOrg.9-46
$C_6F_6HN_2O_4S_3Tl$	$[SNSNSCHC(Tl(OOCCF_3)_2)]$	S:	S-N Comp.4-112/7
$C_6F_6H_2N_2O$	$(CF_3)_2C(OH)CH(CN)_2$	F:	PFHOrg.9-78
$C_6F_6H_2S_3$	$SC_4H_2(SCF_3)_2$	F:	PFHOrg.SVol.2-166
$C_6F_6H_3N_3$	$C_3H_3N_2(CN)(CF_3)_2$	F:	PFHOrg.8-107, 112
$C_6F_6H_4N_2O_2$	$(CF_3)_2CC[C(O)NH_2]_2$	F:	PFHOrg.9-78
$C_6F_6H_4N_2O_3S$	$(O)SN_2C_2(OCH_2CF_3)_2$	S:	S-N Comp.3-250/7
$C_6F_6H_6N_2$	$(CF_3)_2CNNC(CH_3)_2$	F:	PFHOrg.8-125, 126
–	$[N(CF_3)CH_2CHCHCH_2N(CF_3)]$	F:	PFHOrg.8-143/4
$C_6F_6H_6N_2O_2$	$(CF_3N)(CH_3O)CC(OCH_3)(NCF_3)$	F:	PFHOrg.9-153
–	$CH_3C(O)N(CF_3)N(CF_3)C(O)CH_3$	F:	PFHOrg.8-122
$C_6F_6H_6OS$	$(CF_3)_2(CH_3O)C_3H_3S$	F:	PFHOrg.SVol.2-11
$C_6F_6H_6O_4Sn$	$(CH_3)_2Sn(OOCCF_3)_2$	Sn:	Org.Comp.14-87
$C_6F_6H_7N$	$(CF_3)_2CNC_3H_7-n$	F:	PFHOrg.SVol.2-8
$C_6F_6H_7NO$	$(CF_3)_2H_6C_4ONH$	F:	PFHOrg.SVol.2-9
$C_6F_6H_7NO_2$	$(CF_3)_2NCH_2C(O)OC_2H_5$	F:	PFHOrg.9-47

$C_6F_6H_7N_3$	$(CF_3)_2CNNCHN(CH_3)_2$	F:	PFHOrg.8-125, 128
$C_6F_6H_8N_2O$	$C_3H_7CON(CF_3)NHCF_3$	F:	PFHOrg.8-144
$C_6F_6H_8N_4$	$[CF_3NC(NHCH_3)]_2$	F:	PFHOrg.9-154
$C_6F_6H_9NSi$	$(CF_3)_2CNSi(CH_3)_3$	F:	PFHOrg.9-46
$C_6F_6H_9NSn$	$(CH_3)_3SnNC(CF_3)_2$	F:	PFHOrg.9-46
$C_6F_6H_9OP$	$(CF_3)_2PO(C_4H_9-t)$	F:	PFHOrg.SVol.1-122
$C_6F_6H_9OSn$	$(CH_3)_3SnOC(CF_3)_2$ (radical)	Sn:	Org.Comp.11-164
$C_6F_6H_9PS$	$(CF_3)_2PS(C_4H_9-t)$	F:	PFHOrg.SVol.1-122
$C_6F_6H_{10}N_2S_2$	$CF_3SN(CH_3)C_2H_4N(CH_3)SCF_3$	F:	PFHOrg.SVol.3-44
$C_6F_6H_{10}OSn$	$(CH_3)_3SnOCH(CF_3)_2$	Sn:	Org.Comp.11-57, 60
$C_6F_6H_{10}Si$	$(CF_3)_2CHSi(CH_3)_3$	F:	PFHOrg.8-136
$C_6F_6H_{10}Sn$	$(CH_3)_3SnCH(CF_3)_2$	F:	PFHOrg.8-108, 114
$C_6F_6H_{12}NPSi$	$(CH_3)[(CH_3)_3Si]NP(CF_3)_2$	F:	PFHOrg.SVol.1-129
$C_6F_6H_{12}N_2O$	$CF_3NHNHCF_3 \cdot C_2H_5OC_2H_5$	F:	PFHOrg.8-63, 89
$C_6F_6N_2$	$(CF_3)_2CC(CN)_2$	F:	PFHOrg.9-57, 72, 78, 80, 82, 84/5, 87/9
–	$CF_3C(CN)C(CN)CF_3$	F:	PFHOrg.9-57, 72, 85/6
–	$[NCCFCF_2CF(CN)CF_2]$	F:	PFHOrg.9-58/9, 73
–	$[NCCFCF_2CF_2CFCN]$	F:	PFHOrg.9-58/9, 73
$C_6F_6N_2O$	$(CF_3)_2(NC)_2C_2O$	F:	PFHOrg.9-58/9, 73, 90
$C_6F_6N_2O_2$	$[CF_3C(O)CN]_2$	F:	PFHOrg.9-53, 69, 79, 88
$C_6F_6N_4S_3$	$(CF_3C_2SN_2)S(N_2SC_2CF_3)$	F:	PFHOrg.SVol.2-45, 50
C_6F_6OS	$C_6F_5S(O)F$	F:	PFHOrg.SVol.3-12, 35, 42
C_6F_6OSe	$C_6F_5Se(O)F$	F:	PFHOrg.SVol.3-215/6, 223
$C_6F_6O_2S$	$C_6F_5SO_2F$	F:	PFHOrg.SVol.3-158, 168, 171
C_6F_6S	C_6F_5SF	F:	PFHOrg.SVol.2-115
$C_6F_7HO_2S_4$	$SC_4H(SCF_3)_2SO_2F$	F:	PFHOrg.SVol.2-167
$C_6F_7HP^+$	$[C_6F_5PF_2H]^+$	F:	PFHOrg.SVol.1-135
$C_6F_7H_2NO_2$	$[N(C_3F_7)C(O)CH_2C(O)]$	F:	PFHOrg.9-118
$C_6F_7H_4NO_2$	$[N(C_3F_7)C(O)OCH_2CH_2]$	F:	PFHOrg.9-119
$C_6F_7H_5N_2O$	$CF_3NCFC(OC_2H_5)NCF_3$	F:	PFHOrg.9-153
$C_6F_7H_5N_2O_2$	$C_3F_7NHC(O)NHC(O)CH_3$	F:	PFHOrg.9-118
$C_6F_7H_6NO$	$C_3F_7C(NH)OC_2H_5$	F:	PFHOrg.9-81
$C_6F_7H_6NO_2$	$CF_3NHC(O)OC_2H_4CHFCF_3$	F:	PFHOrg.9-116
–	$C_3F_7NHC(O)OC_2H_5$	F:	PFHOrg.9-116/7, 154
$C_6F_7H_6NS$	$C_3F_7C(NH)SC_2H_5$	F:	PFHOrg.9-82
$C_6F_7H_7N_2$	$C_3F_7C(NH)N(CH_3)_2$	F:	PFHOrg.9-79/80
$C_6F_7H_7N_2O$	$(CF_3)_2C(ONC(CH_3)_2)NFH$	F:	PFHOrg.8-221
$C_6F_7H_8NO$	$(CF_3)_2C(OCH(CH_3)_2)NFH$	F:	PFHOrg.8-221
$C_6F_7H_8O_2P$	$C_3F_7PH(O)OC_3H_7-i$	F:	PFHOrg.SVol.1-123
C_6F_7I	$C_6F_5IF_2$	F:	PFHOrg.SVol.3-255/6, 259
C_6F_7IO	$C_6F_5I(O)F_2$	F:	PFHOrg.SVol.3-253, 257
C_6F_7N	$C_6F_5NF_2$	F:	PFHOrg.8-159
C_6F_7NOS	$C_6F_5NS(O)F_2$	F:	PFHOrg.9-8/9, 28
C_6F_7NS	$C_6F_5NSF_2$	F:	PFHOrg.9-4/5, 20, 33
C_6F_7OP	$C_6F_5P(O)F_2$	F:	PFHOrg.SVol.1-98/9
C_6F_7P	$C_6F_5PF_2$	F:	PFHOrg.SVol.1-132/5

$C_6F_{14}HO_2P$	$(C_3F_7)_2P(O)OH$	F:	PFHOrg.SVol.1-97
$C_6F_{14}HgS_2$	$Hg[SCF(CF_3)_2]_2$	F:	PFHOrg.SVol.2-251, 253, 256/7
$C_6F_{14}IP$	$(i-C_3F_7)_2PI$	F:	PFHOrg.SVol.1-106, 117/9
$C_6F_{14}N_2$	$(CF_3)_2CFNNCF(CF_3)_2$	F:	PFHOrg.8-73/4, 97, 101
–	$(CF_3)_2NCFCFN(CF_3)_2$	F:	PFHOrg.9-164, 173
–	$CF_3C(NF)CF(NF_2)(C_3F_7-n)$	F:	PFHOrg.8-166
–	$CF_3C(NF_2)C(NF_2)C_3F_7$	F:	PFHOrg.8-191
–	$C_3F_7NNC_3F_7$	F:	PFHOrg.8-73/4, 98, 144, 203
$C_6F_{14}N_2^-$	$[(CF_3)_2CFNNCF(CF_3)_2]^-$	F:	PFHOrg.8-101
$C_6F_{14}N_2O$	$C_3F_7N(O)NC_3F_7$	F:	PFHOrg.8-55/6, 82
$C_6F_{14}N_2S$	$(CF_3)_2CFNSNCF(CF_3)_2$	F:	PFHOrg.9-7, 24
–	$(CF_3)_2CNC(CF_3)_2NSF_2$	F:	PFHOrg.9-35
$C_6F_{14}N_2S_2$	$(CF_3)_2CFNSNSCF(CF_3)_2$	F:	PFHOrg.SVol.2-99, 110
$C_6F_{14}N_2S_3$	$(CF_3)_2CFSNSNSCF(CF_3)_2$	F:	PFHOrg.SVol.2-99, 104
$C_6F_{14}O_2S$	$C_6F_{13}SO_2F$	F:	PFHOrg.SVol.3-154, 172
$C_6F_{14}S$	$(CF_3)_2CFSCF(CF_3)_2$	F:	PFHOrg.SVol.2-211, 225
–	$C_2F_5CF(CF_3)SC_2F_5$	F:	PFHOrg.SVol.2-211
$C_6F_{14}S_2$	$(CF_3)_2CFSSCF(CF_3)_2$	F:	PFHOrg.SVol.2-178, 191, 200, 203
–	$CF_3S(CF_2)_4SCF_3$	F:	PFHOrg.SVol.2-212, 226, 233
–	$C_2F_5CF(CF_3)SSC_2F_5$	F:	PFHOrg.SVol.2-178, 191, 197
$C_6F_{14}S_3$	$(CF_3)_2CFS_3CF(CF_3)_2$	F:	PFHOrg.SVol.2-181, 193, 197
$C_6F_{14}S_4$	$(CF_3)_2CFS_4CF(CF_3)_2$	F:	PFHOrg.SVol.2-181
$C_6F_{14}S_5$	$(CF_3)_2CFS_5CF(CF_3)_2$	F:	PFHOrg.SVol.2-181
$C_6F_{14}S_6$	$(CF_3S)_2CFSSCF(SCF_3)_2$	F:	PFHOrg.SVol.2-177
$C_6F_{14}Se$	$C_3F_7SeC_3F_7$	F:	PFHOrg.SVol.3-220, 234
$C_6F_{14}Se_2$	$C_3F_7SeSeC_3F_7$	F:	PFHOrg.SVol.3-218/9, 232
–	$C_4F_9SeSeC_2F_5$	F:	PFHOrg.SVol.3-218/9
$C_6F_{14}Se_3$	$C_4F_9SeSeSeC_2F_5$	F:	PFHOrg.SVol.3-218/9
$C_6F_{14}Te$	$C_2F_5TeC_4F_9-n$	F:	PFHOrg.SVol.3-249/50
$C_6F_{15}H_2P_3$	$(CF_3)_2PCH_2P(CF_3)P(CF_3)_2$	F:	PFHOrg.SVol.1-85
$C_6F_{15}I$	$C_6F_{13}IF_2$	F:	PFHOrg.SVol.3-255/6, 259, 264, 265
$C_6F_{15}N$	$(CF_3)_2CFN(CF_3)C_2F_5$	F:	PFHOrg.9-165, 174
–	$(CF_3)_2NC(CF_3)_3$	F:	PFHOrg.9-160, 168
–	$(C_2F_5)_3N$	F:	PFHOrg.9-159, 167, 176/7
–	$(C_3F_7)_2NF$	F:	PFHOrg.8-162, 186
$C_6F_{15}NS$	$(CF_3)_2CFNSF[CF(CF_3)_2]$	F:	PFHOrg.9-4/5, 18
–	$(CF_3)_2CFS(F)NCF(CF_3)_2$	F:	PFHOrg.SVol.3-7/8, 27/8
$C_6F_{15}N_2P$	$[(CF_3)_2CN]_2PF_3$	F:	PFHOrg.9-39
$C_6F_{15}N_3O_2$	$CF_3NC[ON(CF_3)_2]_2$	F:	PFHOrg.9-121
$C_6F_{15}NdO_9S_3$	$Nd(OSO_2C_2F_5)_3$	F:	PFHOrg.SVol.3-61
$C_6F_{15}OP$	$(C_2F_5)_3PO$	F:	PFHOrg.SVol.1-87/90, 95

$C_6FeMo_2N_6O_4$	$(MoO_2)_2Fe(CN)_6$	Mo:	SVol.A3-206
$C_6FeN_6Na_4$	$Na_4[Fe(CN)_6]$ systems		
	$Na_4[Fe(CN)_6]-UO_2(NO_3)_2-H_2O$	U:	SVol.C13-32
$C_6FeN_6O_4U_2$	$(UO_2)_2[Fe(CN)_6]$	U:	SVol.C13-30/2
–	$(UO_2)_2[Fe(CN)_6] \cdot 6 H_2O$	U:	SVol.C13-30, 32
$C_6FeN_6Rb_4$	$Rb_4[Fe(CN)_6]$ systems		
	$Rb_4[Fe(CN)_6]-UO_2(NO_3)_2-H_2O$	U:	SVol.C13-32
C_6FeN_6Rh	$Fe[Rh(CN)_6]$	Rh:	SVol.B1-192
C_6FeN_6U	$U[Fe(CN)_6] \cdot 6 H_2O$	U:	SVol.C13-29/30
$C_6Fe_4H_{10}NO^-$	$[Fe_4CC(O)N(C_2H_5)_2]^-$	Fe:	Org.Comp.C7-232, 235
$C_6GaGeH_{18}N_3$	$Ga(CH_3)_3 \cdot N_3Ge(CH_3)_3$	Ga:	Org.Comp.1-47
C_6GaH_3	$Ga(CCH)_3$	Ga:	Org.Comp.1-97
$C_6GaH_5I_2$	$Ga(C_6H_5)I_2$	Ga:	Org.Comp.1-162/4
C_6GaH_9	$Ga(CHCH_2)_3$	Ga:	Org.Comp.1-25/6, 98/101
$C_6GaH_9O_4S_2$	$Ga(CHCH_2)(OS(O)CHCH_2)_2$	Ga:	Org.Comp.1-222
C_6GaH_{11}	$Ga(C_2H_5)_2CCH$	Ga:	Org.Comp.1-116
$C_6GaH_{11}N_2$	$Ga(CH_3)_2NCHCHNCCH_3$	Ga:	Org.Comp.1-262, 264,
			268, 270
–	$Ga(CH_3)_2N_2C_3H_2CH_3$	Ga:	Org.Comp.1-262/3, 267/8
$C_6GaH_{12}NO_2$	$Ga(CH_3)_2N(COCH_3)_2$	Ga:	Org.Comp.1-182, 186
$C_6GaH_{13}OS$	$Ga(C_2H_5)_2OC(S)CH_3$	Ga:	Org.Comp.1-194/5
$C_6GaH_{13}O_4S$	$Ga(C_2H_5)(OOCCH_3)OS(O)C_2H_5$	Ga:	Org.Comp.1-222
$C_6GaH_{13}S_2$	$Ga(C_2H_5)_2SC(S)CH_3$	Ga:	Org.Comp.1-230, 233/4
$C_6GaH_{14}N$	$Ga(CH_3)_2N(CH_2)_4$	Ga:	Org.Comp.1-260
–	$Ga(C_2H_5)_2N(CH_2)_2$	Ga:	Org.Comp.1-259
C_6GaH_{15}	$Ga(C_2H_5)_3$	Ga:	Org.Comp.1-61/72
$C_6GaH_{15}N_2$	$Ga(CH_3)_2CN \cdot N(CH_3)_3$	Ga:	Org.Comp.1-169
–	$Ga(CH_3)_2N(CH_3)C(CH_3)NCH_3$	Ga:	Org.Comp.1-254/6
$C_6GaH_{15}O$	$Ga(CH_3)_2OC_4H_9$-t	Ga:	Org.Comp.1-176, 178
–	$Ga(CH_3)_3 \cdot OC(CH_3)_2$	Ga:	Org.Comp.1-33, 38
–	$Ga(C_2H_5)_2OC_2H_5$	Ga:	Org.Comp.1-177
$C_6GaH_{15}O_2$	$Ga(CH_3)_2OOC_4H_9$-t	Ga:	Org.Comp.1-200
$C_6GaH_{15}O_2S$	$Ga(C_2H_5)_2OS(O)C_2H_5$	Ga:	Org.Comp.1-207, 211
$C_6GaH_{15}O_3S$	$Ga(C_2H_5)_2OSO_2C_2H_5$	Ga:	Org.Comp.1-207
$C_6GaH_{16}K$	$K[Ga(C_2H_5)_3H]$	Ga:	Org.Comp.1-321
$C_6GaH_{16}NO$	$Ga(CH_3)_2OCH_2CH_2N(CH_3)_2$	Ga:	Org.Comp.1-182, 185
$C_6GaH_{16}Na$	$Na[Ga(CH_3)_2(C_2H_5)_2]$	Ga:	Org.Comp.1-316
–	$Na[Ga(C_2H_5)_3H]$	Ga:	Org.Comp.1-321
$C_6GaH_{16}O_2P$	$Ga(C_2H_5)_2OP(O)(CH_3)_2$	Ga:	Org.Comp.1-208
$C_6GaH_{16}P$	$Ga(CH_3)_2P(C_2H_5)_2$	Mo:	Org.Comp.1-296
$C_6GaH_{18}LiSn$	$Li[Ga(CH_3)_3Sn(CH_3)_3]$	Ga:	Org.Comp.1-418
$C_6GaH_{18}N$	$Ga(CH_3)_3 \cdot N(CH_3)_3$	Ga:	Org.Comp.1-41/2, 49/51
–	$Ga(C_2H_5)_3 \cdot NH_3$	Ga:	Org.Comp.1-73, 75
$C_6GaH_{18}NO$	$Ga(CH_3)_3 \cdot ON(CH_3)_3$	Ga:	Org.Comp.1-33, 38
$C_6GaH_{18}NS$	$Ga(CH_3)_2SCH_3 \cdot N(CH_3)_3$	Ga:	Org.Comp.1-232
$C_6GaH_{18}NSe$	$Ga(CH_3)_2SeCH_3 \cdot N(CH_3)_3$	Ga:	Org.Comp.1-243
$C_6GaH_{18}N_{33}$	$[N(CH_3)_4][Ga(CH_3)_2N_{32}]$	Ga:	Org.Comp.1-338
$C_6GaH_{18}OP$	$Ga(CH_3)_3 \cdot OP(CH_3)_3$	Ga:	Org.Comp.1-33
$C_6GaH_{18}P$	$Ga(CH_3)_3 \cdot P(CH_3)_3$	Ga:	Org.Comp.1-41, 47, 54

$C_6GeH_{18}OSn$	$(CH_3)_3SnOGe(CH_3)_3$	Sn:	Org.Comp.11-154
$C_6GeH_{18}SSn$	$(CH_3)_3SnSGe(CH_3)_3$	Sn:	Org.Comp.9-65
$C_6GeH_{18}SeSn$	$(CH_3)_3SnSeGe(CH_3)_3$	Sn:	Org.Comp.10-261
$C_6Ge_2H_{18}O_4S$	$[(CH_3)_3GeO]_2SO_2$	S:	SVol.3-321
C_6HMnN_{15}	$Mn(NCCN_4H)(NCCN_4)_2 \cdot 2 H_2O$	Mn:	MVol.D4-76
$C_6HN_5O_2S_2$	$O_2NC_6H(N_2S)_2$	S:	S-N Comp.3-222/3
$C_6HO_{14}U^{3-}$	$[UO_2(C_2O_4)_2(HC_2O_4)]^{3-}$	U:	SVol.D2-377/8
$C_6H_2HoO_8S_2^-$	$[Ho(O_2C_6H_2(SO_3)_2)]^-$	Sc:	MVol.D3-41/3
$C_6H_2K_3MnN_6O$	$K_2[Mn(CN)_5(H_2O)] \cdot KCN$	Mn:	MVol.D2-239/40
$C_6H_2LaN_3O_7^{2+}$	$[La(OC_6H_2(NO_2)_3)]^{2+}$	Sc:	MVol.D3-39
$C_6H_2MnN_2O_4$	$Mn[2,3-(OOC)_2-1,4-N_2C_4H_2] \cdot 2 H_2O$	Mn:	MVol.D4-335/6
–	$Mn[2,3-(OOC)_2-1,4-N_2C_4H_2] \cdot 4 H_2O$	Mn:	MVol.D4-335/6
–	$Mn[2,5-(OOC)_2-1,4-N_2C_4H_2] \cdot 2 H_2O$	Mn:	MVol.D4-335/6
$C_6H_2MnN_6O^{3-}$	$[Mn(CN)_6H_2O]^{3-}$	Mn:	MVol.D2-226
$C_6H_2N_2O_6U$	$UO_2[1,4-N_2C_4H_2-2,5-(COO)_2] \cdot 2 H_2O$	U:	SVol.C13-305
$C_6H_2N_4S_2$	$C_6H_2(N_2S)_2$	S:	S-N Comp.3-222/3
$C_6H_2N_4S_4$	$C_6H_2(NSS)(NSNSN)$	S:	S-N Comp.4-119/21
$C_6H_2N_6O_{24}Th_2$	$Th_2(C_6H_2O_6)(NO_3)_6$	Th:	SVol.E-147, 152
$C_6H_2NaNdO_8S_2$	$Na[Nd(O_2C_6H_2(SO_3)_2)]$	Sc:	MVol.D3-42
$C_6H_2NdO_8S_2^-$	$[Nd(O_2C_6H_2(SO_3)_2)]^-$	Sc:	MVol.D3-41/3
$C_6H_2O_6U$	$UO_2(C_6H_2O_4) \cdot x H_2O$	U:	SVol.E2-29, 30/1
$C_6H_2O_8S_2Sm^-$	$[Sm(O_2C_6H_2(SO_3)_2)]^-$	Sc:	MVol.D3-41/3
$C_6H_2O_8S_2Tb^-$	$[Tb(O_2C_6H_2(SO_3)_2)]^-$	Sc:	MVol.D3-41/3
$C_6H_2O_8S_2Tm^-$	$[Tm(O_2C_6H_2(SO_3)_2)]^-$	Sc:	MVol.D3-41/3
$C_6H_2O_8S_2Y^-$	$[Y(O_2C_6H_2(SO_3)_2)]^-$	Sc:	MVol.D3-41/3
$C_6H_2O_8S_2Yb^-$	$[Yb(O_2C_6H_2(SO_3)_2)]^-$	Sc:	MVol.D3-41/3
$C_6H_2O_{10}S_2U^{2-}$	$[UO_2(1,2-(O)_2C_6H_2(SO_3)_2-3,5)]^{2-}$	U:	SVol.D1-237
$C_6H_3HoK_3N_3O_9$	$K_3[Ho(HNC(O)COO)_3] \cdot 2 H_2O$	Sc:	MVol.D2-257
$C_6H_3HoO_6$	$Ho(OOCCHC(COO)CH_2COO)$	Sc:	MVol.D5-217/8
–	$Ho(OOCCHC(COO)CH_2COO) \cdot 2 H_2O$	Sc:	MVol.D5-217/8
$C_6H_3HoO_6S$	$Ho(O_2C_6H_2(OH)SO_3)$	Sc:	MVol.D3-45
$C_6H_3HoO_8S_2$	$Ho[O_2C_6H_2(SO_3)_2H]$	Sc:	MVol.D3-41/3
$C_6H_3K_3N_3NdO_9$	$K_3[Nd(HNC(O)COO)_3] \cdot 3 H_2O$	Sc:	MVol.D2-257
$C_6H_3K_3N_3O_9Pr$	$K_3[Pr(HNC(O)COO)_3] \cdot 3 H_2O$	Sc:	MVol.D2-257
$C_6H_3K_3N_3O_9Sm$	$K_3[Sm(HNC(O)COO)_3] \cdot 3 H_2O$	Sc:	MVol.D2-257
$C_6H_3K_3N_3O_9Y$	$K_3[Y(HNC(O)COO)_3]$	Sc:	MVol.D2-257
$C_6H_3K_3N_3O_9Yb$	$K_3[Yb(HNC(O)COO)_3] \cdot 2 H_2O$	Sc:	MVol.D2-257
$C_6H_3K_5MnN_7$	$K_5[Mn(CN)_6] \cdot NH_3$	Mn:	MVol.D2-202
$C_6H_3LaN_2O_5^{2+}$	$[La(OC_6H_3(NO_2)_2)]^{2+}$	Sc:	MVol.D3-38/9
$C_6H_3LaO_6$	$La(OOCCHC(COO)CH_2COO)$	Sc:	MVol.D5-217/8
–	$La(OOCCHC(COO)CH_2COO) \cdot 2 H_2O$	Sc:	MVol.D5-217/8
–	$La(OOCCHC(COO)CH_2COO) \cdot 3 H_2O$	Sc:	MVol.D5-217/8
$C_6H_3LaO_8S_2$	$La[O_2C_6H_2(SO_3)_2H]$	Sc:	MVol.D3-41/3
$C_6H_3LuO_6$	$Lu(OOCCHC(COO)CH_2COO)$	Sc:	MVol.D5-217/8
–	$Lu(OOCCHC(COO)CH_2COO) \cdot 2 H_2O$	Sc:	MVol.D5-217/8
$C_6H_3LuO_8S_2$	$Lu[O_2C_6H_2(SO_3)_2H]$	Sc:	MVol.D3-41/3
$C_6H_3MnN_4O^+$	$Mn(4-(O)-1,3,5,8-N_4C_6H_3)^+$	Mn:	MVol.D4-56/7
$C_6H_3MnN_6$	$H_3[Mn(CN)_6]$	Mn:	MVol.D2-227
$C_6H_3MnO_6^-$	$[Mn(C(CHCOO)(COO)CH_2COO)]^-$	Mn:	MVol.D2-144
$C_6H_3MnO_{12}$	$H_3[Mn(C_2O_4)_3]$	Mn:	MVol.D2-111/2

C$_6$H$_5$O$_7$Sm	Sm(HOC(CH$_2$COO)$_2$COO)	Sc:	MVol.D5-344/51
–	Sm(HOC(CH$_2$COO)$_2$COO) · 3.5 H$_2$O	Sc:	MVol.D5-352/4
–	Sm(HOC(CH$_2$COO)$_2$COO) · 4 H$_2$O	Sc:	MVol.D5-352/4
C$_6$H$_5$O$_7$Tb	Tb(HOC(CH$_2$COO)$_2$COO)	Sc:	MVol.D5-344/51
–	Tb(HOC(CH$_2$COO)$_2$COO) · 5 H$_2$O	Sc:	MVol.D5-352/4
C$_6$H$_5$O$_7$Tm	Tm(HOC(CH$_2$COO)$_2$COO)	Sc:	MVol.D5-344/51
C$_6$H$_5$O$_7$Y	Y(HOC(CH$_2$COO)$_2$COO)	Sc:	MVol.D5-344/54
–	Y(HOC(CH$_2$COO)$_2$COO) · H$_2$O	Sc:	MVol.D5-352/4
–	Y(HOC(CH$_2$COO)$_2$COO) · 2 H$_2$O	Sc:	MVol.D5-352/4
–	Y(HOC(CH$_2$COO)$_2$COO) · 3 H$_2$O	Sc:	MVol.D5-352/4
–	Y(HOC(CH$_2$COO)$_2$COO) · 4 H$_2$O	Sc:	MVol.D5-352/4
–	Y(HOC(CH$_2$COO)$_2$COO) · 5 H$_2$O	Sc:	MVol.D5-352/4
–	Y(HOC(CH$_2$COO)$_2$COO) · x H$_2$O	Sc:	MVol.D5-352/4
C$_6$H$_5$O$_7$Yb	Yb(HOC(CH$_2$COO)$_2$COO)	Sc:	MVol.D5-344/54
–	Yb(HOC(CH$_2$COO)$_2$COO) · 4 H$_2$O	Sc:	MVol.D5-352/4
C$_6$H$_5$O$_8$Pr	[Pr(CH$_2$(COO)$_2$)$_2$H]	Sc:	MVol.D5-152, 154
C$_6$H$_5$O$_8$SU$^-$	[UO$_2$(OOCCH(SCH$_2$COO)CH$_2$COO)]$^-$	U:	SVol.D1-210
C$_6$H$_5$O$_8$Sm	[Sm(CH$_2$(COO)$_2$)$_2$H]	Sc:	MVol.D5-152, 154
C$_6$H$_5$O$_8$Tb	[Tb(CH$_2$(COO)$_2$)$_2$H]	Sc:	MVol.D5-152, 154
C$_6$H$_5$O$_8$Tm	[Tm(CH$_2$(COO)$_2$)$_2$H]	Sc:	MVol.D5-152, 154
C$_6$H$_5$O$_8$Yb	[Yb(CH$_2$(COO)$_2$)$_2$H]	Sc:	MVol.D5-152, 154
C$_6$H$_5$O$_9$U$^-$	[UO$_2$(OOCCH$_2$C(OH)(COO)CH$_2$COO)]$^-$	U:	SVol.D1-210
		U:	SVol.D2-378
		U:	SVol.D3-393
C$_6$H$_5$SSb	C$_6$H$_5$SbS	Sb:	Org.Comp.2-131
C$_6$H$_5$Sb	(C$_6$H$_5$Sb)$_n$	Sb:	Org.Comp.2-150
C$_6$H$_6$HoNO^{2+}	[Ho(H$_2$NC$_6$H$_4$O)]$^{2+}$	Sc:	MVol.D2-3
C$_6$H$_6$HoNO$_6$	Ho[N(CH$_2$COO)$_3$]	Sc:	MVol.D1-137/42
–	Ho[N(CH$_2$COO)$_3$] · n H$_2$O	Sc:	MVol.D1-143
C$_6$H$_6$HoN$_9$S$_6$	Ho[(S)(H$_2$N)C$_2$N$_2$S]$_3$	Sc:	MVol.D4-103
C$_6$H$_6$HoO$_3$P^{2+}	Ho(C$_6$H$_5$P(O$_2$)OH)$^{2+}$	Sc:	MVol.D4-144
C$_6$H$_6$HoO$_9$$^{3-}$	[Ho(OH)$_2$(OC(CH$_2$COO)$_2$COO)]$^{3-}$	Sc:	MVol.D5-349
C$_6$H$_6$Ho$_2$O$_7$$^{4+}$	[Ho$_2$(HOC(CH$_2$COO)$_2$COOH)]$^{4+}$	Sc:	MVol.D5-349
C$_6$H$_6$I$_2$MnN$_2$O$_2$	Mn(ONC$_3$H$_3$)$_2$I$_2$ · H$_2$O	Mn:	MVol.D4-213/5
C$_6$H$_6$I$_2$NSb	NH$_2$C$_6$H$_4$SbI$_2$ · HI	Sb:	Org.Comp.2-117
C$_6$H$_6$I$_3$LaO$_6$	La(CH$_2$ICOO)$_3$ · 3 H$_2$O	Sc:	MVol.D5-66
C$_6$H$_6$I$_3$NdO$_6$	Nd(CH$_2$ICOO)$_3$	Sc:	MVol.D5-55/6
–	Nd(CH$_2$ICOO)$_3$ · 3 H$_2$O	Sc:	MVol.D5-66
C$_6$H$_6$I$_3$O$_6$Pr	Pr(CH$_2$ICOO)$_3$	Sc:	MVol.D5-55/6
–	Pr(CH$_2$ICOO)$_3$ · 3 H$_2$O	Sc:	MVol.D5-66
C$_6$H$_6$I$_3$O$_6$Sm	Sm(CH$_2$ICOO)$_3$ · 3 H$_2$O	Sc:	MVol.D5-66
C$_6$H$_6$I$_3$O$_6$Y	Y(CH$_2$ICOO)$_3$ · 3 H$_2$O	Sc:	MVol.D5-66
C$_6$H$_6$KLaO$_8$	K[La(HOC(CH$_2$COO)$_2$COO)(OH)] · 2 H$_2$O	Sc:	MVol.D5-355
C$_6$H$_6$KNdO$_8$	K[Nd(HOC(CH$_2$COO)$_2$COO)(OH)] · 3 H$_2$O	Sc:	MVol.D5-355
C$_6$H$_6$KO$_8$Pr	K[Pr(HOC(CH$_2$COO)$_2$COO)(OH)] · 2.5 H$_2$O	Sc:	MVol.D5-355
C$_6$H$_6$KO$_8$Sm	K[Sm(HOC(CH$_2$COO)$_2$COO)(OH)] · 3 H$_2$O	Sc:	MVol.D5-355
C$_6$H$_6$KO$_8$Tb	K[Tb(HOC(CH$_2$COO)$_2$COO)(OH)] · 2 H$_2$O	Sc:	MVol.D5-355
C$_6$H$_6$KO$_8$Y	K[Y(HOC(CH$_2$COO)$_2$COO)(OH)] · 1.5 H$_2$O	Sc:	MVol.D5-355
C$_6$H$_6$KO$_8$Yb	K[Yb(HOC(CH$_2$COO)$_2$COO)(OH)] · 2 H$_2$O	Sc:	MVol.D5-355
C$_6$H$_6$K$_2$O$_{11}$U	K$_2$[UO$_2$(CH$_2$(COO)$_2$)$_2$(H$_2$O)]	U:	SVol.C13-237/8, 240

$C_6H_6K_3N_3O_9Rh$	$K_3[Rh(O_2CCONH_2)_3] \cdot 3 H_2O$	Rh:	SVol.B2-70
$C_6H_6K_3O_{12}Sc$	$K_3Sc(HCOO)_6$	Sc:	MVol.D5-4
$C_6H_6LaNO^{2+}$	$[La(H_2NC_6H_4O)]^{2+}$	Sc:	MVol.D2-3
$C_6H_6LaNO_6$	$La[N(CH_2COO)_3]$	Sc:	MVol.D1-137/42
–	$La[N(CH_2COO)_3] \cdot 2 H_2O$	Sc:	MVol.D1-143
–	$La[N(CH_2COO)_3] \cdot 3 H_2O$	Sc:	MVol.D1-143/5
–	$La[N(CH_2COO)_3] \cdot 5 H_2O$	Sc:	MVol.D1-143
$C_6H_6LaO_3P^{2+}$	$La(C_6H_5P(O_2)OH)^{2+}$	Sc:	MVol.D4-144
$C_6H_6LaO_6^+$	$[LaH(CH(CH_2COO)_2COO)]^+$	Sc:	MVol.D5-216
$C_6H_6LaO_7^+$	$[La(HOC(CH_2COO)_2COOH)]^+$	Sc:	MVol.D5-344/51
$C_6H_6LaO_8^+$	$[La(CH_2(COO)_2H)_2]^+$	Sc:	MVol.D5-152
$C_6H_6La_2N_6O_{12}$	$La_2(ONHC(O)C(O)NHO)_3 \cdot 5 H_2O$	Sc:	MVol.D2-260
$C_6H_6La_2O_7^{4+}$	$[La_2(HOC(CH_2COO)_2COOH)]^{4+}$	Sc:	MVol.D5-349
$C_6H_6Li_3O_{12}Sc$	$Li_3Sc(HCOO)_6$	Sc:	MVol.D5-4
$C_6H_6LuNO_6$	$Lu[N(CH_2COO)_3]$	Sc:	MVol.D1-137/42
–	$Lu[N(CH_2COO)_3] \cdot n H_2O$	Sc:	MVol.D1-143
$C_6H_6LuO_3P^{2+}$	$Lu(C_6H_5P(O_2)OH)^{2+}$	Sc:	MVol.D4-144
$C_6H_6LuO_7^+$	$[Lu(HOC(CH_2COO)_2COOH)]^+$	Sc:	MVol.D5-344/51
$C_6H_6LuO_8^+$	$[Lu(CH_2(COO)_2H)_2]^+$	Sc:	MVol.D5-152
$C_6H_6Lu_2O_7^{4+}$	$[Lu_2(HOC(CH_2COO)_2COOH)]^{4+}$	Sc:	MVol.D5-349
$C_6H_6MnNO^+$	$Mn(2-NH_2C_6H_4O)^+$	Mn:	MVol.D4-236
$C_6H_6MnNO_6$	$Mn(N(CH_2COO)_3)$	Mn:	MVol.D5-20/1
–	$Mn(N(CH_2COO)_3) \cdot H_2O$	Mn:	MVol.D5-20/1
$C_6H_6MnNO_6^-$	$Mn(N(CH_2COO)_3)^-$	Mn:	MVol.D5-15/8
$C_6H_6MnN_2O_4$	$Mn(NCCH_2N(CH_2COO)_2)$	Mn:	MVol.D5-7/13
$C_6H_6MnN_3O^+$	$Mn(NC_5H_4C(O)NNH_2)^+$	Mn:	MVol.D5-190
$C_6H_6MnN_6O_{10}S$	$[MnSO_4(NH(CONH)_2CO)_2] \cdot H_2O$	Mn:	MVol.D4-73
$C_6H_6MnN_8S_2$	$Mn(1,2,4-C_2H_2N_3H)_2(NCS)_2$	Mn:	MVol.D4-63/5
$C_6H_6MnO_4$	$Mn(CH_2CHCOO)_2$	Mn:	MVol.D2-82
–	$[Mn(CH_2CHCOO)_2]_n$	Mn:	MVol.D2-144/5
$C_6H_6MnO_7$	$MnH(HOC_3H_4(COO)_3)$	Mn:	MVol.D2-180/1, 185/6
–	$MnH(HOC_3H_4(COO)_3) \cdot 0.5 H_2O$	Mn:	MVol.D2-182
$C_6H_6Mo_3N_2O_{10}$	$C_6H_5N_2[HMo_3O_{10}]$	Mo:	SVol.B4-158
$C_6H_6NNdO^{2+}$	$[Nd(H_2NC_6H_4O)]^{2+}$	Sc:	MVol.D2-3
$C_6H_6NNdO_6$	$Nd[N(CH_2COO)_3]$	Sc:	MVol.D1-137/42
–	$Nd[N(CH_2COO)_3] \cdot 3 H_2O$	Sc:	MVol.D1-143/5
–	$Nd[N(CH_2COO)_3] \cdot 5 H_2O$	Sc:	MVol.D1-143
C_6H_6NOSb	$[(NH_2)(HO)C_6H_3Sb]_n$	Sb:	Org.Comp.2-151
–	$NH_2C_6H_4SbO$	Sb:	Org.Comp.2-122
$C_6H_6NOTb^{2+}$	$[Tb(H_2NC_6H_4O)]^{2+}$	Sc:	MVol.D2-3
$C_6H_6NOTm^{2+}$	$[Tm(H_2NC_6H_4O)]^{2+}$	Sc:	MVol.D2-3
$C_6H_6NOYb^{2+}$	$[Yb(H_2NC_6H_4O)]^{2+}$	Sc:	MVol.D2-3
$C_6H_6NO_3SSb$	$NH_2SO_2C_6H_4SbO$	Sb:	Org.Comp.2-123
$C_6H_6NO_4U^+$	$[UO_2(ONCHCH_2C_4H_3O)]^+$	U:	SVol.D1-252
$C_6H_6NO_6PU$	$UO_2(HOC_5H_3NCH_2PO_3)$	U:	SVol.D1-247
$C_6H_6NO_6Pr$	$Pr[N(CH_2COO)_3]$	Sc:	MVol.D1-137/42
–	$Pr[N(CH_2COO)_3] \cdot 3 H_2O$	Sc:	MVol.D1-143/5
–	$Pr[N(CH_2COO)_3] \cdot 5 H_2O$	Sc:	MVol.D1-143
$C_6H_6NO_6Sc$	$Sc[N(CH_2COO)_3]$	Sc:	MVol.D1-137/42
–	$Sc[N(CH_2COO)_3] \cdot 3 H_2O$	Sc:	MVol.D1-143/5

$C_6H_6O_8Sm^+$	$[Sm(CH_2(COO)_2H)_2]^+$	Sc:	MVol.D5-152
$C_6H_6O_8Tb^+$	$[Tb(CH_2(COO)_2H)_2]^+$	Sc:	MVol.D5-152
$C_6H_6O_8U$	$UO[HOC(CH_2COO)_2COOH]$	U:	SVol.C13-262/4
–	$UO_2(CH_3COCOO)_2$	U:	SVol.D1-204
$C_6H_6O_8Yb^+$	$[Yb(CH_2(COO)_2H)_2]^+$	Sc:	MVol.D5-152
$C_6H_6O_9Tm^{3-}$	$[Tm(OH)_2(OC(CH_2COO)_2COO)]^{3-}$	Sc:	MVol.D5-349
$C_6H_6O_9U$	$UO_2H(HOC(CH_2COO)_2COO)$	U:	SVol.D1-210
–	$UO_2[HOC(CH_2COO)_2COOH]\cdot 4\,H_2O$	U:	SVol.C13-262/4
$C_6H_6O_{11}Rb_2U$	$Rb_2[UO_2(CH_2(COO)_2)_2(H_2O)]$	U:	SVol.C13-238/40
$C_6H_6O_{12}U_2$	$(UO_2)_2(C_2O_4)(CH_3COO)_2\cdot 2\,H_2O$	U:	SVol.C13-235/6
$C_6H_6Ti^{2+}$	$[C_6H_6Ti]^{2+}$	Ti:	Org.Comp.4-76
C_6H_6Yb	$Yb(CCCH_3)_2$	Sc:	MVol.D6-141
$C_6H_7HoN_3O^{2+}$	$Ho[4-H_2NNHC(O)C_5H_4N]^{2+}$	Sc:	MVol.D2-248
$C_6H_7HoO_6{}^{2+}$	$[Ho((HOCH_2CHOH)(HO)C_4HO_3)]^{2+}$	Sc:	MVol.D3-286
$C_6H_7I_2MnNO$	$Mn(4-CH_3C_5H_4NO)I_2$	Mn:	MVol.D3-157
$C_6H_7I_3NSb$	$NH_2C_6H_4SbI_2\cdot HI$	Sb:	Org.Comp.2-117
$C_6H_7LaO_6{}^{2+}$	$[La((HOCH_2CHOH)(HO)C_4HO_3)]^{2+}$	Sc:	MVol.D3-286
$C_6H_7LaO_7$	$LaO((HOCH_2CHOH)(HO)C_4HO_3)\cdot 2\,H_2O$	Sc:	MVol.D3-286
$C_6H_7LaO_7{}^{2+}$	$[La(HOC(CH_2COOH)_2COO)]^{2+}$	Sc:	MVol.D5-344/51
$C_6H_7LaO_8$	$La(C_6H_7O_8)\cdot 6\,H_2O = [La(C_6H_7O_8)(H_2O)_3]\cdot 3\,H_2O$	Sc:	MVol.D5-340/1
$C_6H_7LuO_6{}^{2+}$	$[Lu((HOCH_2CHOH)(HO)C_4HO_3)]^{2+}$	Sc:	MVol.D3-286
$C_6H_7MnN^{2+}$	$Mn(2-CH_3C_5H_4N)^{2+}$	Mn:	MVol.D3-112
–	$Mn(3-CH_3C_5H_4N)^{2+}$	Mn:	MVol.D3-112
–	$Mn(4-CH_3C_5H_4N)^{2+}$	Mn:	MVol.D3-112
$C_6H_7MnNO^{2+}$	$Mn(NC_5H_4CH_2OH)^{2+}$	Mn:	MVol.D3-130
$C_6H_7MnNO_4$	$Mn[(OOC)_2C_4H_6NH]$	Mn:	MVol.D4-306
$C_6H_7MnNO_4S$	$Mn(C_6H_5NH_2)SO_4\cdot C_2H_5OH\cdot 1.5\,H_2O$	Mn:	MVol.D3-30
$C_6H_7MnN_3O^{2+}$	$Mn(NC_5H_4C(O)NHNH_2)^{2+}$	Mn:	MVol.D5-190
$C_6H_7MnN_3O_5S$	$Mn(NC_5H_4C(O)NHNH_2)SO_4$	Mn:	MVol.D5-188/9
–	$Mn(NC_5H_4C(O)NHNH_2)SO_4\cdot 3\,H_2O$	Mn:	MVol.D5-193
$C_6H_7MnO_7{}^+$	$[Mn(OC_3H_4(COOH)_3)]^+$	Mn:	MVol.D2-180/1
$C_6H_7MnO_8{}^-$	$[Mn(HOC_3H_4(COO)_3)(H_2O)]^-$	Mn:	MVol.D2-183
–	$[Mn(OCOCH(O)(CHOH)_3COO)]^-$	Mn:	MVol.D2-179
$C_6H_7NNd^{3+}$	$[Nd(CH_3C_5H_4N)]^{3+}$	Sc:	MVol.D1-36/7
$C_6H_7NO_8Tc^{2-}$	$[TcO(OH)(N(CH_2COO)_3)]^{2-}$	Tc:	SVol.2-127, 221
$C_6H_7NO_8U$	$UO_2[HOOCCH_2N(CH_2COO)_2]$	U:	SVol.C13-262/4
–	$UO_2[HOOCCH_2N(CH_2COO)_2]\cdot 2\,H_2O$	U:	SVol.C13-262/4
–	$UO_2[HOOCCH_2N(CH_2COO)_2]\cdot 5\,H_2O$	U:	SVol.C13-262/4
$C_6H_7NPr^{3+}$	$[Pr(CH_3C_5H_4N)]^{3+}$	Sc:	MVol.D1-36/7
$C_6H_7N_3NdO^{2+}$	$Nd[4-H_2NNHC(O)C_5H_4N]^{2+}$	Sc:	MVol.D2-248
$C_6H_7N_3OPr^{2+}$	$Pr[4-H_2NNHC(O)C_5H_4N]^{2+}$	Sc:	MVol.D2-248
$C_6H_7N_3OS$	$[(O)SC_5H_4N_3CH_3]$	S:	S-N Comp.3-86/7
$C_6H_7N_3OSm^{2+}$	$Sm[4-H_2NNHC(O)C_5H_4N]^{2+}$	Sc:	MVol.D2-248
$C_6H_7N_3OTb^{2+}$	$Tb[4-H_2NNHC(O)C_5H_4N]^{2+}$	Sc:	MVol.D2-248
$C_6H_7N_3O_2S$	$SN_2C_2(CONH_2)OCH_2CHCH_2$	S:	S-N Comp.3-93/8, 176
$C_6H_7N_3O_9U$	$UO_2(NO_3)_2\cdot CH_3C_5H_4NO$	U:	SVol.D2-56
$C_6H_7NdO_6{}^{2+}$	$[Nd((HOCH_2CHOH)(HO)C_4HO_3)]^{2+}$	Sc:	MVol.D3-286
$C_6H_7NdO_7$	$NdO((HOCH_2CHOH)(HO)C_4HO_3)\cdot 2\,H_2O$	Sc:	MVol.D3-286
$C_6H_7NdO_7{}^{2+}$	$[Nd(HOC(CH_2COOH)_2COO)]^{2+}$	Sc:	MVol.D5-344/51

C$_6$H$_8$MnO$_6$.........	Mn[C$_2$H$_4$(OCH$_2$COO)$_2$]	Mn:	MVol.D2-138
C$_6$H$_8$MnO$_6$$^-$	[Mn(OCH(CH$_3$)COO)$_2$]$^-$	Mn:	MVol.D2-151
C$_6$H$_8$MnO$_{10}$$^-$	[Mn(CH$_2$(COO)$_2$)$_2$(H$_2$O)$_2$]$^-$	Mn:	MVol.D2-128/30
C$_6$H$_8$MnO$_{10}$$^{2-}$	[Mn(CH$_2$(COO)$_2$)$_2$(H$_2$O)$_2$]$^{2-}$	Mn:	MVol.D2-129
C$_6$H$_8$MnO$_{10}$Rb	Rb[Mn(CH$_2$(COO)$_2$)$_2$(H$_2$O)$_2$].........	Mn:	MVol.D2-133, 135
C$_6$H$_8$MnO$_{10}$Tl	Tl[Mn(CH$_2$(COO)$_2$)$_2$(H$_2$O)$_2$].	Mn:	MVol.D2-132, 133, 135
C$_6$H$_8$Mn$_2$N$_8$........	(NH$_4$)$_2$Mn[Mn(CN)$_6$]	Mn:	MVol.D2-215
C$_6$H$_8$NO$_4$U$^+$	UO$_2$(CH$_3$CHCHCHCHCONHO)$^+$	U:	SVol.D1-253
C$_6$H$_8$NO$_8$U$^-$	[UO$_2$(NH$_2$C$_2$H$_4$COO)(OOCCH$_2$COO)]$^-$	U:	SVol.D1-257
C$_6$H$_8$NO$_9$U$^-$	[UO$_2$(OOCCH$_2$OCH$_2$COO)(NH$_2$CH$_2$COO)]$^-$	U:	SVol.D1-259
C$_6$H$_8$N$_2$Na$_2$O$_{14}$U	Na$_2$(NH$_4$)$_2$[UO$_2$(C$_2$O$_4$)$_3$] · 3 H$_2$O	U:	SVol.C13-201
C$_6$H$_8$N$_2$O$_2$S........	SN$_2$C$_2$HCH$_2$COOC$_2$H$_5$...............	S:	S-N Comp.3-93/8, 124
C$_6$H$_8$N$_2$O$_6$OsS$_4$.....	[OsO$_2$(S$_2$CNHCH$_2$COOH)$_2$].........	Os:	SVol.1-290
C$_6$H$_8$N$_2$O$_8$Os.......	Os(C$_2$H$_4$(NH$_2$)$_2$)(C$_2$O$_4$)$_2$	Os:	SVol.1-252
C$_6$H$_8$N$_2$O$_{16}$U$_2$	(NH$_4$)$_2$[(UO$_2$)$_2$(C$_2$O$_4$)$_3$]	U:	SVol.A6-37
		U:	SVol.C13-203/6
–	(NH$_4$)$_2$[(UO$_2$)$_2$(C$_2$O$_4$)$_3$] · 3 H$_2$O	U:	SVol.C13-203/5
C$_6$H$_8$N$_2$S	SN$_2$C$_2$(CH$_2$)$_4$.....................	S:	S-N Comp.3-191
C$_6$H$_8$N$_3$NdO$_2$$^{2+}$	[Nd(N$_2$C$_3$H$_3$CH$_2$CH(NH$_2$)COO)]$^{2+}$	Sc:	MVol.D1-111/2
C$_6$H$_8$N$_3$O$_2$Pr^{2+}	[Pr(N$_2$C$_3$H$_3$CH$_2$CH(NH$_2$)COO)]$^{2+}$	Sc:	MVol.D1-111/2
C$_6$H$_8$N$_3$O$_2$Sm^{2+}	[Sm(N$_2$C$_3$H$_3$CH$_2$CH(NH$_2$)COO)]$^{2+}$	Sc:	MVol.D1-111/2
C$_6$H$_8$N$_3$O$_2$Yb^{2+}	[Yb(N$_2$C$_3$H$_3$CH$_2$CH(NH$_2$)COO)]$^{2+}$	Sc:	MVol.D1-111/2
C$_6$H$_8$N$_3$O$_4$U$^+$.....	[UO$_2$(OOCCH(NH$_2$)CH$_2$N$_2$C$_3$H$_3$)]$^+$	U:	SVol.D1-219
C$_6$H$_8$N$_4$Nd$_2$O$_{12}$	Nd$_2$(N$_2$H$_2$)$_2$(C$_2$O$_4$)$_3$	Sc:	MVol.D1-14
C$_6$H$_8$N$_4$O$_3$S	[(O)SNC(O)NCH$_3$C(O)CHCNCH$_3$NH]......	S:	S-N Comp.3-90
C$_6$H$_8$N$_4$O$_{10}$Th	Th(C$_2$O$_4$)$_2$ · 2 (NH$_2$)$_2$CO	Th:	SVol.E-39, 42
–	Th(C$_2$O$_4$)$_2$ · 2 (NH$_2$)$_2$CO · 2 H$_2$O	Th:	SVol.E-39, 42
C$_6$H$_8$N$_6$NaRhS$_4$.....	Na[Rh(NH$_2$C$_2$H$_4$NH$_2$)(SCN)$_4$]	Rh:	SVol.B2-184
C$_6$H$_8$N$_7$O$_6$Rh.......	[Rh(NHCHNCHCH)$_2$(NO$_2$)$_2$]NO$_2$	Rh:	SVol.B2-241
C$_6$H$_8$N$_8$S$_6$Tc	(NH$_4$)$_2$[Tc(NCS)$_6$]	Tc:	SVol.1-329
		Tc:	SVol.2-167, 200/1
C$_6$H$_8$NdO$_4$$^+$	[Nd((CH$_2$CH$_2$COO)$_2$)]$^+$	Sc:	MVol.D5-177/8
C$_6$H$_8$NdO$_4$S$^+$.......	Nd[S(C$_2$H$_4$COO)$_2$]$^+$..............	Sc:	MVol.D4-65
C$_6$H$_8$NdO$_6$$^+$	[Nd(C$_2$H$_4$(OCH$_2$COO)$_2$)]$^+$...........	Sc:	MVol.D5-193/5
C$_6$H$_8$O$_4$S$_2$Sm$^+$	[Sm(C$_2$H$_4$(SCH$_2$COO)$_2$)]$^+$...........	Sc:	MVol.D4-67
C$_6$H$_8$O$_4$Sm$^+$	[Sm((CH$_2$CH$_2$COO)$_2$)]$^+$	Sc:	MVol.D5-177/8
C$_6$H$_8$O$_4$Sn	(CH$_3$)$_2$SnOOCCHCHCOO	Sn:	Org.Comp.14-103
C$_6$H$_8$O$_4$U^{2+}	[U(OOC(CH$_2$)$_4$COO)]$^{2+}$	U:	SVol.D1-208
C$_6$H$_8$O$_6$Pr$^+$	[Pr(C$_2$H$_4$(OCH$_2$COO)$_2$)]$^+$	Sc:	MVol.D5-193/5
C$_6$H$_8$O$_6$SU	UO$_2$((OOCC$_2$H$_4$)$_2$S)................	U:	SVol.D1-210
C$_6$H$_8$O$_6$Sm$^+$	[Sm(C$_2$H$_4$(OCH$_2$COO)$_2$)]$^+$	Sc:	MVol.D5-193/5
C$_6$H$_8$O$_6$Tb$^+$	[Tb(C$_2$H$_4$(OCH$_2$COO)$_2$)]$^+$	Sc:	MVol.D5-193/5
C$_6$H$_8$O$_6$Tm$^+$	[Tm(C$_2$H$_4$(OCH$_2$COO)$_2$)]$^+$.	Sc:	MVol.D5-193/5
C$_6$H$_8$O$_6$U	UO$_2$(OOC(CH$_2$)$_4$COO)............	U:	SVol.D1-208
–	UO$_2$(OOCCH(CH$_3$)CH(CH$_3$)COO)	U:	SVol.D1-205, 207/8
–	UO$_2$(OOCCH$_2$C(CH$_3$)$_2$COO)	U:	SVol.D1-205/7
C$_6$H$_8$O$_6$Yb$^+$	[Yb(C$_2$H$_4$(OCH$_2$COO)$_2$)]$^+$	Sc:	MVol.D5-193/5
C$_6$H$_8$O$_{10}$U$^-$	[UO$_2$(OOCCH$_2$OCH$_2$COO)(HOCH$_2$COOH)]$^-$	U:	SVol.D1-259
C$_6$H$_8$S$_2$Sn	(CH$_2$CH)$_2$Sn(SCH)$_2$.................	Sn:	Org.Comp.10-134
C$_6$H$_8$Sm...........	Sm(C$_3$H$_4$)$_2$....................	Sc:	MVol.D6-161

C$_6$H$_8$Yb	Yb(C$_3$H$_4$)$_2$	Sc:	MVol.D6-161
C$_6$H$_9$HoNO$_5$$^+$	[Ho(HOC$_2$H$_4$N(CH$_2$COO)$_2$)]$^+$	Sc:	MVol.D1-128/9
C$_6$H$_9$HoN$_2$O$_8$	Ho[HOC$_2$H$_4$N(CH$_2$COO)$_2$](NO$_3$) · 2 H$_2$O . . .	Sc:	MVol.D1-130
C$_6$H$_9$HoO$_3$$^{2+}$	[Ho((CH$_2$)$_4$C(OH)COO)]$^{2+}$	Sc:	MVol.D5-278/9
C$_6$H$_9$HoO$_6$	Ho(CH$_3$COO)$_3$	Sc:	MVol.D5-24, 27/9, 32
−	Ho(CH$_3$COO)$_3$ · 2 H$_2$O	Sc:	MVol.D5-34/5
−	Ho(CH$_3$COO)$_3$ · 3 H$_2$O	Sc:	MVol.D5-43
−	Ho(CH$_3$COO)$_3$ · 4 H$_2$O	Sc:	MVol.D5-34/9, 44
C$_6$H$_9$HoO$_6$S$_3$	Ho(OOCCH$_2$SH)$_3$ · 3 H$_2$O	Sc:	MVol.D4-54/5
C$_6$H$_9$HoO$_7$	Ho[HOCH$_2$(CHOH)$_2$(CH(O))$_2$COO]	Sc:	MVol.D5-275/8
C$_6$H$_9$HoO$_9$	Ho(HOCH$_2$COO)$_3$	Sc:	MVol.D5-222/30, 238/9
−	Ho(HOCH$_2$COO)$_3$ · 2 H$_2$O	Sc:	MVol.D5-233/4, 236, 238/9
C$_6$H$_9$HoO$_{12}$	Ho((HO)$_2$CHCOO)$_3$	Sc:	MVol.D5-249/50
C$_6$H$_9$ISn	(CH$_2$CH)$_3$SnI	Sn:	Org.Comp.8-35/6
C$_6$H$_9$I$_2$Sb	(CH$_2$CH)$_3$SbI$_2$	Sb:	Org.Comp.4-96
C$_6$H$_9$KO$_8$Os	K[OsO$_2$(CH$_3$COO)$_3$]	Os:	SVol.1-184
C$_6$H$_9$KO$_8$U	KUO$_2$(CH$_3$COO)$_3$	U:	SVol.A6-51
		U:	SVol.C13-126, 133
−	KUO$_2$(CH$_3$COO)$_3$ · H$_2$O	U:	SVol.C13-126, 133
C$_6$H$_9$K$_5$O$_{19}$U$_2$	K$_5$[(UO$_2$)$_2$(CH$_2$(COO)$_2$)$_2$(OO)$_2$(OH)(H$_2$O)$_2$] . .	U:	SVol.C13-238, 240/1
C$_6$H$_9$LaNO$_5$$^+$	[La(HOC$_2$H$_4$N(CH$_2$COO)$_2$)]$^+$	Sc:	MVol.D1-128/9
C$_6$H$_9$LaN$_2$O$_8$	La[HOC$_2$H$_4$N(CH$_2$COO)$_2$](NO$_3$) · H$_2$O	Sc:	MVol.D1-130
C$_6$H$_9$LaN$_3$O$_2$$^{3+}$	[La(N$_2$C$_3$H$_3$CH$_2$CH(NH$_2$)COOH)]$^{3+}$	Sc:	MVol.D1-112
C$_6$H$_9$LaO$_2$$^{2+}$	[La(CH$_3$COCHCOCH$_2$CH$_3$)]$^{2+}$	Sc:	MVol.D3-116
C$_6$H$_9$LaO$_2$S^{2+}	[La(CH$_3$C(S)CHCOOC$_2$H$_5$)]$^{2+}$	Sc:	MVol.D4-100
C$_6$H$_9$LaO$_3$$^{2+}$	[La((CH$_2$)$_4$C(OH)COO)]$^{2+}$	Sc:	MVol.D5-278/9
−	[La(CH$_3$COCHCOOC$_2$H$_5$)]$^{2+}$	Sc:	MVol.D5-363
C$_6$H$_9$LaO$_6$	La(CH$_3$COO)$_3$	Sc:	MVol.D5-23/8, 31/3, 43
−	La(CD$_3$COO)$_3$	Sc:	MVol.D5-32
−	La(CH$_3$COO)$_3$ · H$_2$O	Sc:	MVol.D5-35, 43/4
−	La(CH$_3$COO)$_3$ · 1.5 H$_2$O	Sc:	MVol.D5-34/6, 39/40, 43
−	La(CH$_3$COO)$_3$ · 2 H$_2$O	Sc:	MVol.D5-34/5
−	La(CH$_3$COO)$_3$ · 5 H$_2$O	Sc:	MVol.D5-35, 43
−	La(CH$_3$COO)$_3$ systems		
	La(CH$_3$COO)$_3$–CH$_3$COOH–H$_2$O	Sc:	MVol.D5-35
C$_6$H$_9$LaO$_6$S$_3$	La(OOCCH$_2$SH)$_3$	Sc:	MVol.D4-53/4
−	La(OOCCH$_2$SH)$_3$ · 3 H$_2$O	Sc:	MVol.D4-54/5
C$_6$H$_9$LaO$_7$	La[HOCH$_2$(CHOH)$_2$(CH(O))$_2$COO]	Sc:	MVol.D5-275/8
C$_6$H$_9$LaO$_9$	La(HOCH$_2$COO)$_3$	Sc:	MVol.D5-222/30, 234/5, 238/9
C$_6$H$_9$LaO$_{12}$	La((HO)$_2$CHCOO)$_3$	Sc:	MVol.D5-249/50
C$_6$H$_9$LuNO$_5$$^+$	[Lu(HOC$_2$H$_4$N(CH$_2$COO)$_2$)]$^+$	Sc:	MVol.D1-128/9
C$_6$H$_9$LuN$_2$O$_8$	Lu[HOC$_2$H$_4$N(CH$_2$COO)$_2$](NO$_3$) · 2 H$_2$O . . .	Sc:	MVol.D1-130
C$_6$H$_9$LuO$_3$$^{2+}$	[Lu((CH$_2$)$_4$C(OH)COO)]$^{2+}$	Sc:	MVol.D5-278/9
C$_6$H$_9$LuO$_6$	Lu(CH$_3$COO)$_3$	Sc:	MVol.D5-25, 32, 43
−	Lu(CH$_3$COO)$_3$ · 3 H$_2$O	Sc:	MVol.D5-44
−	Lu(CH$_3$COO)$_3$ · 4 H$_2$O	Sc:	MVol.D5-34/6, 39, 44
C$_6$H$_9$LuO$_6$S$_3$	Lu(OOCCH$_2$SH)$_3$ · 3 H$_2$O	Sc:	MVol.D4-54/5
C$_6$H$_9$LuO$_7$	Lu[HOCH$_2$(CHOH)$_2$(CH(O))$_2$COO]	Sc:	MVol.D5-275/8

$C_6H_9LuO_9$	$Lu(HOCH_2COO)_3$	Sc:	MVol.D5-222/7, 238/9
–	$Lu(HOCH_2COO)_3 \cdot 2\ H_2O$	Sc:	MVol.D5-233/4, 236, 238/9
$C_6H_9LuO_{12}$	$Lu((HO)_2CHCOO)_3$	Sc:	MVol.D5-249/50
$C_6H_9MnNO_5$	$Mn(HOCH_2CH_2N(CH_2COO)_2)$	Mn:	MVol.D5-7/13
–	$[Mn(HOCH_2CH_2N(CH_2COO)_2)(H_2O)] \cdot H_2O$	Mn:	MVol.D5-14/5
$C_6H_9MnN_2O_2{}^+$	$Mn(ONC_6H_8NOH)^+$	Mn:	MVol.D5-266/7
$C_6H_9MnN_3O$	$Mn(OCCH_3N(CH_3)_2)(CN)_2$	Mn:	MVol.D5-102
$C_6H_9MnN_3O_2{}^{2+}$	$Mn[4-OOCCH(NH_3)CH_2-1,3-C_3H_2N_2H]^{2+}$ ..	Mn:	MVol.D4-275/7
$C_6H_9MnN_5O_8$	$[Mn(NO_3)_2(NC_5H_4C(O)NHNH_2)(H_2O)]$	Mn:	MVol.D5-190/1
$C_6H_9MnN_6O_6{}^-$	$[Mn(ONCCH_3NO)_3]^-$	Mn:	MVol.D5-277
$C_6H_9MnN_6O_6{}^{2-}$	$[Mn(ONCCH_3NO)_3]^{2-}$	Mn:	MVol.D5-277
$C_6H_9MnN_7O_4$	$Mn[NHNCHC(NO_2)CH]_2 \cdot NH_3 \cdot H_2O$	Mn:	MVol.D3-279
$C_6H_9MnN_{11}O_6$.....	$Mn(1,2,4-C_2H_2N_3H)_3(NO_3)_2 \cdot 1.5\ H_2O$.....	Mn:	MVol.D4-63
$C_6H_9MnO_3{}^+$	$[Mn(CH_3COCHCOOC_2H_5)]^+$	Mn:	MVol.D2-189
$C_6H_9MnO_6$........	$Mn(CH_3COO)_3$	Mn:	MVol.A1-165
		Mn:	MVol.D2-29/30, 55/8, 60
–	$Mn(CH_3COO)_3 \cdot 2\ H_2O$	Mn:	MVol.D2-29/30, 58/60
$C_6H_9MnO_6{}^-$	$[Mn(CH_3COO)_3]^-$	Mn:	MVol.D2-30/1
$C_6H_9MnO_{10}$.......	$H[Mn(CH_2(COO)_2)_2(H_2O)_2] \cdot H_2O$	Mn:	MVol.D2-133
–	$H[Mn(CH_2(COO)_2)_2(H_2O)_2] \cdot 4\ H_2O$.	Mn:	MVol.D2-133
$C_6H_9Mn_2O_8$........	$Mn(CH_3COO)_3 \cdot MnO_2$............	Mn:	MVol.D2-61
$C_6H_9NNdO_5{}^+$	$[Nd(HOC_2H_4N(CH_2COO)_2)]^+$	Sc:	MVol.D1-128/9
$C_6H_9NO_5Pm^+$	$[Pm(HOC_2H_4N(CH_2COO)_2)]^+$	Sc:	MVol.D1-128/9
$C_6H_9NO_5Pr^+$	$[Pr(HOC_2H_4N(CH_2COO)_2)]^+$	Sc:	MVol.D1-128/9
$C_6H_9NO_5Sm^+$	$[Sm(HOC_2H_4N(CH_2COO)_2)]^+$	Sc:	MVol.D1-128/9
$C_6H_9NO_5Tb^+$	$[Tb(HOC_2H_4N(CH_2COO)_2)]^+$.........	Sc:	MVol.D1-128/9
$C_6H_9NO_5Tm^+$	$[Tm(HOC_2H_4N(CH_2COO)_2)]^+$	Sc:	MVol.D1-128/9
$C_6H_9NO_5Y^+$	$[Y(HOC_2H_4N(CH_2COO)_2)]^+$..........	Sc:	MVol.D1-128/9
$C_6H_9NO_5Yb^+$	$[Yb(HOC_2H_4N(CH_2COO)_2)]^+$........	Sc:	MVol.D1-128/9
$C_6H_9NO_7U$.........	$UO_2((OOCCH_2)_2NC_2H_4OH)$	U:	SVol.D1-220/1
C_6H_9NSSn........	$(CH_2CH)_2(CH_3)SnNCS$	Sn:	Org.Comp.8-183
$C_6H_9N_2NdO_8$.......	$Nd[HOC_2H_4N(CH_2COO)_2](NO_3) \cdot H_2O$....	Sc:	MVol.D1-130
$C_6H_9N_2O_8Pr$	$Pr[HOC_2H_4N(CH_2COO)_2](NO_3) \cdot H_2O$	Sc:	MVol.D1-130
$C_6H_9N_2O_8Sm$	$Sm[HOC_2H_4N(CH_2COO)_2](NO_3) \cdot 3\ H_2O$...	Sc:	MVol.D1-130
$C_6H_9N_2O_8Tb$	$Tb[HOC_2H_4N(CH_2COO)_2](NO_3) \cdot 2\ H_2O$ ...	Sc:	MVol.D1-130
$C_6H_9N_2O_8Tm$	$Tm[HOC_2H_4N(CH_2COO)_2](NO_3) \cdot H_2O$	Sc:	MVol.D1-130
$C_6H_9N_2O_8Yb$.......	$Yb[HOC_2H_4N(CH_2COO)_2](NO_3) \cdot 2\ H_2O$ ...	Sc:	MVol.D1-130
$C_6H_9N_3NdO_2{}^{3+}$	$[Nd(N_2C_3H_3CH_2CH(NH_2)COOH)]^{3+}$	Sc:	MVol.D1-112
$C_6H_9N_3O_2Pr^{3+}$	$[Pr(N_2C_3H_3CH_2CH(NH_2)COOH)]^{3+}$	Sc:	MVol.D1-112
$C_6H_9N_3O_2S$........	$SN_2C_2OHNC_4H_8O$	S:	S-N Comp.3-93/8, 154
$C_6H_9N_3O_2Sm^{3+}$	$[Sm(N_2C_3H_3CH_2CH(NH_2)COOH)]^{3+}$	Sc:	MVol.D1-112
$C_6H_9N_3O_2Yb^{3+}$	$[Yb(N_2C_3H_3CH_2CH(NH_2)COOH)]^{3+}$	Sc:	MVol.D1-112
$C_6H_9N_3O_3S$........	$(O)SN_2C_2(OH)(N(C_2H_4)_2O)$	S:	S-N Comp.3-250/7
$C_6H_9N_3O_9U$.......	$UO_2(NO_3)_2(H_2O)(CH_3C_5H_4N)$	U:	SVol.A5-217/8
$C_6H_9NaO_8U$........	$NaUO_2(CH_3COO)_3$..................	U:	SVol.A5-125, 129, 134, 137/9, 216, 220, 222
		U:	SVol.A6-32, 43, 49/56, 59/61, 64
		U:	SVol.C13-125/6, 131/3

$C_6H_9O_6S_3Tm$	$Tm(OOCCH_2SH)_3 \cdot 3 H_2O$	Sc:	MVol.D4-54/5
$C_6H_9O_6S_3Y$	$Y(OOCCH_2SH)_3 \cdot 3 H_2O$	Sc:	MVol.D4-54/5
$C_6H_9O_6S_3Yb$	$Yb(OOCCH_2SH)_3 \cdot 3 H_2O$	Sc:	MVol.D4-54/5
$C_6H_9O_6Sc$	$Sc(CH_3COO)_3$	Sc:	MVol.D5-21
–	$Sc(CH_3COO)_3 \cdot H_2O$	Sc:	MVol.D5-21
$C_6H_9O_6Sm$	$Sm(CH_3COO)_3$	Sc:	MVol.D5-24/31, 43
–	$Sm(CH_3COO)_3 \cdot 0.5 H_2O$	Sc:	MVol.D5-32, 44
–	$Sm(CH_3COO)_3 \cdot H_2O$	Sc:	MVol.D5-43
–	$Sm(CH_3COO)_3 \cdot 2 H_2O$	Sc:	MVol.D5-34/5
–	$Sm(CH_3COO)_3 \cdot 4 H_2O$	Sc:	MVol.D5-34/6, 39/40, 43/4
$C_6H_9O_6Tb$	$Tb(CH_3COO)_3$	Sc:	MVol.D5-24, 32/3, 43
–	$Tb(CH_3COO)_3 \cdot 0.5 H_2O$	Sc:	MVol.D5-43
–	$Tb(CH_3COO)_3 \cdot 2 H_2O$	Sc:	MVol.D5-34/5
–	$Tb(CH_3COO)_3 \cdot 4 H_2O$	Sc:	MVol.D5-34/6, 39, 43/4
$C_6H_9O_6Tm$	$Tm(CH_3COO)_3$	Sc:	MVol.D5-25, 32
–	$Tm(CH_3COO)_3 \cdot 3.5 H_2O$	Sc:	MVol.D5-44
–	$Tm(CH_3COO)_3 \cdot 4 H_2O$	Sc:	MVol.D5-36/9, 41/2
$C_6H_9O_6U$	$U(CH_3COO)_3$	U:	SVol.D1-200
$C_6H_9O_6U^+$	$[U(CH_3COO)_3]^+$	U:	SVol.D1-200
–	$[UO_2H(OOC(CH_2)_4COO)]^+$	U:	SVol.D1-208
–	$[UO_2H(OOCCH(CH_3)CH(CH_3)COO)]^+$	U:	SVol.D1-207/8
–	$[UO_2H(OOCCH_2C(CH_3)_2COO)]^+$	U:	SVol.D1-207
$C_6H_9O_6Y$	$Y(CH_3COO)_3$	Sc:	MVol.D5-23, 27/8, 32/3, 43
–	$Y(CH_3COO)_3 \cdot 2 H_2O$	Sc:	MVol.D5-43
–	$Y(CH_3COO)_3 \cdot 4 H_2O$	Sc:	MVol.D5-34/6, 43/4
–	$Y(CH_3COO)_3$ systems $Y(CH_3COO)_3-CH_3COOH-H_2O$	Sc:	MVol.D5-35
$C_6H_9O_6Yb$	$Yb(CH_3COO)_3$	Sc:	MVol.D5-25/9, 32
–	$Yb(CH_3COO)_3 \cdot 4 H_2O$	Sc:	MVol.D5-34/6, 39, 43/4
$C_6H_9O_7Pr$	$Pr[HOCH_2(CHOH)_2(CH(O))_2COO]$	Sc:	MVol.D5-275/8
$C_6H_9O_7Sm$	$Sm[HOCH_2(CHOH)_2(CH(O))_2COO]$	Sc:	MVol.D5-275/8
$C_6H_9O_7Tb$	$Tb[HOCH_2(CHOH)_2(CH(O))_2COO]$	Sc:	MVol.D5-275/8
$C_6H_9O_7Tm$	$Tm[HOCH_2(CHOH)_2(CH(O))_2COO]$	Sc:	MVol.D5-275/8
$C_6H_9O_7Y$	$Y[HOCH_2(CHOH)_2(CH(O))_2COO]$	Sc:	MVol.D5-275/8
$C_6H_9O_7Y^{2+}$	$[Y(OCH(CHOH)_4COO)]^{2+}$	Sc:	MVol.D5-281
$C_6H_9O_7Yb$	$Yb[HOCH_2(CHOH)_2(CH(O))_2COO]$	Sc:	MVol.D5-275/8
$C_6H_9O_8RbU$	$RbUO_2(CH_3COO)_3$	U:	SVol.A5-125, 134, 138, 220
		U:	SVol.A6-49/51, 55, 64
		U:	SVol.C13-127, 133/4
$C_6H_9O_8S_3U^-$	$[UO_2(HSCH_2COO)_3]^-$	U:	SVol.D1-202
$C_6H_9O_8TlU$	$Tl[UO_2(CH_3COO)_3]$	U:	SVol.C13-128, 133
$C_6H_9O_8U^-$	$[UO_2(CH_3COO)_3]^-$	U:	SVol.C13-131/2, 134
		U:	SVol.D1-200
		U:	SVol.D2-377
		U:	SVol.D3-383
–	$[UO_2(CD_3COO)_3]^-$	U:	SVol.C13-126, 131/3

$C_6H_{10}LaO_6{}^+$	$[La(H_3CCHOHCOO)_2]^+$	Sc:	MVol.D5-250/5
$C_6H_{10}LaO_7{}^+$	$[La(OCH_2(CHOH)_4COO)]^+$	Sc:	MVol.D5-275/8
$C_6H_{10}La_3O_{14}{}^{3+}$	$[La_3(HCOO)_6 \cdot 2 H_2O]^{3+}$	Sc:	MVol.D5-4/5
$C_6H_{10}LuN_2O_4{}^+$	$Lu[C_2H_4(NH(CH_2COO))_2]^+$	Sc:	MVol.D1-148/9
$C_6H_{10}LuN_4O_4{}^+$	$[Lu(CH_3C(NO)CHNOH)_2]^+$	Sc:	MVol.D2-116/7
$C_6H_{10}LuO_4{}^+$	$[Lu(C_2H_5COO)_2]^+$	Sc:	MVol.D5-71/4
$C_6H_{10}LuO_6{}^+$	$[Lu(CH_3OCH_2COO)_2]^+$	Sc:	MVol.D5-69/70
—	$[Lu(HOC_2H_4COO)_2]^+$	Sc:	MVol.D5-260/1
—	$[Lu(H_3CCHOHCOO)_2]^+$	Sc:	MVol.D5-250/5
$C_6H_{10}LuO_7{}^+$	$[Lu(OCH_2(CHOH)_4COO)]^+$	Sc:	MVol.D5-275/8
$C_6H_{10}MnNO_8{}^-$	$Mn(N(CH_2COO)_3)(H_2O)_2{}^-$	Mn:	MVol.D5-15/8
$C_6H_{10}MnN_2O_4$	$Mn(C_2H_4(NHCH_2COO)_2)$	Mn:	MVol.D5-23/5
—	$Mn(H_2NCH_2CH_2N(CH_2COO)_2)$	Mn:	MVol.D5-7/13
$C_6H_{10}MnN_2O_4S_2$	$Mn[(OOCCH(NH_2)CH_2)_2S_2]$	Mn:	MVol.D4-292/3
$C_6H_{10}MnN_2O_4S_2{}^{2-}$	$Mn[SCH_2CH(NH_2)COO]_2{}^{2-}$	Mn:	MVol.D4-290
$C_6H_{10}MnN_2O_6S_2$	$Mn(C_6H_4(NH_2)_2)(HSO_3)_2$	Mn:	MVol.D3-45
$C_6H_{10}MnN_3O_2{}^{3+}$	$Mn[4-OOCCH(NH_3)CH_2-1,3-C_3H_2N_2H_2]^{3+}$	Mn:	MVol.D4-276/7
$C_6H_{10}MnN_3O_4{}^+$	$Mn[H_2NCH_2C(O)NHCH_2C(O)NHCH_2COO]^+$	Mn:	MVol.D4-346
$C_6H_{10}MnN_6O_2S_2$	$Mn(C_2H_4(C(O)NHNH_2)_2)(NCS)_2 \cdot 3 H_2O$	Mn:	MVol.D5-204/5
$C_6H_{10}MnO_4$	$Mn(C_2H_5COO)_2$	Mn:	MVol.D2-65/6
—	$Mn(C_2H_5COO)_2 \cdot 2 H_2O$	Mn:	MVol.D2-66/7
$C_6H_{10}MnO_6$	$Mn(CH_3CHOHCOO)_2$	Mn:	MVol.D2-151/2
—	$Mn(CH_3CHOHCOO)_2 \cdot 3 H_2O$	Mn:	MVol.D2-152
—	$Mn(CH_3COO)_2 \cdot CH_3COOH \cdot H_2O$	Mn:	MVol.D2-29, 51/2
—	$Mn(CH_3COO)_2 \cdot CH_3COOH \cdot 1.5 H_2O$	Mn:	MVol.D2-29, 50/2
$C_6H_{10}NO_8U^-$	$[UO_2(OH)((OOCCH_2)_2NC_2H_4OH)]^-$	U:	SVol.D1-220/1
$C_6H_{10}N_2NdO_4{}^+$	$Nd[C_2H_4(NHCH_2COO)_2]^+$	Sc:	MVol.D1-148/9
$C_6H_{10}N_2NdO_4S_2{}^-$	$[Nd(SCH_2CH(NH_2)COO)_2]^-$	Sc:	MVol.D4-59/60
$C_6H_{10}N_2OS$	$SN_2C_2OHC_4H_9-n$	S:	S-N Comp.3-93/8, 150/1
—	$SN_2C_2OHC_4H_9-i$	S:	S-N Comp.3-150/1
$C_6H_{10}N_2O_2S$	$SN_2C_2(OC_2H_5)_2$	S:	S-N Comp.3-93/8, 159
$C_6H_{10}N_2O_3S$	$(O)SN_2C_2(OC_2H_5)_2$	S:	S-N Comp.3-250/7
$C_6H_{10}N_2O_4Pr^+$	$Pr[C_2H_4(NHCH_2COO)_2]^+$	Sc:	MVol.D1-148/9
$C_6H_{10}N_2O_4PrS_2{}^-$	$[Pr(SCH_2CH(NH_2)COO)_2]^-$	Sc:	MVol.D4-59/60
$C_6H_{10}N_2O_4S_2Sm^-$	$[Sm(SCH_2CH(NH_2)COO)_2]^-$	Sc:	MVol.D4-59/60
$C_6H_{10}N_2O_4S_2Tb^-$	$[Tb(SCH_2CH(NH_2)COO)_2]^-$	Sc:	MVol.D4-59/60
$C_6H_{10}N_2O_4S_2Tm^-$	$[Tm(SCH_2CH(NH_2)COO)_2]^-$	Sc:	MVol.D4-59/60
$C_6H_{10}N_2O_4Sm^+$	$Sm[C_2H_4(NHCH_2COO)_2]^+$	Sc:	MVol.D1-148/9
$C_6H_{10}N_2O_4Tb^+$	$Tb[C_2H_4(NHCH_2COO)_2]^+$	Sc:	MVol.D1-148/9
$C_6H_{10}N_2O_4Tm^+$	$Tm[C_2H_4(NHCH_2COO)_2]^+$	Sc:	MVol.D1-148/9
$C_6H_{10}N_2O_4Y^+$	$Y[C_2H_4(NHCH_2COO)_2]^+$	Sc:	MVol.D1-148/9
$C_6H_{10}N_2O_4Yb^+$	$Yb[C_2H_4(NHCH_2COO)_2]^+$	Sc:	MVol.D1-148/9
$C_6H_{10}N_2O_6U$	$UO_2((OOCCH_2NH)_2C_2H_4)$	U:	SVol.D1-221
$C_6H_{10}N_2O_{10}U$	$[NH_3C_2H_4NH_3][UO_2(C_2O_4)_2] \cdot 2 H_2O$	U:	SVol.C13-196/7
$C_6H_{10}N_2O_{16}S_2Sc_2$	$C_2H_8N_2H_2[Sc(SO_4)(C_2O_4)]_2$	Sc:	MVol.D5-118
—	$C_2H_8N_2H_2[Sc(SO_4)(C_2O_4)]_2 \cdot 4 H_2O$	Sc:	MVol.D5-118
$C_6H_{10}N_2S_2Sn$	$(C_2H_5)_2Sn(NCS)_2$	Sn:	Org.Comp.8-186/7
$C_6H_{10}N_2Sn$	$(C_2H_5)_2Sn(CN)_2$	Sn:	Org.Comp.8-157
$C_6H_{10}N_3NdO_2S_2$	$Nd(HN(CH_2CH_2O)_2H)(NCS)_2$	Sc:	MVol.D2-2
$C_6H_{10}N_3NdO_4{}^{2+}$	$Nd(H_2N(CH_2C(O)NH)_2CH_2COO)^{2+}$	Sc:	MVol.D4-282/3

$C_6H_{10}O_6U$	$UO_2(C_2H_5COO)_2$	U:	SVol.D1-201
–	$UO_2(C_2H_5COO)_2 \cdot 2\ H_2O$	U:	SVol.C13-157/8
$C_6H_{10}O_6Y^+$	$[Y(CH_3OCH_2COO)_2]^+$	Sc:	MVol.D5-69/70
–	$[Y(H_3CCHOHCOO)_2]^+$	Sc:	MVol.D5-250/5
$C_6H_{10}O_6Yb^+$	$[Yb(CH_3OCH_2COO)_2]^+$	Sc:	MVol.D5-69/70
–	$[Yb(HOC_2H_4COO)_2]^+$	Sc:	MVol.D5-260/1
–	$[Yb(H_3CCHOHCOO)_2]^+$	Sc:	MVol.D5-250/5
$C_6H_{10}O_7Pr^+$	$[Pr(OCH_2(CHOH)_4COO)]^+$	Sc:	MVol.D5-275/8
$C_6H_{10}O_7Sc^-$	$Sc(OH)(CH_3COO)_3^-$	Sc:	MVol.D5-21
$C_6H_{10}O_7Sm^+$	$[Sm(OCH_2(CHOH)_4COO)]^+$	Sc:	MVol.D5-275/8
$C_6H_{10}O_7Tb^+$	$[Tb(OCH_2(CHOH)_4COO)]^+$	Sc:	MVol.D5-275/8
$C_6H_{10}O_7Tm^+$	$[Tm(OCH_2(CHOH)_4COO)]^+$	Sc:	MVol.D5-275/8
$C_6H_{10}O_7U$	$U(OH)(CH_3COO)_3$	U:	SVol.C13-111, 113
$C_6H_{10}O_7Y^+$	$[Y(OCH_2(CHOH)_4COO)]^+$	Sc:	MVol.D5-275/8
$C_6H_{10}O_7Yb^+$	$[Yb(OCH_2(CHOH)_4COO)]^+$	Sc:	MVol.D5-275/8
$C_6H_{10}O_8U$	$H[UO_2(CH_3COO)_3] = UO_2(CH_3COO)_2$ $\cdot\ CH_3COOH$	U:	SVol.C13-125, 131
–	$UO_2[CH_3CH(OH)COO]_2$	U:	SVol.C13-164
		U:	SVol.D1-203
–	$UO_2[CH_3CH(OH)COO]_2 \cdot 2\ H_2O$	U:	SVol.C13-164
–	$UO_2[CH_3CH(OH)COO]_2 \cdot 5\ H_2O$	U:	SVol.C13-164
–	$[UO_2(CH_3OCH_2COO)_2]_n \cdot x\ H_2O$	U:	SVol.C13-147
–	$UO_2(HOCH_2CH_2COO)_2$	U:	SVol.D1-203
$C_6H_{10}O_9STh$	$Th(OOCH)_4 \cdot (CH_3)_2SO$	Th:	SVol.E-100, 103
$C_6H_{10}S_2$	$2\ H_2S \cdot C_6H_6 \cdot 17\ H_2O$	S:	SVol.4a/b-326
$C_6H_{10}Si_2$	$Si_2H_5C_6H_5$	Si:	SVol.B1-175
$C_6H_{10}Sm$	$Sm(C_6H_{10})$	Sc:	MVol.D6-159, 162
$C_6H_{10}Yb$	$HYbCCC_4H_9$	Sc:	MVol.D6-163
–	$Yb(C_6H_{10})$	Sc:	MVol.D6-159, 162
$C_6H_{11}HoO_2^{2+}$	$[Ho(CH_3(CH_2)_4COO)]^{2+}$	Sc:	MVol.D5-87
–	$[Ho(C_2H_5C(CH_3)_2COO)]^{2+}$	Sc:	MVol.D5-85
$C_6H_{11}HoO_3^{2+}$	$[Ho(i-C_3H_7C(OH)(CH_3)COO)]^{2+}$	Sc:	MVol.D5-271
–	$[Ho(HOC(C_2H_5)_2COO)]^{2+}$	Sc:	MVol.D5-270/1
$C_6H_{11}HoO_4^{2+}$	$[Ho(H_3C(C(OH)CH_3)_2COO)]^{2+}$	Sc:	MVol.D5-272
$C_6H_{11}HoO_7^{2+}$	$[Ho(HOCH_2(CHOH)_4COO)]^{2+}$	Sc:	MVol.D5-275/8
$C_6H_{11}I_2N_4O_4Rh$	$H[Rh(CH_3CNOCHNOH)_2I_2] \cdot 2\ H_2O$	Rh:	SVol.B2-219
$C_6H_{11}LaO_2^{2+}$	$[La(CH_3(CH_2)_4COO)]^{2+}$	Sc:	MVol.D5-87
–	$[La(C_2H_5C(CH_3)_2COO)]^{2+}$	Sc:	MVol.D5-85
$C_6H_{11}LaO_2S^{2+}$	$[La(SCH_2COOC_4H_9)]^{2+}$	Sc:	MVol.D4-58
$C_6H_{11}LaO_3^{2+}$	$[La(n-C_3H_7C(OH)(CH_3)COO)]^{2+}$	Sc:	MVol.D5-273
–	$[La(HOC(C_2H_5)_2COO)]^{2+}$	Sc:	MVol.D5-270/1
$C_6H_{11}LaO_3S$	$La(SCHC(OCH_3)O)(OC_3H_7-i)$	Sc:	MVol.D4-58/9
$C_6H_{11}LaO_4^{2+}$	$[La(H_3C(C(OH)CH_3)_2COO)]^{2+}$	Sc:	MVol.D5-272
$C_6H_{11}LaO_7^{2+}$	$[La(HOCH_2(CHOH)_4COO)]^{2+}$	Sc:	MVol.D5-275/8
$C_6H_{11}LaO_{10}$	$H[La(HOC(CH_2COO)_2COO)(OH)(H_2O)_2]$	Sc:	MVol.D5-354
–	$[La(HOC(CH_2COO)_2COO)(H_2O)_3]$ $= La(HOC(CH_2COO)_2COO) \cdot 3\ H_2O$	Sc:	MVol.D5-352/4
$C_6H_{11}LuO_2^{2+}$	$[Lu(CH_3(CH_2)_4COO)]^{2+}$	Sc:	MVol.D5-87
–	$[Lu(C_2H_5C(CH_3)_2COO)]^{2+}$	Sc:	MVol.D5-85
$C_6H_{11}LuO_4^{2+}$	$[Lu(H_3C(C(OH)CH_3)_2COO)]^{2+}$	Sc:	MVol.D5-272

$C_6H_{12}N_4O_{10}Rh_2$ .... $Rh_2(O_2CH)_4((NH_2)_2CO)_2$ .... Rh: SVol.B2-10
$C_6H_{12}N_5O_7Rh$ ...... $Rh(CH_3CNOCHNOH)_2(NO_2)(H_2O)$ .... Rh: SVol.B2-219
$C_6H_{12}N_6O_6U$ .... $UO_2(OOCCH_2NHC(NH)NH_2)_2$ .... U: SVol.D1-252
$C_6H_{12}N_6O_{10}U$ ...... $[C(NH_2)_3]_2[UO_2(C_2O_4)_2]$ .... U: SVol.C13-193, 196/8
− .... $[C(NH_2)_3]_2[UO_2(C_2O_4)_2]$ · 1.25 H₂O ...... U: SVol.C13-196/8
− .... $[C(NH_2)_3]_2[UO_2(C_2O_4)_2]$ · 1.9 H₂O .... U: SVol.C13-196/8
− .... $[C(NH_2)_3]_2[UO_2(C_2O_4)_2]$ · 2.1 H₂O .... U: SVol.C13-196/8
− .... $[C(NH_2)_3]_2[UO_2(C_2O_4)_2]$ · 2.8 H₂O .... U: SVol.C13-196/8
$C_6H_{12}N_6O_{12}S_3Th$ ... $Th(NO_3)_4$ · $((CH_3)_2NCS)_2S$ .... Th: SVol.C5-33/4
$C_6H_{12}N_6O_{12}S_4Th$ ... $Th(NO_3)_4$ · $((CH_3)_2NCS)_2S_2$ .... Th: SVol.C5-33/4
$C_6H_{12}N_6O_{12}Tm_2$ .... $Tm_2(N_2H_4)_3(C_2O_4)_3$ · 2 H₂O .... Sc: MVol.D1-13
$C_6H_{12}N_6O_{12}Yb_2$ .... $Yb_2(N_2H_4)_3(C_2O_4)_3$ · 3 H₂O .... Sc: MVol.D1-13
− .... $Yb_2(C_2O_4)_3$ · 3 N₂H₄ · 8 H₂O .... Sc: MVol.D5-131/2
$C_6H_{12}N_6S_2$ .... $S_2N_4C_2(N(CH_3)_2)_2$ .... S: S-N Comp.4-136/43
$C_6H_{12}N_9Rh$ .... $(NH_4)_3[Rh(CN)_6]$ .... Rh: SVol.B1-191
$C_6H_{12}Na_4O_{20}U_2$ .... $Na_4[(UO_2)_2(CH_2(COO)_2)_2(OO)_2(H_2O)_4]$ · 7 H₂O
                                                                    U: SVol.C13-238/41
$C_6H_{12}O_2SSn$ .... $(CH_3)_3SnOS(O)CH_2CCH$ .... Sn: Org.Comp.11-140/4
− .... $(C_2H_5)_2Sn(SCH_2COO)$ .... Sn: Org.Comp.10-231
$C_6H_{12}O_2S_2Sn$ ...... $(CH_3)_2Sn(OC(S)CH_3)_2$ .... Sn: Org.Comp.14-110
− .... $(CH_3)_2Sn(SC(O)CH_3)_2$ .... Sn: Org.Comp.10-16
$C_6H_{12}O_2Sn$ .... $(CH_3)_3SnOOCCHCH_2$ .... Sn: Org.Comp.11-108
$C_6H_{12}O_4Sn$ .... $(CH_3)_2Sn(OOCCH_3)_2$ .... Sn: Org.Comp.14-83, 84/5
− .... $(C_2H_5)_2Sn(OOCH)_2$ .... Sn: Org.Comp.14-170, 174
$C_6H_{12}O_6U$ .... $U(OH)_2(CH_3CH_2COO)_2$ .... U: SVol.C13-157
$C_6H_{12}O_8U$ .... $[UO_2(OOC(CH_2)_4COO)(H_2O)_2]_n$ .... U: SVol.C13-254/5
$C_6H_{12}O_9S$ .... SO_3 · 3 CH₃COOH .... S: SVol.3-315
$C_6H_{12}O_{12}U_2$ .... $[UO_2(OH)(CH_3CHOHCOO)]_2$ .... U: SVol.D1-203
$C_6H_{12}S_2Sn$ .... $(C_2H_5)_2Sn(SCH)_2$ .... Sn: Org.Comp.10-39
$C_6H_{12}Sb_2$ .... $(CH_3)_2SbCCSb(CH_3)_2$ .... Sb: Org.Comp.1-177
$C_6H_{12}Si_3$ .... $Si_3H_7C_6H_5$ .... Si: SVol.B1-175
$C_6H_{13}HoN_2O_{12}P_4^{4-}$ $Ho[H(C_2H_4(N(CH_2PO_3)_2)_2)]^{4-}$ .... Sc: MVol.D4-154/5
$C_6H_{13}ISn$ .... $CH_2(CH_2CH_2)_2Sn(CH_3)I$ .... Sn: Org.Comp.8-86
$C_6H_{13}I_3MnN_2$ ...... $[Mn(N(C_2H_4)_3NH)I_3]$ .... Mn: MVol.D4-51
$C_6H_{13}LaN_2O_{12}P_4^{4-}$ $La[H(C_2H_4(N(CH_2PO_3)_2)_2)]^{4-}$ .... Sc: MVol.D4-154/5
$C_6H_{13}LaO_3S$ .... $La(C_3H_6O_2S)(OC_3H_7-i)$ .... Sc: MVol.D4-49
$C_6H_{13}LaO_4S_2$ ..... $La(C_3H_6O_2S)(C_3H_7O_2S)$ .... Sc: MVol.D4-49
$C_6H_{13}LaO_{11}$ .... $[La(C_6H_7O_8)(H_2O)_3]$ · 3 H₂O
                                = La(C₆H₇O₈) · 6 H₂O ...... Sc: MVol.D5-340/1
$C_6H_{13}MnNO_5^{2+}$ .... $Mn[NH_2C_5H_5O(OH)_3CH_2OH]^{2+}$ .... Mn: MVol.D4-242
$C_6H_{13}MnN_2O_2^+$ .... $Mn[H_2NC_4H_8CH(NH_2)COO]^+$ .... Mn: MVol.D4-288/9
$C_6H_{13}MnN_2O_4^+$ .... $[MnH(OOCCH(CH_3)NH_2)_2]^+$ .... Mn: MVol.D4-266/7
$C_6H_{13}MnN_4O_2^+$ .... $Mn[H_2NC(NH)NHC_3H_6CH(NH_2)COO]^+$ ..... Mn: MVol.D4-288/9
$C_6H_{13}NNdO_7^+$ ...... $[Nd(HOCH_2(CHOH)_3CH(NH_2)COO)(OH)]^+$ .. Sc: MVol.D4-281/2
$C_6H_{13}NO_2S$ .... $[OS(O)N(C_4H_9-t)CH_2CH_2]$ .... S: S-N Comp.3-294/303
− .... $[OS(O)N(C_3H_7-i)CH_2CH_2CH_2]$ .... S: S-N Comp.4-86/95
$C_6H_{13}NO_2Sn$ ...... $(CH_3)_3SnOOCN(CH_2)_2$ .... Sn: Org.Comp.11-118/9
$C_6H_{13}NO_3Sn$ ...... $(CH_3)_3SnOOCCH_2NHCHO$ .... Sn: Org.Comp.11-96, 105
$C_6H_{13}NO_8U$ .... $NH_4UO_2(CH_3COO)_3$ .... U: SVol.A5-125, 134, 138
                                                                    U: SVol.A6-49/51, 55/6

$C_6H_{13}NO_8U$	$NH_4UO_2(CH_3COO)_3$	U:	SVol.C13-126, 133/4
		U:	SVol.D2-249
–	$NH_4UO_2(CH_3COO)_3 \cdot 6 H_2O$	U:	SVol.A6-185
		U:	SVol.C13-126
$C_6H_{13}NO_{10}U$	$NH_4[UO_2(OH)_2(C_6H_7O_6)]$	U:	SVol.C13-95
$C_6H_{13}NO_{11}Rh$	$[Rh(NH_3)(H_2O)_3((O_2C)_2CH_2)]((O_2C)_2CH_2)$	Rh:	SVol.B2-154
$C_6H_{13}N_2NdO_5$	$Nd[CH_3CH(NH_2)COO]_2(OH) \cdot 3 H_2O$	Sc:	MVol.D1-108/9
$C_6H_{13}N_2NdO_{12}P_4^{4-}$	$Nd[H(C_2H_4(N(CH_2PO_3)_2)_2)]^{4-}$	Sc:	MVol.D4-154/5
$C_6H_{13}N_2O_2Y^{2+}$	$Y[NH_2(CH_2)_4CH(NH_2)COO]^{2+}$	Sc:	MVol.D1-116
$C_6H_{13}N_2O_4U^+$	$[UO_2(OOCCH(NH_2)(CH_2)_4NH_2)]^+$	U:	SVol.D1-218
$C_6H_{13}N_2O_5Pr$	$Pr[CH_3CH(NH_2)COO]_2(OH) \cdot 4 H_2O$	Sc:	MVol.D1-108/9
$C_6H_{13}N_2O_{12}P_4Sm^{4-}$	$Sm[H(C_2H_4(N(CH_2PO_3)_2)_2)]^{4-}$	Sc:	MVol.D4-154/5
$C_6H_{13}N_2O_{12}P_4Tm^{4-}$	$Tm[H(C_2H_4(N(CH_2PO_3)_2)_2)]^{4-}$	Sc:	MVol.D4-154/5
$C_6H_{13}N_3SSn$	$(CH_3)_3SnSC(NCH_3)NHCN$	Sn:	Org.Comp.9-59, 60
$C_6H_{13}N_4OPS_2$	$SN_2C_2OP(S)H(N(CH_3)_2)_2$	S:	S-N Comp.3-93/8, 160
$C_6H_{13}N_4O_2PS$	$SN_2C_2OP(O)H(N(CH_3)_2)_2$	S:	S-N Comp.3-93/8, 160
$C_6H_{13}N_4O_2Y^{2+}$	$Y[NHC(NH_2)NH(CH_2)_3CH(NH_2)COO]^{2+}$	Sc:	MVol.D1-116
$C_6H_{13}N_5O_7U$	$UO_2[H_2NC(NH)NHCH_3H_6CH(NH_2)COO]NO_3$		
	$\cdot 2 H_2O$	U:	SVol.C13-172/3
$C_6H_{13}N_6O_{16}PU_2$	$(CN_3H_6)_2[(UO_2)_2(C_2O_4)_2(HPO_4)] \cdot 4 H_2O$	U:	SVol.C14-129
$C_6H_{13}NdO_3S$	$Nd(C_3H_6O_2S)(OC_3H_7-i)$	Sc:	MVol.D4-49
$C_6H_{13}NdO_4S_2$	$Nd(C_3H_6O_2S)(C_3H_7O_2S)$	Sc:	MVol.D4-49
$C_6H_{13}OSb$	$(C_2H_5)_2SbCH_2CHO$	Sb:	Org.Comp.1-119
$C_6H_{13}O_3PrS$	$Pr(C_3H_6O_2S)(OC_3H_7-i)$	Sc:	MVol.D4-49
$C_6H_{13}O_3SSm$	$Sm(C_3H_6O_2S)(OC_3H_7-i)$	Sc:	MVol.D4-49
$C_6H_{13}O_3Sm$	$Sm(OC_3H_7)(OCH_2)_2CH_2$	Sc:	MVol.D3-31/2
$C_6H_{13}O_4PrS_2$	$Pr(C_3H_6O_2S)(C_3H_7O_2S)$	Sc:	MVol.D4-49
$C_6H_{13}O_4S_2Sm$	$Sm(C_3H_6O_2S)(C_3H_7O_2S)$	Sc:	MVol.D4-49
$C_6H_{13}O_4Sm$	$Sm(OC_3H_6O)(OC_3H_6OH)$	Sc:	MVol.D3-31
$C_6H_{13}O_5U$	$U(1,2-C_2H_4O_2)_2(OC_2H_5)$	U:	SVol.C13-91/2
$C_6H_{13}O_6U$	$U(O_2C_2H_4)_2(OC_2H_4OH)$	U:	SVol.C13-91/2
		U:	SVol.D2-50/1
$C_6H_{13}O_{11}Sm$	$[Sm(C_6H_7O_8)(H_2O)_3] \cdot 3 H_2O$		
	$= Sm(C_6H_7O_8) \cdot 6 H_2O$	Sc:	MVol.D5-340/1
$C_6H_{13}O_{11}Yb$	$[Yb(C_6H_7O_8)(H_2O)_3] \cdot 3 H_2O$		
	$= Yb(C_6H_7O_8) \cdot 6 H_2O$	Sc:	MVol.D5-340/1
$C_6H_{13}Sb$	$CH_3Sb(CH_2)_5$	Sb:	Org.Comp.1-148
$C_6H_{14}HoN_2O_{12}P_4^{3-}$	$Ho[H_2(C_2H_4(N(CH_2PO_3)_2)_2)]^{3-}$	Sc:	MVol.D4-154/5
$C_6H_{14}ISb$	$(CH_3)_2ISb(CH_2)_4$	Sb:	Org.Comp.3-166
$C_6H_{14}I_2MnN_2O_4$	$Mn[OOCCH(CH_3)NH_3]_2I_2$	Mn:	MVol.D4-268
$C_6H_{14}I_2Sn$	$(CH_3)_2(I(CH_2)_4)SnI$	Sn:	Org.Comp.8-52
–	$(C_3H_7)_2SnI_2$	Sn:	Org.Comp.8-102/3
–	$(i-C_3H_7)_2SnI_2$	Sn:	Org.Comp.8-103/4
$C_6H_{14}LaN_2O_{12}P_4^{3-}$	$La[H_2(C_2H_4(N(CH_2PO_3)_2)_2)]^{3-}$	Sc:	MVol.D4-154/5
$C_6H_{14}LaNaO_{10}$	$Na[La(HOCH_2CHOH(CH(O))_3COO)(H_2O)_3]$	Sc:	MVol.D5-278
$C_6H_{14}LaO_4P^{2+}$	$La[(C_3H_7O)_2PO_2]^{2+}$	Sc:	MVol.D4-177
$C_6H_{14}LaS_2^+$	$La(SC_3H_7-i)_2^+$	Sc:	MVol.D4-48
$C_6H_{14}MnN_2^{2+}$	$Mn(C_6H_{10}(NH_2)_2)^{2+}$	Mn:	MVol.D3-43
$C_6H_{14}MnN_2O_3^{2+}$	$Mn[H_2NCH_2CH(OH)C_2H_4CH(NH_3)COO]^{2+}$	Mn:	MVol.D4-288/9
$C_6H_{14}MnN_2O_4^{2+}$	$Mn[OOCCH(CH_3)NH_3]_2^{2+}$	Mn:	MVol.D4-266/7

C$_6$H$_{14}$MnN$_4$O$_2$$^{2+}$....	Mn(C$_2$H$_4$(CH$_2$C(O)NHNH$_2$)$_2$)$^{2+}$..........	Mn:	MVol.D5-205
−...............	Mn[H$_2$NC(NH)NHC$_3$H$_6$CH(NH$_3$)COO]$^{2+}$....	Mn:	MVol.D4-288/9
C$_6$H$_{14}$MnN$_4$O$_6$.....	Mn(OC(NH$_2$)$_2$)$_2$(CH$_3$COO)$_2$............	Mn:	MVol.D5-146/7
C$_6$H$_{14}$MnN$_4$O$_6$S.....	Mn(C$_2$H$_4$(CH$_2$C(O)NHNH$_2$)$_2$)SO$_4$ · 3 H$_2$O ..	Mn:	MVol.D5-206
C$_6$H$_{14}$MnN$_6$O$_6$$^{2-}$....	Mn(H(ONCH$_2$)$_3$)$_2$$^{2-}$..................	Mn:	MVol.D5-232
C$_6$H$_{14}$NNd$_2$O$_8$$^{3+}$....	[Nd$_2$(HOCH$_2$(CHOH)$_3$CH(NH$_2$)COO)(OH)$_2$]$^{3+}$		
		Sc:	MVol.D4-281/2
C$_6$H$_{14}$N$_2$NdO$_{12}$P$_4$$^{3-}$	Nd[H$_2$(C$_2$H$_4$(N(CH$_2$PO$_3$)$_2$)$_2$)]$^{3-}$..........	Sc:	MVol.D4-154/5
C$_6$H$_{14}$N$_2$OS........	OS(NC$_2$H$_5$CH$_2$)$_2$	S:	S-N Comp.3-264/6
C$_6$H$_{14}$N$_2$O$_4$PtS	(C$_6$H$_{10}$(NH$_2$)$_2$)PtSO$_4$		
	Cytostatic activity	Pt:	SVol.A1-332/4
C$_6$H$_{14}$N$_2$O$_4$Sn	(CH$_3$)$_2$Sn(ONHCOCH$_3$)$_2$	Sn:	Org.Comp.14-113, 117/9
−...........	(CH$_3$)$_2$Sn(ONHCOCH$_3$)$_2$ · H$_2$O......	Sn:	Org.Comp.14-117/9
−...........	(CH$_3$)$_2$Sn(OOCCH$_2$NH$_2$)$_2$	Sn:	Org.Comp.14-88, 91
C$_6$H$_{14}$N$_2$O$_6$Sn	(C$_3$H$_7$)$_2$Sn(ONO$_2$)$_2$	Sn:	Org.Comp.14-207/8
C$_6$H$_{14}$N$_2$O$_6$U^{2+}.....	[UO$_2$H$_2$(OOCCH(NH$_2$)CH$_3$)$_2$]$^{2+}$........	U:	SVol.D1-216
−........	[UO$_2$H$_2$(OOCCH$_2$CH$_2$NH$_2$)$_2$]$^{2+}$........	U:	SVol.D1-216
C$_6$H$_{14}$N$_2$O$_8$Rh$_2$	Rh$_2$(O$_2$CH)$_4$(NH$_2$CH$_3$)$_2$	Rh:	SVol.B2-10
C$_6$H$_{14}$N$_2$O$_8$SU	UO$_2$SO$_4$ · 2 (CH$_3$)$_2$NCHO...........	U:	SVol.A5-139
		U:	SVol.C10-163
C$_6$H$_{14}$N$_2$O$_9$U	UO$_2$(NO$_3$)$_2$ · (C$_3$H$_7$)$_2$O · 3 H$_2$O.........	U:	SVol.C7-168
C$_6$H$_{14}$N$_2$O$_{10}$U	(NH$_4$)$_2$[UO$_2$(C$_2$O$_4$)(CH$_3$COO)$_2$].......	U:	SVol.C13-235/6
C$_6$H$_{14}$N$_2$O$_{11}$S$_5$	OS(N(SO$_2$CH$_3$)CH(SO$_3$CH$_3$))$_2$.........	S:	S-N Comp.3-266/7
C$_6$H$_{14}$N$_2$O$_{11}$U	(NH$_4$)$_2$[UO$_2$(CH$_2$(COO)$_2$)$_2$(H$_2$O)]..........	U:	SVol.A6-36
		U:	SVol.C13-238/9
C$_6$H$_{14}$N$_2$O$_{12}$P$_4$Sm^{3-}	Sm[H$_2$(C$_2$H$_4$(N(CH$_2$PO$_3$)$_2$)$_2$)]$^{3-}$.........	Sc:	MVol.D4-154/5
C$_6$H$_{14}$N$_2$O$_{12}$P$_4$Tm^{3-}	Tm[H$_2$(C$_2$H$_4$(N(CH$_2$PO$_3$)$_2$)$_2$)]$^{3-}$.........	Sc:	MVol.D4-154/5
C$_6$H$_{14}$N$_4$O$_{10}$S$_2$U	[(CH$_2$)$_6$N$_4$H$_2$][UO$_2$(SO$_4$)$_2$]	U:	SVol.C10-192
−...............	[(CH$_2$)$_6$N$_4$H$_2$][UO$_2$(SO$_4$)$_2$] · 2 H$_2$O........	U:	SVol.C10-192
C$_6$H$_{14}$N$_4$O$_{10}$Se$_2$U ...	(CH$_2$)$_6$N$_4$H$_2$[UO$_2$(SeO$_4$)$_2$]...........	U:	SVol.C11-80
C$_6$H$_{14}$N$_4$O$_{12}$U	[UO$_2$(NO$_3$)$_2$(C$_2$H$_5$OCONH$_2$)$_2$]	U:	SVol.A6-32
C$_6$H$_{14}$N$_6$O$_8$Rh$_2$	Rh$_2$(O$_2$CH)$_4$(NHC(NH$_2$)$_2$)$_2$..............	Rh:	SVol.B2-11
−...............	Rh$_2$(O$_2$CH)$_4$(NHC(NH$_2$)$_2$)$_2$ · 2 H$_2$O	Rh:	SVol.B2-11
C$_6$H$_{14}$N$_6$O$_8$Rh$_2$S$_2$ ...	Rh$_2$(O$_2$CH)$_4$(NH$_2$CSNHNH$_2$)$_2$	Rh:	SVol.B2-13
C$_6$H$_{14}$N$_6$O$_{11}$U	[C(NH$_2$)$_3$]$_2$[UO$_2$(C$_2$O$_4$)$_2$(H$_2$O)] · 2 H$_2$O	U:	SVol.C13-193, 196/9
C$_6$H$_{14}$N$_6$Sn	(C$_3$H$_7$)$_2$Sn(N$_3$)$_2$	Sn:	Org.Comp.8-205
C$_6$H$_{14}$N$_7$O$_8$Rh	NH$_4$[Rh(CH$_3$CNOCHNOH)$_2$(NO$_2$)$_2$]	Rh:	SVol.B2-219
C$_6$H$_{14}$OSSn........	(C$_2$H$_5$)$_2$Sn(SCH$_2$CH$_2$O)	Sn:	Org.Comp.10-230/1
C$_6$H$_{14}$OS$_2$Sn.......	(CH$_3$)$_2$Sn(SCH$_2$CH$_2$)$_2$O	Sn:	Org.Comp.10-34
−........	(CH$_3$)$_3$SnSC(S)OC$_2$H$_5$	Sn:	Org.Comp.9-55
C$_6$H$_{14}$OSn.........	(CH$_3$)$_2$SnC$_4$H$_8$O...........	Sn:	Org.Comp.13-276
−........	(CH$_3$)$_3$SnOC(CH$_3$)CH$_2$.............	Sn:	Org.Comp.11-63/5
−........	(CH$_3$)$_3$SnOCH$_2$CHCH$_2$.............	Sn:	Org.Comp.11-63
C$_6$H$_{14}$O$_2$SSn.......	(CH$_3$)$_2$SnC$_4$H$_8$S(O)O	Sn:	Org.Comp.13-276
−........	(CH$_3$)$_2$Sn(OCH$_2$CH$_2$)$_2$S	Sn:	Org.Comp.14-51, 59
−........	(CH$_3$)$_3$SnOS(O)CH$_2$CHCH$_2$	Sn:	Org.Comp.11-140
−........	(CH$_3$)$_3$SnOS(O)C$_3$H$_5$-c	Sn:	Org.Comp.11-139
−........	(CH$_3$)$_3$SnSCH$_2$CO$_2$CH$_3$	Sn:	Org.Comp.9-32
C$_6$H$_{14}$O$_2$Sn........	(CH$_3$)$_2$SnO$_2$C$_2$H$_2$(CH$_3$)$_2$	Sn:	Org.Comp.14-47
−........	(CH$_3$)$_3$SnOOCC$_2$H$_5$	Sn:	Org.Comp.11-89/90

$C_6H_{14}O_2Sn$	$(C_2H_5)_2SnOCH_2CH_2O$	Sn:	Org.Comp.14-161, 164
$C_6H_{14}O_3SSn$	$(C_3H_7)_2SnOSO_2$	Sn:	Org.Comp.14-210
$C_6H_{14}O_3Sn$	$(C_2H_5)_2Sn(OCHCH_3OO)$	Sn:	Org.Comp.14-185/6
$C_6H_{14}O_4PYb^{2+}$	$Yb[(C_3H_7O)_2PO_2]^{2+}$	Sc:	MVol.D4-177
$C_6H_{14}O_4SSn$	$(C_3H_7)_2SnOSO_3$	Sn:	Org.Comp.14-210
–	$((CH_3)_2CH)_2SnOSO_3$	Sn:	Org.Comp.14-212
$C_6H_{14}O_4U$	$UO_2(OC_3H_7\text{-}n)_2$	U:	SVol.C13-85/6
$C_6H_{14}O_6Os$	$OsO_2(OH)_2(O_2C_2(CH_3)_4)$	Os:	SVol.1-190
$C_6H_{14}O_6S_2U$	$UO_2(SCH_2CHOHCH_2OH)_2$	U:	SVol.D1-251
$C_6H_{14}O_8Tc^{2-}$	$[Tc(OH)_2((OCH_2)_2CHOH)_2]^{2-}$	Tc:	SVol.2-204
$C_6H_{14}O_{11}Rh_2S$	$Rh_2(O_2CH)_4((CH_3)_2SO)(H_2O)_2$	Rh:	SVol.B2-13
$C_6H_{14}O_{16}P_6U^{6-}$	$[UO_2((O_2POCH_2)_2P(O)CH_3)_2]^{6-}$	U:	SVol.D1-253
$C_6H_{14}PrS_2^+$	$Pr(SC_3H_7\text{-}i)_2^+$	Sc:	MVol.D4-48
$C_6H_{14}SSn$	$(CH_3)_2SnSC_4H_8$	Sn:	Org.Comp.9-229
$C_6H_{14}S_2Sm^+$	$Sm(SC_3H_7\text{-}i)_2^+$	Sc:	MVol.D4-48
$C_6H_{14}S_2Sn$	$(CH_3)_2Sn(SCH_2CH_2)_2$	Sn:	Org.Comp.10-34
–	$(C_2H_5)_2Sn(SCH_2)_2$	Sn:	Org.Comp.10-39, 42/3
$C_6H_{14}Si_4$	$(SiH_3)_2SiHSiH_2C_6H_5$	Si:	SVol.B1-175
$C_6H_{15}HoN_2O_{12}P_4^{2-}$	$Ho[H_3(C_2H_4(N(CH_2PO_3)_2)_2)]^{2-}$	Sc:	MVol.D4-154/5
$C_6H_{15}HoO_3$	$Ho(OC_2H_5)_3$	Sc:	MVol.D3-19/20
$C_6H_{15}HoO_{12}S_3$	$Ho(C_2H_5SO_4)_3 \cdot 9\,H_2O$	Sc:	MVol.C8-362/414
$C_6H_{15}IN_3Rh$	$Rh(NHCH_2CH_2)_3I$	Rh:	SVol.B2-155
$C_6H_{15}IO_3Sn$	$(C_2H_5)_3SnOIO_2$	Sn:	Org.Comp.11-259
$C_6H_{15}ISn$	$(CH_3)(C_2H_5)(C_3H_7)SnI$	Sn:	Org.Comp.8-82, 83
–	$(CH_3)_2(C_4H_9)SnI$	Sn:	Org.Comp.8-49/50, 56
–	$(CH_3)_2(i\text{-}C_4H_9)SnI$	Sn:	Org.Comp.8-50
–	$(CH_3)_2(s\text{-}C_4H_9)SnI$	Sn:	Org.Comp.8-50
–	$(C_2H_5)_3SnI$	Sn:	Org.Comp.8-20/5
–	$(C_2H_5)_3Sn^{129}I$	Sn:	Org.Comp.8-20
–	$(C_3H_7)_2SnIH$	Sn:	Org.Comp.8-142
–	$(i\text{-}C_3H_7)_2SnIH$	Sn:	Org.Comp.8-142
$C_6H_{15}I_2Sb$	$(C_2H_5)_3SbI_2$	Sb:	Org.Comp.4-94/5
$C_6H_{15}I_3LaN_3O_6$	$La(NH_2CH_2COOH)_3I_3 \cdot 3\,H_2O$	Sc:	MVol.D1-104
$C_6H_{15}InO_3S$	$C_2H_5SO_3In(C_2H_5)_2$	S:	SVol.3-322
$C_6H_{15}LaN_2O_{12}P_4^{2-}$	$La[H_3(C_2H_4(N(CH_2PO_3)_2)_2)]^{2-}$	Sc:	MVol.D4-154/5
$C_6H_{15}LaO_{10}$	$[La(HOCH_2(CHOH)_2(CH(O))_2COO)(H_2O)_3]$	Sc:	MVol.D5-278
$C_6H_{15}LaO_{12}S_3$	$La(C_2H_5SO_4)_3 \cdot 9\,H_2O$	Sc:	MVol.C8-362/414
$C_6H_{15}LuO_{12}S_3$	$Lu(C_2H_5SO_4)_3 \cdot 9\,H_2O$	Sc:	MVol.C8-362/414
$C_6H_{15}MnNO_3^{2+}$	$[Mn(N(C_2H_4OH)_3)]^{2+}$	Mn:	MVol.D4-227/8
$C_6H_{15}MnNO_7S$	$Mn[N(C_2H_4OH)_3]SO_4$	Mn:	MVol.D4-232
$C_6H_{15}MnN_3^{2+}$	$Mn[HN(C_2H_4NH)_2C_2H_4]^{2+}$	Mn:	MVol.D4-74
$C_6H_{15}NOSSn$	$(CH_3)_3SnSC(O)N(CH_3)_2$	Sn:	Org.Comp.9-54
$C_6H_{15}NOSeSn$	$(CH_3)_3SnSeC(O)N(CH_3)_2$	Sn:	Org.Comp.10-260
$C_6H_{15}NOSn$	$(CH_3)_3SnONC(CH_3)_2$	Sn:	Org.Comp.11-128, 134
$C_6H_{15}NO_2SSn$	$(CH_3)_3SnSCH_2CH(NH_2)CO_2H$	Sn:	Org.Comp.9-32
$C_6H_{15}NO_2Sn$	$(CH_3)_3SnON(CH_3)COCH_3$	Sn:	Org.Comp.11-130, 135/6
–	$(CH_3)_3SnON(O)C(CH_3)_2$	Sn:	Org.Comp.11-131/7
–	$(CH_3)_3SnOOCN(CH_3)_2$	Sn:	Org.Comp.11-118/9
–	$(CH_3)_3SnOOCCH(CH_3)NH_2$	Sn:	Org.Comp.11-95, 104
–	$(CH_3)_3SnOOCCH_2CH_2NH_2$	Sn:	Org.Comp.11-95

$C_6H_{16}N_{11}O_{13}Y$	$Y(NO_3)_3 \cdot 2 [NH_2NHC(O)]_2CH_2 \cdot 2 H_2O$ ...	Sc:	MVol.D2-244/5
$C_6H_{16}N_{13}O_6Rh$	$[Rh((N_5C_2H_6)_2C_2H_4)(NO_2)_2](NO_2)$.........	Rh:	SVol.B2-224
$C_6H_{16}OS_3Sn$	$(CH_3)_2Sn(SCH_2)_2 \cdot (CH_3)_2SO$	Sn:	Org.Comp.10-31
$C_6H_{16}OSn$	$(CH_3)(C_2H_5)(C_3H_7)SnOH$...............	Sn:	Org.Comp.13-266, 272
–	$(CH_3)_3SnOC_3H_7$.	Sn:	Org.Comp.11-54
–	$(CH_3)_3SnOC_3H_7$-i	Sn:	Org.Comp.11-54
–	$(C_2H_5)_3SnOH$.	Sn:	Org.Comp.11-170/7
$C_6H_{16}O_2Sn$	$(CH_3)_2Sn(OC_2H_5)_2$	Sn:	Org.Comp.14-27, 29
–	$(C_2H_5)_2Sn(OCH_3)_2$	Sn:	Org.Comp.14-149, 155/7
–	$(C_2H_5)_3SnOOH$	Sn:	Org.Comp.11-255/6
–	$(C_3H_7)_2Sn(OH)_2$.	Sn:	Org.Comp.14-198
$C_6H_{16}O_6OsS_2$	$H_2[OsO_2(SO_2C_3H_7)_2] \cdot 3 C_2H_5OH$	Os:	SVol.1-293
$C_6H_{16}O_6S_2Sn$	$(CH_3)_2Sn(OSO_2C_2H_5)_2$	Sn:	Org.Comp.14-126
–	$(C_2H_5)_2Sn(OSO_2CH_3)_2$	Sn:	Org.Comp.14-190
$C_6H_{16}O_8S_2Sn$	$(C_3H_7)_2Sn(OSO_3H)_2$	Sn:	Org.Comp.14-210
$C_6H_{16}PSb$	$(CH_3)_2SbCHP(CH_3)_3$	Sb:	Org.Comp.1-111
$C_6H_{16}SSn$	$(CH_3)_3SnSC_3H_7$.	Sn:	Org.Comp.9-29/30
–	$(CH_3)_3SnSC_3H_7$-i	Sn:	Org.Comp.9-30
–	$(C_2H_5)_3SnSH$.	Sn:	Org.Comp.9-71
$C_6H_{16}S_2Sn$	$(CH_3)_2Sn(SC_2H_5)_2$.	Sn:	Org.Comp.10-3/4
–	$(C_2H_5)_2Sn(SCH_3)_2$.	Sn:	Org.Comp.10-37
$C_6H_{16}Sb^+$	$(CH_3)_2(C_2H_5)_2Sb^+$	Sb:	Org.Comp.3-38
$C_6H_{16}Si_5$	$(SiH_3)_3SiSiH_2C_6H_5$	Si:	SVol.B1-175
$C_6H_{17}HoN_2O_{12}P_4$ ..	$Ho[H_5(C_2H_4(N(CH_2PO_3)_2)_2)]$............	Sc:	MVol.D4-154/5
–	$Ho[H_5(C_2H_4(N(CH_2PO_3)_2)_2)] \cdot 3 H_2O$	Sc:	MVol.D4-155
$C_6H_{17}LaN_2O_{12}P_4$ ..	$La[H_5(C_2H_4(N(CH_2PO_3)_2)_2)]$............	Sc:	MVol.D4-154/5
–	$La[H_5(C_2H_4(N(CH_2PO_3)_2)_2)] \cdot 4 H_2O$	Sc:	MVol.D4-155
$C_6H_{17}MnNO_5$	$[Mn(OH)_2(N(C_2H_4OH)_3)]$..............	Mn:	MVol.D4-227/8
$C_6H_{17}MnN_{10}O_2$.....	$[Mn(OH)(C_2H_4(C_2N_5H_5)_2)(H_2O)] \cdot 0.5 H_2O$	Mn:	MVol.D5-164
$C_6H_{17}N_3NdO_{12}P_4$ ..	$Nd[H_5(C_2H_4(N(CH_2PO_3)_2)_2)]$............	Sc:	MVol.D4-154/5
$C_6H_{17}N_2O_{12}P_4Pr$....	$Pr[H_5(C_2H_4(N(CH_2PO_3)_2)_2)]$	Sc:	MVol.D4-154/5
–	$Pr[H_5(C_2H_4(N(CH_2PO_3)_2)_2)] \cdot 4 H_2O$	Sc:	MVol.D4-155
$C_6H_{17}N_2O_{12}P_4Sm$ ..	$Sm[H_5(C_2H_4(N(CH_2PO_3)_2)_2)]$	Sc:	MVol.D4-154/5
–	$Sm[H_5(C_2H_4(N(CH_2PO_3)_2)_2)] \cdot 3.5 H_2O$....	Sc:	MVol.D4-155
$C_6H_{17}N_2O_{12}P_4Tm$ ..	$Tm[H_5(C_2H_4(N(CH_2PO_3)_2)_2)]$	Sc:	MVol.D4-154/5
–	$Tm[H_5(C_2H_4(N(CH_2PO_3)_2)_2)] \cdot 3 H_2O$	Sc:	MVol.D4-155
$C_6H_{17}N_2O_{12}P_4Y$	$Y[H_5(C_2H_4(N(CH_2PO_3)_2)_2)] \cdot 6 H_2O$	Sc:	MVol.D4-155
$C_6H_{17}N_2O_{12}P_4Yb$ ...	$Yb[H_5(C_2H_4(N(CH_2PO_3)_2)_2)]$............	Sc:	MVol.D4-154/5
–	$Yb[H_5(C_2H_4(N(CH_2PO_3)_2)_2)] \cdot 4.5 H_2O$	Sc:	MVol.D4-155
$C_6H_{17}N_5NiS_4Si_2$	$[Ni((S_2N_2Si(CH_3)_2)_2NC_2H_5)]$	S:	S-N Comp.2-291, 295, 296
$C_6H_{17}OSb$	$(CH_3)_4SbOC_2H_5$.	Sb:	Org.Comp.3-102
$C_6H_{17}O_2Sb$	$(C_2H_5)_3Sb(OH)_2$.	Sb:	Org.Comp.4-116
$C_6H_{17}O_3PSSn$	$(CH_3)_3SnOP(O)(OC_2H_5)SCH_3$	Sn:	Org.Comp.11-156
$C_6H_{17}O_3Sn$	$(C_2H_5)_3SnOOH \cdot 0.5 H_2O_2$.	Sn:	Org.Comp.11-255/6
$C_6H_{17}O_4PSn$	$(C_2H_5)_3SnOPO(OH)_2$.	Sn:	Org.Comp.11-265
$C_6H_{17}SSb$	$(CH_3)_4SbSC_2H_5$.	Sb:	Org.Comp.3-140
$C_6H_{17}Sb$	$(CH_3)_4SbC_2H_5$.	Sb:	Org.Comp.3-18
$C_6H_{18}HoN_3O_{12}S_3$ ...	$Ho(NO_3)_3 \cdot 3 (CH_3)_2SO$	Sc:	MVol.D4-5, 7/8, 10
$C_6H_{18}HoO_{12}P_3$	$Ho[(CH_3O)_2PO_2]_3$	Sc:	MVol.D4-180

$C_6H_{18}Ho_2O_{15}P_6$	$Ho_2[((O)_2PCH_3)_2O]_3$	Sc:	MVol.D4-159
$C_6H_{18}I_3N_4Rh$.......	$[Rh(N(C_2H_4NH_2)_3)I_2]I$	Rh:	SVol.B2-199
–	$[Rh((NH_2C_2H_4NH)_2C_2H_4)I_2]I$	Rh:	SVol.B2-197
$C_6H_{18}I_3O_3RhS_3$....	$RhI_3[(CH_3)_2SO]_3$	Rh:	SVol.B1-144
$C_6H_{18}I_3RhS_3$.....	$Rh[S(CH_3)_2]_3I_3$	Rh:	SVol.B3-3/4
$C_6H_{18}I_3RhTe_3$.....	$Rh[Te(CH_3)_2]_3I_3$	Rh:	SVol.B3-64
$C_6H_{18}InSb$........	$In(CH_3)_3[Sb(CH_3)_3]$	Sb:	Org.Comp.1-15
$C_6H_{18}KO_6U$.......	$K[U(OCH_3)_6]$	U:	SVol.C13-74
$C_6H_{18}LaN_7O_9$.....	$[La((NH_2C_2H_4)_3N)(NO_3)_3]$	Sc:	MVol.D1-32/3
$C_6H_{18}LaP_3S_6$.......	$La[(CH_3)_2PS_2]_3 \cdot 2 H_2O$........	Sc:	MVol.D4-247/8
$C_6H_{18}La_2O_{15}$.....	$[La_2(OH)_2(OC(CH_2COO)_2COO)(H_2O)_6]$.....	Sc:	MVol.D5-349
$C_6H_{18}La_2O_{15}P_6$....	$La_2[((O)_2PCH_3)_2O]_3$	Sc:	MVol.D4-159
$C_6H_{18}LiO_6U$........	$Li[U(OCH_3)_6]$.....................	U:	SVol.C13-74
$C_6H_{18}Li_2O_6U$.......	$Li_2[U(OCH_3)_6]$	U:	SVol.C13-62
$C_6H_{18}Li_3Lu$........	$Li_3[Lu(CH_3)_6] \cdot 3 (N(CH_3)_2CH_2)_2$........	Sc:	MVol.D6-152, 155/6
$C_6H_{18}Li_3Nd$........	$Li_3[Nd(CH_3)_6] \cdot 3 (N(CH_3)_2CH_2)_2$.	Sc:	MVol.D6-152, 155/6
$C_6H_{18}Li_3Pr$	$Li_3[Pr(CH_3)_6] \cdot 3 (N(CH_3)_2CH_2)_2$	Sc:	MVol.D6-152, 155/6
$C_6H_{18}Li_3Sm$	$Li_3[Sm(CH_3)_6] \cdot 3 (N(CH_3)_2CH_2)_2$	Sc:	MVol.D6-152, 155/6
$C_6H_{18}Li_3Tm$.......	$Li_3[Tm(CH_3)_6] \cdot 3 (N(CH_3)_2CH_2)_2$	Sc:	MVol.D6-152, 155/6
$C_6H_{18}Li_3Yb$	$Li_3[Yb(CH_3)_6] \cdot 3 (N(CH_3)_2CH_2)_2$.	Sc:	MVol.D6-152, 155/6
$C_6H_{18}LuN_3O_{12}S_3$...	$Lu(NO_3)_3 \cdot 3 (CH_3)_2SO$	Sc:	MVol.D4-5, 7/9
$C_6H_{18}Lu_2O_{15}P_6$....	$Lu_2[((O)_2PCH_3)_2O]_3$	Sc:	MVol.D4-159
$C_6H_{18}MnNO_5{}^+$.....	$[Mn(OH)(N(C_2H_4OH)_3)(H_2O)]^+$	Mn:	MVol.D4-227/8
$C_6H_{18}MnN_4{}^{2+}$......	$[Mn(H_2N(CH_2)_2NHCH_2)_2]^{2+}$	Mn:	MVol.D3-56/7
–	$Mn[N(CH_2CH_2NH_2)_3]^{2+}$	Mn:	MVol.D3-59
$C_6H_{18}MnN_6O_6{}^{2+}$....	$Mn[1,3,5-(OH)_3-1,3,5-C_3H_6N_3]_2{}^{2+}$........	Mn:	MVol.D4-73
		Mn:	MVol.D5-232
$C_6H_{18}NOSb$........	$(CH_3)_4SbON(CH_3)_2$	Sb:	Org.Comp.3-115
$C_6H_{18}NSbSi$	$(CH_3)_2SbN(CH_3)Si(CH_3)_3$	Sb:	Org.Comp.2-47
$C_6H_{18}N_2OSSi_2$	$OS(N(CH_3)Si(CH_3)_2)_2$	S:	S-N Comp.3-6/7
$C_6H_{18}N_2O_4Os$......	$OsO_2(OH)_2[C_2H_4(N(CH_3)_2)_2]$.	Os:	SVol.1-254
$C_6H_{18}N_2O_{10}P_2U$...	$UO_2(NO_3)_2((CH_3)_3PO)_2$	U:	SVol.A6-65
$C_6H_{18}N_2O_{15}P_2Th$ ...	$ThO(NO_3)_2 \cdot 2 (CH_3O)_3PO$	Th:	SVol.E-93, 94
$C_6H_{18}N_2S_2Sn$	$(CH_3)_2Sn(SCH_2CH_2NH_2)_2$	Sn:	Org.Comp.10-5
$C_6H_{18}N_3NdO_3$......	$Nd(H_2NCH_2CH_2O)_3$	Sc:	MVol.D2-1
$C_6H_{18}N_3O_3Pr$	$Pr(H_2NCH_2CH_2O)_3$	Sc:	MVol.D2-1
$C_6H_{18}N_3O_3Sm$......	$Sm(H_2NCH_2CH_2O)_3$.	Sc:	MVol.D2-1
$C_6H_{18}N_3O_6PSU$.....	$UO_2SO_3 \cdot ((CH_3)_2N)_3PO$	U:	SVol.C10-139
$C_6H_{18}N_3O_6Sc$	$Sc(CH_3COO)_3 \cdot 3 NH_3$	Sc:	MVol.D1-11
$C_6H_{18}N_3O_6Y$	$Y(CH_3COO)_3 \cdot 3 NH_3$	Sc:	MVol.D1-11
$C_6H_{18}N_3O_{12}S_3Tb$ ...	$Tb(NO_3)_3 \cdot 3 (CH_3)_2SO$	Sc:	MVol.D4-5, 7/8
$C_6H_{18}N_3O_{12}S_3Tm$ ...	$Tm(NO_3)_3 \cdot 3 (CH_3)_2SO$	Sc:	MVol.D4-5, 7/8
$C_6H_{18}N_3O_{12}S_3Y$ ...	$Y(NO_3)_3 \cdot 3 (CH_3)_2SO$	Sc:	MVol.D4-5, 7/8, 10
$C_6H_{18}N_3O_{12}S_3Yb$ ...	$Yb(NO_3)_3 \cdot 3 (CH_3)_2SO$	Sc:	MVol.D4-5, 7/10
$C_6H_{18}N_4NiS_4Si_2$ ...	$[Ni(S_2N_2Si(CH_3)_3)_2]$.	S:	S-N Comp.2-294, 296
$C_6H_{18}N_4O_2S_2Si_2$....	$S(NSi(CH_3)_2)_2NSO_2N(CH_3)_2$	S:	S-N Comp.3-7/8
$C_6H_{18}N_4O_6Pt$......	$(i\text{-}C_3H_7NH_2)_2Pt(NO_3)_2$		
	Cytostatic activity	Pt:	SVol.A1-332/4
$C_6H_{18}N_4O_{15}S_3Th$ ...	$Th(NO_3)_4 \cdot 3 (CH_3)_2SO$	Th:	SVol.E-100, 102, 106
–	$Th(NO_3)_4 \cdot 3 (CH_3)_2SO \cdot 2 H_2O$	Th:	SVol.E-100, 102

$C_6H_{18}N_6NiS_4Si_2$	$[Ni(S_2N_2Si(CH_3)_2NCH_3)_2]$	S:	S–N Comp.2-291, 295, 296
$C_6H_{18}N_7NdO_9$	$[Nd((NH_2C_2H_4)_3N)(NO_3)_3]$	Sc:	MVol.D1-32/3
$C_6H_{18}N_7O_7Rh$	$[Rh(N(C_2H_4NH_2)_3)(NO_2)_2]NO_3$	Rh:	SVol.B2-200
$C_6H_{18}N_7O_9Pr$	$[Pr((NH_2C_2H_4)_3N)(NO_3)_3]$	Sc:	MVol.D1-32/3
$C_6H_{18}N_7O_9Sm$	$[Sm((NH_2C_2H_4)_3N)(NO_3)_3]$	Sc:	MVol.D1-32/3
$C_6H_{18}N_7O_9Yb$	$[Yb((NH_2C_2H_4)_3N)(NO_3)_3]$	Sc:	MVol.D1-32/3
$O_6H_{18}N_{13}Rh$	$[Rh((NH_2C_2H_4NH)_2C_2H_4)(N_3)_2]N_3$	Rh:	SVol.B2-197
$C_6H_{18}NaO_6U$	$Na[U(OCH_3)_6]$	U:	SVol.C13-73/4
$C_6H_{18}NdP_3S_6$	$Nd[(CH_3)_2PS_2]_3$	Sc:	MVol.D4-247/8
$C_6H_{18}Nd_2O_{15}$	$[Nd_2(OH)_2(OC(CH_2COO)_2COO)(H_2O)_6]$	Sc:	MVol.D5-349
$C_6H_{18}Nd_2O_{15}P_6$	$Nd_2[((O)_2PCH_3)_2O]_3$	Sc:	MVol.D4-159
$C_6H_{18}OPSSb$	$(CH_3)_4SbOP(S)(CH_3)_2$	Sb:	Org.Comp.3-138
$C_6H_{18}OSiSn$	$(CH_3)_3SnOSi(CH_3)_3$	Sn:	Org.Comp.11-154/9
$C_6H_{18}O_2PSb$	$(CH_3)_4SbOP(O)(CH_3)_2$	Sb:	Org.Comp.3-136/7
$C_6H_{18}O_4P_2S_4Sn$	$(CH_3)_2Sn(SP(S)(OCH_3)_2)_2$	Sn:	Org.Comp.10-26
$C_6H_{18}O_4SSi_2$	$(CH_3)_3SiOSO_2OSi(CH_3)_3$	S:	SVol.3-320/1
$C_6H_{18}O_6P_2S_2Sn$	$(CH_3)_2Sn(OP(S)(OCH_3)_2)_2$	Sn:	Org.Comp.14-135, 138
$C_6H_{18}O_6P_2Sn$	$(CH_3)_2Sn(OP(O)(CH_3)OCH_3)_2$	Sn:	Org.Comp.14-135, 137
–	$(CH_3)_2Sn(P(O)(OCH_3)_2)_2$	Sn:	Org.Comp.8-94
$C_6H_{18}O_6RbU$	$Rb[U(OCH_3)_6]$	U:	SVol.C13-74
$C_6H_{18}O_6Te$	$Te(OCH_3)_6$ solutions		
	$Te(OCH_3)_6$–HF	F:	SVol.3-195
$C_6H_{18}O_6U$	$U(OCH_3)_6$	U:	SVol.A6-115
		U:	SVol.C13-76/80
$C_6H_{18}O_7S_2Si_2$	$(CH_3)_3Si(OSO_2)_2OSi(CH_3)_3$	S:	SVol.3-320
$C_6H_{18}O_8P_2Sn$	$(CH_3)_2Sn(OP(O)(OCH_3)_2)_2$	Sn:	Org.Comp.8-94
$C_6H_{18}O_8S_2U$	$[UO_2(SCH_2CHOHCH_2OH)_2(H_2O)_2]$	U:	SVol.C13-308/10
$C_6H_{18}O_{10}S_3Si_2$	$(CH_3)_3Si(OSO_2)_3OSi(CH_3)_3$	S:	SVol.3-320
$C_6H_{18}O_{12}P_3Tb$	$Tb[(CH_3O)_2PO_2]_3$	Sc:	MVol.D4-180
$C_6H_{18}O_{12}P_3Yb$	$Yb[(CH_3O)_2PO_2]_3$	Sc:	MVol.D4-180
$C_6H_{18}O_{15}P_6Pr_2$	$Pr_2[((O)_2PCH_3)_2O]_3$	Sc:	MVol.D4-159
$C_6H_{18}O_{15}P_6Sm_2$	$Sm_2[((O)_2PCH_3)_2O]_3$	Sc:	MVol.D4-159
$C_6H_{18}O_{15}P_6Tb_2$	$Tb_2[((O)_2PCH_3)_2O]_3$	Sc:	MVol.D4-159
$C_6H_{18}O_{15}P_6Tm_2$	$Tm_2[((O)_2PCH_3)_2O]_3$	Sc:	MVol.D4-159
$C_6H_{18}O_{15}P_6Yb_2$	$Yb_2[((O)_2PCH_3)_2O]_3$	Sc:	MVol.D4-159
$C_6H_{18}O_{15}Pr_2$	$[Pr_2(OH)_2(OC(CH_2COO)_2COO)(H_2O)_6]$	Sc:	MVol.D5-349
$C_6H_{18}O_{15}Sm_2$	$[Sm_2(OH)_2(OC(CH_2COO)_2COO)(H_2O)_6]$	Sc:	MVol.D5-349
$C_6H_{18}P_2S_4Sn$	$(CH_3)_2Sn(SP(S)(CH_3)_2)_2$	Sn:	Org.Comp.10-26
$C_6H_{18}P_3PrS_6$	$Pr[(CH_3)_2PS_2]_3$	Sc:	MVol.D4-247/8
$C_6H_{18}SSiSn$	$(CH_3)_3SnSSi(CH_3)_3$	Sn:	Org.Comp.9-65
$C_6H_{18}SSn_2$	$[(CH_3)_3Sn]_2S$	Sn:	Org.Comp.9-26
$C_6H_{18}Si_6$	$SiH_3SiH_2Si(SiH_3)_2SiH_2C_6H_5$	Si:	SVol.B1-175
$C_6H_{19}IOSn_2$	$(CH_3)_3SnOH \cdot (CH_3)_3SnI \cdot H_2O$	Sn:	Org.Comp.8-18
$C_6H_{19}MnNO_7^-$	$[Mn(OH)_4(N(C_2H_4OH)_3)]^-$	Mn:	MVol.D4-228/30
$C_6H_{19}MnNO_7^{2-}$	$[Mn(OH)_4(N(C_2H_4OH)_3)]^{2-}$	Mn:	MVol.D4-227/8
$C_6H_{19}MnN_4O^+$	$Mn(H_2N(CH_2)_2NHCH_2)_2(OH)^+$	Mn:	MVol.D3-57
$C_6H_{19}NS_2$	$H_2S \cdot [(C_2H_5)_3NH]SH$	S:	SVol.4a/b-368
$C_6H_{19}N_3OPbSn$	$(CH_3)_3SnOH \cdot (CH_3)_3PbN_3$	Sn:	Org.Comp.11-46
$C_6H_{19}N_3OSn_2$	$(CH_3)_3SnN_3 \cdot (CH_3)_3SnOH$	Sn:	Org.Comp.8-194, 197

$C_6H_{19}N_3OSn_2$	$(CH_3)_3SnN_3 \cdot (CH_3)_3SnOH$	Sn:	Org.Comp.11-46
$C_6H_{20}HgI_4N_3O_3Rh$	$[Rh(NHCH_2CH_2)_3(H_2O)_2(OH)](HgI_4)$	Rh:	SVol.B2-159
$C_6H_{20}I_2N_5O_2Rh$	$[Rh(NH_2C_2H_4NH_2)_2(OCOCH_2NH_2)]I_2$	Rh:	SVol.B2-171
$C_6H_{20}KN_6Rh$	$K[Rh(NHC_2H_4NH_2)_2((NH)_2C_2H_4)]$	Rh:	SVol.B2-186
$C_6H_{20}MnN_4O_2$	$Mn(H_2N(CH_2)_2NHCH_2)_2(OH)_2$	Mn:	MVol.D3-57
$C_6H_{20}Mo_3N_2O_{10}$	$[(CH_3)_3NH]_2O \cdot 3\ MoO_3$	Mo:	SVol.B4-136
–	$[(CH_3)_3NH]_2O \cdot 3\ MoO_3 \cdot H_2O$	Mo:	SVol.B4-15, 135/6
$C_6H_{20}Mo_4N_2O_{13}$	$[(CH_3)_3NH]_2O \cdot 4\ MoO_3$	Mo:	SVol.B4-136
$C_6H_{20}N_2O_9Rh_2$	$Rh_2(O_2CCH_3)_2(HOH \cdot O_2CCH_3)(OH)(H_2O)(NH_3)_2$		
		Rh:	SVol.B2-35
$C_6H_{20}N_4O_2Rh^+$	$[Rh(N(C_2H_4NH_2)_3)(OH)_2]^+$	Rh:	SVol.B2-199
$C_6H_{20}N_4O_{16}U_2$	$(NH_4)_4[(UO_2)_2(CH_2(COO)_2)_2(OO)_2] \cdot 2\ H_2O$	U:	SVol.C13-238/41
$C_6H_{20}N_5O_8RhS_2$	$[Rh(C_2H_8N_2)_2(OCOCH_2NH_2)]S_2O_6$	Rh:	SVol.B2-171
$C_6H_{20}N_5PS_2Si_2$	$(SN)_2P(NHSi(CH_3)_3)_2N$	S:	S-N Comp.3-14/9
$C_6H_{20}N_6O_{18}S_3U_2$	$(NH_4)_4[(UO_2)_2(C_2O_4)_2(NCS)_2(SO_4)(H_2O)_2]$	U:	SVol.C13-233/4
$C_6H_{20}N_7NdO_9$	$[Nd(NH_2C_3H_6NH_2)_2(NO_3)_3]$	Sc:	MVol.D1-26
$C_6H_{20}N_7O_9Sm$	$[Sm(NH_2C_3H_6NH_2)_2(NO_3)_3]$	Sc:	MVol.D1-26
$C_6H_{20}N_7O_9Yb$	$[Yb(NH_2C_3H_6NH_2)_2(NO_3)_3]$	Sc:	MVol.D1-26
$C_6H_{20}N_8O_{12}Th$	$Th(NO_3)_4 \cdot 2\ CH_3CH(NH_2)CH_2NH_2 \cdot 2\ H_2O$		
		Th:	SVol.E-8
$C_6H_{20}N_8O_{14}U$	$(N_2H_5)_4[UO_2(C_2O_4)_3] \cdot H_2O$	U:	SVol.C13-200/2
$C_6H_{21}IN_6Os$	$[Os(C_2H_4(NH_2)NH)_3]I$	Os:	SVol.1-252
$C_6H_{21}I_3N_{15}Rh$	$[Rh((NH_2CNH)_2NH)_3]I_3$	Rh:	SVol.B2-220/1
$C_6H_{21}LaN_6O_6$	$La(N_2H_4)_3(CH_3COO)_3 \cdot 4\ H_2O$	Sc:	MVol.D1-13
$C_6H_{21}LaO_9P_3$	$La(HP(O)(OH)OC_2H_5)_3$	Sc:	MVol.D4-157/8
$C_6H_{21}NS_3$	$2\ H_2S \cdot [(C_2H_5)_3NH]SH$	S:	SVol.4a/b-368, 495
$C_6H_{21}N_6O_6Pr$	$Pr(N_2H_4)_3(CH_3COO)_3 \cdot 3\ H_2O$	Sc:	MVol.D1-13
$C_6H_{21}N_6O_6Sm$	$Sm(N_2H_4)_3(CH_3COO)_3 \cdot 3\ H_2O$	Sc:	MVol.D1-13
$C_6H_{21}N_6O_6Y$	$Y(N_2H_4)_3(CH_3COO)_3 \cdot 3\ H_2O$	Sc:	MVol.D1-13
$C_6H_{21}N_6Os$	$Os(C_2H_4(NH_2)NH)_3$	Os:	SVol.1-252
$C_6H_{21}N_6Rh$	$Rh(NHC_2H_4NH_2)_3$	Rh:	SVol.B2-186
$C_6H_{21}N_{15}Rh^{3+}$	$[Rh((NH_2CNH)_2NH)_3]^{3+}$	Rh:	SVol.B2-220/1
$C_6H_{21}N_{18}O_9Rh$	$[Rh((NH_2CNH)_2NH)_3](NO_3)_3$	Rh:	SVol.B2-220/1
$C_6H_{22}IN_6Os$	$[Os(C_2H_4(NH_2)NH)_2(C_2H_4(NH_2)_2)]I$	Os:	SVol.1-253
$C_6H_{22}IN_6Rh$	$[Rh(NHC_2H_4NH_2)_2((NH_2)_2C_2H_4)]I$	Rh:	SVol.B2-186
$C_6H_{22}I_2N_6Os$	$[Os(C_2H_4(NH_2)NH)_2(C_2H_4(NH_2)_2)]I_2$	Os:	SVol.1-253
$C_6H_{22}N_4O_2Rh^{3+}$	$[Rh(N(C_2H_4NH_2)_3)(OH_2)_2]^{3+}$	Rh:	SVol.B2-199
$C_6H_{22}N_{10}O_2Rh^{3+}$	$[Rh(NH_2C(NH)NHC(NH)NHCH_3)_2(H_2O)_2]^{3+}$	Rh:	SVol.B2-223
$C_6H_{23}I_2N_6Os_3$	$[Os_3(C_2H_4(NH_2)NH)(C_2H_4(NH_2)_2)_2]I_2$	Os:	SVol.1-253
$C_6H_{23}I_2N_6Rh$	$[Rh(NHC_2H_4NH_2)((NH_2)_2C_2H_4)_2]I_2$	Rh:	SVol.B2-186
$C_6H_{23}I_3N_6Os$	$[Os(C_2H_4(NH_2)NH)(C_2H_4(NH_2)_2)_2]I_3$	Os:	SVol.1-253
$C_6H_{23}N_4ORh^{2+}$	$[Rh(NH_2CH_2CH(CH_3)NH_2)_2H(OH_2)]^{2+}$	Rh:	SVol.B2-188
$C_6H_{24}HoN_6^{3+}$	$[Ho(C_2H_8N_2)_3]^{3+}$	Sc:	MVol.D1-19/21
$C_6H_{24}HoN_9O_9$	$Ho(C_2H_8N_2)_3(NO_3)_3$	Sc:	MVol.D1-21/2
$C_6H_{24}Ho_2N_{12}O_{18}S_3$	$Ho_2(SO_4)_3 \cdot 6\ CO(NH_2)_2$	Sc:	MVol.C8-185
		Sc:	MVol.D2-222/3
$C_6H_{24}I_2MnN_{12}O_6$	$Mn(OC(NH_2)_2)_6I_2$	Mn:	MVol.D5-142/3
$C_6H_{24}I_3N_6Os$	$[Os(C_2H_4(NH_2)_2)_3]I_3 \cdot 2\ H_2O$	Os:	SVol.1-250/1
$C_6H_{24}I_3N_6Rh$	$[Rh(NH_2C_2H_4NH_2)_3]I_3$	Rh:	SVol.B2-162, 164/6
–	$[Rh(NH_2C_2H_4NH_2)_3]I_3 \cdot H_2O$	Rh:	SVol.B2-162, 164/5

$C_6H_{24}I_4N_{12}O_6Th$	$ThI_4 \cdot 6\ (NH_2)_2CO \cdot 2\ H_2O$	Th:	SVol.E-37, 40
$C_6H_{24}I_6N_6Os$	$[Os(C_2H_4(NH_2)_2)_3I_2]I_4$	Os:	SVol.1-252
$C_6H_{24}I_9N_6Os$	$[Os(C_2H_4(NH_2)_2)_3](I_3)_3$	Os:	SVol.1-250/1
$C_6H_{24}LaN_6{}^{3+}$	$[La(C_2H_8N_2)_3]^{3+}$	Sc:	MVol.D1-19/21
$C_6H_{24}LuN_6{}^{3+}$	$[Lu(C_2H_8N_2)_3]^{3+}$	Sc:	MVol.D1-19/21
$C_6H_{24}MgN_{18}O_{24}Th$ ..	$MgTh(NO_3)_6 \cdot 6\ (NH_2)_2CO \cdot 4\ H_2O$	Th:	SVol.E-38, 41
$C_6H_{24}MnN_6{}^{2+}$	$[Mn(C_2H_8N_2)_3]^{2+}$	Mn:	MVol.D3-37/9
$C_6H_{24}MnN_6O_3SSe$	$Mn(C_2H_8N_2)_3SeSO_3$	Mn:	MVol.D3-40/1
$C_6H_{24}MnN_6O_3S_2$	$Mn(C_2H_8N_2)_3S_2O_3$.	Mn:	MVol.D3-40/1
$C_6H_{24}MnN_8O_6$	$Mn(C_2H_8N_2)_3(NO_3)_2$ solid solutions		
	$(Mn,Zn)(C_2H_8N_2)_3(NO_3)_2$	Mn:	MVol.D3-40
$C_6H_{24}MnN_{12}O_6{}^{2+}$ ...	$[Mn(OC(NH_2)_2)_6]^{2+}$	Mn:	MVol.D5-140
$C_6H_{24}MnN_{12}O_6{}^{3+}$ ...	$[Mn(OC(NH_2)_2)_6]^{3+}$	Mn:	MVol.D5-140/1
$C_6H_{24}MnN_{12}O_{10}S$ ...	$Mn(OC(NH_2)_2)_6SO_4$.	Mn:	MVol.D5-143/4
–	$Mn(OC(NH_2)_2)_6SO_4 \cdot H_2O$	Mn:	MVol.D5-143/4
$C_6H_{24}N_4O_{18}U_2$	$(NH_4)_4[(UO_2)_2(CH_2(COO)_2)_2(OO)_2(H_2O)_2]$ ..	U:	SVol.C13-238/41
$C_6H_{24}N_6Nd^{3+}$	$[Nd(C_2H_8N_2)_3]^{3+}$	Sc:	MVol.D1-19/21
$C_6H_{24}N_6O_{20}U_2$	$[C(NH_2)_3]_2[(UO_2)_2(C_2O_4)(CO_3)_2(H_2O)_6]$	U:	SVol.C13-228/9
$C_6H_{24}N_6O_{28}S_4U_2$ ...	$(NH_4)_6U_2(C_2O_4)_3(SO_4)_4$.	U:	SVol.C13-225/7
–	$(NH_4)_6U_2(C_2O_4)_3(SO_4)_4 \cdot 2\ H_2O$	U:	SVol.C13-225/7
–	$(NH_4)_6U_2(C_2O_4)_3(SO_4)_4 \cdot 4\ H_2O$	U:	SVol.C13-225/7
$C_6H_{24}N_6Os^{3+}$	$[Os(C_2H_4(NH_2)_2)_3]^{3+}$	Os:	SVol.1-251
$C_6H_{24}N_6Os^{4+}$	$[Os(C_2H_4(NH_2)_2)_3]^{4+}$	Os:	SVol.1-251
$C_6H_{24}N_6Pr^{3+}$	$[Pr(C_2H_8N_2)_3]^{3+}$	Sc:	MVol.D1-19/21
$C_6H_{24}N_6Rh^{3+}$	$[Rh(NH_2C_2H_4NH_2)_3]^{3+}$	Rh:	SVol.B2-162/6
$C_6H_{24}N_6Sm^{3+}$	$[Sm(C_2H_8N_2)_3]^{3+}$	Sc:	MVol.D1-19/21
$C_6H_{24}N_6Tb^{3+}$	$[Tb(C_2H_8N_2)_3]^{3+}$	Sc:	MVol.D1-19/21
$C_6H_{24}N_6Y^{3+}$	$[Y(C_2H_8N_2)_3]^{3+}$	Sc:	MVol.D1-19/21
$C_6H_{24}N_6Yb^{3+}$	$[Yb(C_2H_8N_2)_3]^{3+}$	Sc:	MVol.D1-19/21
$C_6H_{24}N_8O_6Zn$	$Zn(C_2H_8N_2)_3(NO_3)_2$ solid solutions		
	$(Zn,Mn)(C_2H_8N_2)_3(NO_3)_2$	Mn:	MVol.D3-40
$C_6H_{24}N_9O_9Tb$	$Tb(C_2H_8N_2)_3(NO_3)_3$	Sc:	MVol.D1-21/2
$C_6H_{24}N_{12}O_{10}U$	$(CN_3H_6)_4[UO_2O_2(CO_3)_2] \cdot 3\ H_2O$	U:	SVol.C13-24/5
$C_6H_{24}N_{12}O_{14}S_2Th$ ...	$Th(SO_4)_2 \cdot 6\ (NH_2)_2CO$	Th:	SVol.E-38, 41/2
$C_6H_{24}N_{12}O_{18}Pr_2S_3$ ..	$Pr_2(SO_4)_3 \cdot 6\ CO(NH_2)_2$.	Sc:	MVol.C8-133
		Sc:	MVol.D2-222/3
$C_6H_{24}N_{12}O_{18}S_3Tm_2$	$Tm_2(SO_4)_3 \cdot 6\ OC(NH_2)_2$	Sc:	MVol.D2-222/3
$C_6H_{24}N_{12}O_{18}S_3Yb_2$.	$Yb_2(SO_4)_3 \cdot 6\ OC(NH_2)_2$	Sc:	MVol.D2-222/3
$C_6H_{24}N_{12}OsS_6{}^{2+}$ ...	$[Os(SC(NH_2)_2)_6]^{2+}$	Os:	SVol.1-288
$C_6H_{24}N_{12}OsS_6{}^{3+}$ ...	$[Os(SC(NH_2)_2)_6]^{3+}$	Os:	SVol.1-288
$C_6H_{24}N_{12}RhS_6{}^{3+}$ ...	$[Rh(SC(NH_2)_2)_6]^{3+}$	Rh:	SVol.B3-9
$C_6H_{24}N_{15}O_{15}Pr$	$Pr(NO_3)_3 \cdot 6\ OC(NH_2)_2$	Sc:	MVol.D2-220
$C_6H_{24}N_{15}O_{15}Sc$	$Sc(NO_3)_3 \cdot 6\ OC(NH_2)_2$	Sc:	MVol.D2-219
$C_6H_{24}N_{16}O_{18}Th$	$Th(NO_3)_4 \cdot 6\ (NH_2)_2CO$	Th:	SVol.E-38, 40
–	$Th(NO_3)_4 \cdot 6\ (NH_2)_2CO \cdot 2\ H_2O$	Th:	SVol.E-38, 41
–	$Th(NO_3)_4 \cdot 6\ (NH_2)_2CO \cdot 4\ H_2O$	Th:	SVol.E-38, 41
$C_6H_{26}I_4N_{12}O_{12}U_2$ ...	$[UO_2(OH)(CO(NH_2)_2)_3]_2I_4$	U:	SVol.A6-40, 44
$C_6H_{28}N_{16}O_{20}Th$	$[Th((NH_2)_2CO)_6(H_2O)_2](NO_3)_4$	Th:	SVol.E-41
$C_6H_{30}I_2MnN_6$	$Mn(CH_3NH_2)_6I_2$	Mn:	MVol.D3-21/2

$C_6K_4MnN_6$........	$K_4[Mn(CN)_6]$	Mn: MVol.D2-210
–	$K_4[Mn(CN)_6]$ · $3 H_2O$	Mn: MVol.D2-210/5
$C_6K_4MnN_6S_2Se_4$....	$K_4[Mn(NCS)_2(NCSe)_4]$ · $0.5 (CH_3)_2CO$	Mn: MVol.D2-296/7
$C_6K_4MnN_6S_4Se_2$....	$K_4[Mn(NCS)_4(NCSe)_2]$ · $(CH_3)_2CO$	Mn: MVol.D2-296/7
$C_6K_4MnN_6S_6$......	$K_4[Mn(NCS)_6]$ · $3 H_2O$	Mn: MVol.D2-287
–	$K_4[Mn(NCS)_6]$ · $4 H_2O$	Mn: MVol.D2-287
$C_6K_4N_6Os$	$K_4[Os(CN)_6]$	Os: SVol.1-174/6, 178
–	$K_4[Os(CN)_6]$ · $3 H_2O$	Os: SVol.1-174/6
–	$K_4[Os(CN)_6]$ · $3 D_2O$	Os: SVol.1-174
$C_6K_5MnN_6$........	$K_5[Mn(CN)_6]$	Mn: MVol.D2-199/202
$C_6K_5N_6Tc$	$K_5[Tc(CN)_6]$.....................	Tc: SVol.2-125, 200
$C_6LaO_{12}{}^{3-}$	$[La(C_2O_4)_3]^{3-}$	Sc: MVol.D5-120/4
$C_6LaO_{14}{}^{5-}$	$[La(CO_3)_2(C_2O_4)_2]^{5-}$	Sc: MVol.D5-145/6
$C_6La_2O_{12}$	$La_2(C_2O_4)_3$	Sc: MVol.D5-132/5
–	$La_2(C_2O_4)_3$ · $8 H_2O$	Sc: MVol.D5-125
–	$La_2(C_2O_4)_3$ · $10 H_2O$	Sc: MVol.D5-124/36
–	$La_2(C_2O_4)_3$ · $10.5 H_2O$	Sc: MVol.D5-128
$C_6Li_2O_{16}U_2$	$Li_2[(UO_2)_2(C_2O_4)_3]$	U: SVol.C13-203/5
–	$Li_2[(UO_2)_2(C_2O_4)_3]$ · $7 H_2O$	U: SVol.C13-203/6
$C_6Li_3MnN_6$	$Li_3[Mn(CN)_6]$....................	Mn: MVol.D2-244
$C_6Li_3N_6Rh$.......	$Li_3[Rh(CN)_6]$	Rh: SVol.B1-190
$C_6Li_4N_6Os$........	$Li_4[Os(CN)_6]$ · x H_2O	Os: SVol.1-174
$C_6LuO_{12}{}^{3-}$	$[Lu(C_2O_4)_3]^{3-}$	Sc: MVol.D5-120/4
$C_6LuO_{14}{}^{5-}$	$[Lu(CO_3)_2(C_2O_4)_2]^{5-}$	Sc: MVol.D5-145/6
$C_6Lu_2O_{12}$	$Lu_2(C_2O_4)_3$	Sc: MVol.D5-133
–	$Lu_2(C_2O_4)_3$ · $6 H_2O$	Sc: MVol.D5-124/35
$C_6MgMnNaO_{12}$.....	$NaMgMn(C_2O_4)_3$ solid solutions	
	$NaMg(Mn,Al)(C_2O_4)_3$ · $9 H_2O$	Mn: MVol.D2-115
C_6MnNO_5	$[Mn(CO)_5CN]$.....................	Mn: MVol.D2-198
$C_6MnN_5O^{4-}$	$[Mn(CN)_5CO]^{4-}$	Mn: MVol.D2-198
$C_6MnN_6{}^{2-}$........	$[Mn(CN)_6]^{2-}$	Mn: MVol.D2-224/5, 227,
		248/9
$C_6MnN_6{}^{3-}$........	$[Mn(CN)_6]^{3-}$	Mn: MVol.D2-219/28
$C_6MnN_6{}^{4-}$........	$[Mn(CN)_6]^{4-}$	Mn: MVol.D2-203/8
$C_6MnN_6{}^{5-}$........	$[Mn(CN)_6]^{5-}$	Mn: MVol.D2-197/8
$C_6MnN_6Na_3$........	$Na_3[Mn(CN)_6]$ · $2 H_2O$	Mn: MVol.D2-228
–	$Na_3[Mn(CN)_6]$ · $3 H_2O$	Mn: MVol.D2-228/9
$C_6MnN_6Na_4$........	$Na_4[Mn(CN)_6]$ · $8 H_2O$	Mn: MVol.D2-209
–	$Na_4[Mn(CN)_6]$ · $10 H_2O$	Mn: MVol.D2-208/9
–	$Na_4[Mn(CN)_6]$ · n H_2O	Mn: MVol.D2-209
$C_6MnN_6Na_4S_6$......	$Na_4[Mn(NCS)_6]$ · $12 H_2O$	Mn: MVol.D2-287
$C_6MnN_6Na_5$........	$Na_5[Mn(CN)_6]$	Mn: MVol.D2-199
$C_6MnN_6Rb_3$........	$Rb_3[Mn(CN)_6]$	Mn: MVol.D2-243/4
$C_6MnN_6S_6{}^{4-}$......	$[Mn(NCS)_6]^{4-}$	Mn: MVol.D2-282/3, 286
$C_6MnN_6Se_6{}^{4-}$......	$[Mn(NCSe)_6]^{4-}$	Mn: MVol.D2-293/4
C_6MnN_6Zn........	$Zn[Mn(CN)_6]$	Mn: MVol.D2-251/2
$C_6MnNa_3O_{12}$.......	$Na_3[Mn(C_2O_4)_3]$ · n H_2O	Mn: MVol.D2-112
$C_6MnO_{12}{}^{3-}$........	$[Mn(C_2O_4)_3]^{3-}$	Mn: MVol.D2-107/11
$C_6MnO_{12}{}^{4-}$........	$[Mn(C_2O_4)_3]^{4-}$	Mn: MVol.D2-86/9
$C_6MnO_{12}Tl_3$	$Tl_3[Mn(C_2O_4)_3]$	Mn: MVol.D2-115

$C_6N_{12}O_{42}Th_4$	$Th_4(C_6O_6)(NO_3)_{12}$	Th:	SVol.E-147, 152
$C_6Na_2O_{16}U_2$	$Na_2[(UO_2)_2(C_2O_4)_3]$	U:	SVol.C13-203/5
–	$Na_2[(UO_2)_2(C_2O_4)_3] \cdot 5\,H_2O$	U:	SVol.C13-203/6
$C_6Na_3O_{12}Rh$	$Na_3[Rh(C_2O_4)_3] \cdot n\,H_2O$	Rh:	SVol.B2-2
$C_6Na_3O_{12}Sc$	$Na_3[Sc(C_2O_4)_3]$	Sc:	MVol.D5-116/7
–	$Na_3[Sc(C_2O_4)_3] \cdot 4\,H_2O$	Sc:	MVol.D5-117
–	$Na_3[Sc(C_2O_4)_3] \cdot 6\,H_2O$	Sc:	MVol.D5-116/7
C_6NdO_6	$Nd(CO)_6$	Sc:	MVol.D6-166
$C_6NdO_{12}{}^{3-}$	$[Nd(C_2O_4)_3]^{3-}$	Sc:	MVol.D5-120/4
$C_6NdO_{14}{}^{5-}$	$[Nd(CO_3)_2(C_2O_4)_2]^{5-}$	Sc:	MVol.D5-145/6
$C_6Nd_2O_{12}$	$Nd_2(C_2O_4)_3$	Sc:	MVol.D5-132/5
–	$Nd_2(C_2O_4)_3 \cdot 6\,H_2O$	Sc:	MVol.D5-129/30
–	$Nd_2(C_2O_4)_3 \cdot 6\,D_2O$	Sc:	MVol.D5-130
–	$Nd_2(C_2O_4)_3 \cdot 10\,H_2O$	Sc:	MVol.D5-124/6, 130/5
–	$Nd_2(C_2O_4)_3 \cdot 10\,D_2O$	Sc:	MVol.D5-130
–	$Nd_2(C_2O_4)_3 \cdot 10.5\,H_2O$	Sc:	MVol.D5-126/8
–	$Nd_2(C_2O_4)_3 \cdot n\,H_2O$	Sc:	MVol.D5-133/7
C_6O_6Pr	$Pr(CO)_6$	Sc:	MVol.D6-166
$C_6O_6RhS_6{}^{3-}$	$[Rh(S_2C_2O_2)_3]^{3-}$	Rh:	SVol.B3-24/5
$C_6O_6Tc^+$	$Tc(CO)_6{}^+$	Tc:	SVol.1-316/7
		Tc:	SVol.2-130, 301
–	$^{99m}Tc(CO)_6{}^+$	Tc:	SVol.1-255/60
C_6O_6U	$U(CO)_6$	U:	SVol.E2-171/2
C_6O_6Yb	$Yb(CO)_6$	Sc:	MVol.D6-166
C_6O_8U	$UO_2(C_6O_6)$	U:	SVol.D1-241
$C_6O_8U^{2-}$	$[UO_2(C_6O_6)]^{2-}$	U:	SVol.D1-241
$C_6O_{12}Pm^{3-}$	$[Pm(C_2O_4)_3]^{3-}$	Sc:	MVol.D5-120/4
$C_6O_{12}Pm_2$	$Pm_2(C_2O_4)_3 \cdot 3\,H_2O$	Sc:	MVol.D5-133
–	$Pm_2(C_2O_4)_3 \cdot 10\,H_2O$	Sc:	MVol.D5-125/6, 133
$C_6O_{12}Pr_2$	$Pr_2(C_2O_4)_3$	Sc:	MVol.D5-132/5
–	$Pr_2(C_2O_4)_3 \cdot 10\,H_2O$	Sc:	MVol.D5-124/6, 130/6
–	$Pr_2(C_2O_4)_3 \cdot 10.5\,H_2O$	Sc:	MVol.D5-128
$C_6O_{12}Rh^{3-}$	$[Rh(C_2O_4)_3]^{3-}$	Rh:	SVol.B2-2/6
$C_6O_{12}RhTl_3$	$Tl_3[Rh(C_2O_4)_3] \cdot 2\,H_2O$	Rh:	SVol.B2-6
$C_6O_{12}Sc^{3-}$	$[Sc(C_2O_4)_3]^{3-}$	Sc:	MVol.D5-112/3
$C_6O_{12}Sc_2$	$Sc_2(C_2O_4)_3$	Sc:	MVol.D5-113/5
–	$Sc_2(C_2O_4)_3 \cdot 6\,H_2O$	Sc:	MVol.D5-129/30
–	$Sc_2(C_2O_4)_3 \cdot x\,H_2O$ (x = 1 to 11.8)	Sc:	MVol.D5-113/5
$C_6O_{12}Sm_2$	$Sm_2(C_2O_4)_3$	Sc:	MVol.D5-133/5
–	$Sm_2(C_2O_4)_3 \cdot 5\,H_2O$	Sc:	MVol.D5-130
–	$Sm_2(C_2O_4)_3 \cdot 10\,H_2O$	Sc:	MVol.D5-124/6, 130/6
$C_6O_{12}Tb^{3-}$	$[Tb(C_2O_4)_3]^{3-}$	Sc:	MVol.D5-120/4
$C_6O_{12}Tb_2$	$Tb_2(C_2O_4)_3$	Sc:	MVol.D5-133/4
–	$Tb_2(C_2O_4)_3 \cdot 10\,H_2O$	Sc:	MVol.D5-125/36
$C_6O_{12}Tm^{3-}$	$[Tm(C_2O_4)_3]^{3-}$	Sc:	MVol.D5-120/4
$C_6O_{12}Tm_2$	$Tm_2(C_2O_4)_3 \cdot 5\,H_2O$	Sc:	MVol.D5-130, 133
–	$Tm_2(C_2O_4)_3 \cdot 6\,H_2O$	Sc:	MVol.D5-124/5, 128/34
–	$Tm_2(C_2O_4)_3 \cdot 18\,H_2O$	Sc:	MVol.D5-125
$C_6O_{12}Y^{3-}$	$[Y(C_2O_4)_3]^{3-}$	Sc:	MVol.D5-120/4
$C_6O_{12}Y_2$	$Y_2(C_2O_4)_3$	Sc:	MVol.D5-132/3, 135/6

$C_6O_{12}Y_2$	$Y_2(C_2O_4)_3$ · 6 H_2O	Sc:	MVol.D5-125, 134
–	$Y_2(C_2O_4)_3$ · 9 H_2O	Sc:	MVol.D5-124/5, 132/3
–	$Y_2(C_2O_4)_3$ · 10 H_2O	Sc:	MVol.D5-125, 133/6
–	$Y_2(C_2O_4)_3$ · 17 H_2O	Sc:	MVol.D5-125
–	$Y_2(C_2O_4)_3$ · 18 H_2O	Sc:	MVol.D5-125
$C_6O_{12}Yb^{3-}$	$[Yb(C_2O_4)_3]^{3-}$	Sc:	MVol.D5-120/4
$C_6O_{12}Yb_2$	$Yb_2(C_2O_4)_3$	Sc:	MVol.D5-135
–	$Yb_2(C_2O_4)_3$ · 5 H_2O	Sc:	MVol.D5-125, 130, 133
–	$Yb_2(C_2O_4)_3$ · 6 H_2O	Sc:	MVol.D5-124/5, 128/36
–	$Yb_2(C_2O_4)_3$ · 9 H_2O	Sc:	MVol.D5-125
–	$Yb_2(C_2O_4)_3$ · 18 H_2O	Sc:	MVol.D5-125
$C_6O_{14}Pr^{5-}$	$[Pr(CO_3)_2(C_2O_4)_2]^{5-}$	Sc:	MVol.D5-145/6
$C_6O_{14}Sm^{5-}$	$[Sm(CO_3)_2(C_2O_4)_2]^{5-}$	Sc:	MVol.D5-145/6
$C_6O_{14}Tb^{5-}$	$[Tb(CO_3)_2(C_2O_4)_2]^{5-}$	Sc:	MVol.D5-145/6
$C_6O_{14}Tl_4U$	$Tl_4[UO_2(C_2O_4)_3]$	U:	SVol.C13-201
$C_6O_{14}Tm^{5-}$	$[Tm(CO_3)_2(C_2O_4)_2]^{5-}$	Sc:	MVol.D5-145/6
$C_6O_{14}U^{4-}$	$[UO_2(C_2O_4)_3]^{4-}$	U:	SVol.C13-201/2
$C_6O_{14}Y^{5-}$	$[Y(CO_3)_2(C_2O_4)_2]^{5-}$	Sc:	MVol.D5-145/6
$C_6O_{14}Yb^{5-}$	$[Yb(CO_3)_2(C_2O_4)_2]^{5-}$	Sc:	MVol.D5-145/6
$C_6O_{16}Rb_2U_2$	$Rb_2[(UO_2)_2(C_2O_4)_3]$	U:	SVol.C13-204
–	$Rb_2[(UO_2)_2(C_2O_4)_3]$ · 3 H_2O	U:	SVol.C13-204/6
$C_6O_{16}SU_2$	$U_2(C_2O_4)_3(SO_4)$	U:	SVol.C13-225/7
–	$U_2(C_2O_4)_3(SO_4)$ · 2 H_2O	U:	SVol.C13-225/7
–	$U_2(C_2O_4)_3(SO_4)$ · 4 H_2O	U:	SVol.C13-225/7
–	$U_2(C_2O_4)_3(SO_4)$ · 8 H_2O	U:	SVol.C13-225/7
–	$U_2(C_2O_4)_3(SO_4)$ · 12 H_2O	U:	SVol.C13-225/7
$C_6O_{16}Tl_2U_2$	$Tl_2[(UO_2)_2(C_2O_4)_3]$	U:	SVol.C13-204/6
$C_6O_{16}U_2{}^{2-}$	$[(UO_2)_2(C_2O_4)_3$ · 4 $H_2O]^{2-}$	U:	SVol.C13-205/6
–	$[(UO_2)_2(C_2O_4)_3]_n{}^{2n-}$	U:	SVol.C13-204/6
$C_6O_{24}Rb_4S_3U_2$	$Rb_4U_2(C_2O_4)_3(SO_4)_3$	U:	SVol.C13-225/7
–	$Rb_4U_2(C_2O_4)_3(SO_4)_3$ · 4 H_2O	U:	SVol.C13-225/7
–	$Rb_4U_2(C_2O_4)_3(SO_4)_3$ · 6 H_2O	U:	SVol.C13-225/7
$C_6O_{24}U_3{}^{6-}$	$[(UO_2)_3(CO_3)_6]^{6-}$	U:	SVol.D1-170/4
$C_6O_{28}Rb_6S_4U_2$	$Rb_6U_2(C_2O_4)_3(SO_4)_4$	U:	SVol.C13-225/7
–	$Rb_6U_2(C_2O_4)_3(SO_4)_4$ · 2 H_2O	U:	SVol.C13-225/7
–	$Rb_6U_2(C_2O_4)_3(SO_4)_4$ · 4 H_2O	U:	SVol.C13-225/7
$C_6O_{30}U_6$	$(UO_2)_6(CO_3)_6$	U:	SVol.C13-4
C_6U	UC_6	U:	SVol.A6-180
		U:	SVol.C12-54/5
C_6W_{23}	$W_{23}C_6$ solid solutions		
	$(Mn,W)_{23}C_6$	Mn:	MVol.C7-170/1
$C_{6.22}CeCl_{2.98}H_{18.66}O_{3.11}$			
	$CeCl_{2.98}$ · 3.11 C_2H_5OH	Sc:	MVol.D3-18/9
$C_{6.5}Cl_3H_{6.5}N_{1.3}Sm$ . .	$SmCl_3$ · 1.3 C_5H_5N	Sc:	MVol.D1-34
$C_{6.6}CeCl_{3.01}H_{17.6}O_{2.2}$			
	$CeCl_{3.01}$ · 2.2 i-C_3H_7OH	Sc:	MVol.D3-21
$C_{6.99}H_{20.97}N_4O_{14.33}P_{2.33}Th$			
	$Th(NO_3)_4$ · 2.33 $(CH_3)_3PO$	Th:	SVol.E-77, 80

Key to the Gmelin System
of Elements and Compounds

System Number	Symbol	Element
1		Noble Gases
2	H	Hydrogen
3	O	Oxygen
4	N	Nitrogen
5	F	Fluorine
6	**Cl**	**Chlorine**
7	Br	Bromine
8	I	Iodine
8a	At	Astatine
9	S	Sulfur
10	Se	Selenium
11	Te	Tellurium
12	Po	Polonium
13	B	Boron
14	C	Carbon
15	Si	Silicon
16	P	Phosphorus
17	As	Arsenic
18	Sb	Antimony
19	Bi	Bismuth
20	Li	Lithium
21	Na	Sodium
22	K	Potassium
23	NH_4	Ammonium
24	Rb	Rubidium
25	Cs	Caesium
25a	Fr	Francium
26	Be	Beryllium
27	Mg	Magnesium
28	Ca	Calcium
29	Sr	Strontium
30	Ba	Barium
31	Ra	Radium
32	**Zn**	**Zinc**
33	Cd	Cadmium
34	Hg	Mercury
35	Al	Aluminium
36	Ga	Gallium

HCl

$ZnCl_2$

System Number	Symbol	Element
37	In	Indium
38	Tl	Thallium
39	Sc, Y La—Lu	Rare Earth Elements
40	Ac	Actinium
41	Ti	Titanium
42	Zr	Zirconium
43	Hf	Hafnium
44	Th	Thorium
45	Ge	Germanium
46	Sn	Tin
47	Pb	Lead
48	V	Vanadium
49	Nb	Niobium
50	Ta	Tantalum
51	Pa	Protactinium
52	**Cr**	**Chromium**
53	Mo	Molybdenum
54	W	Tungsten
55	U	Uranium
56	Mn	Manganese
57	Ni	Nickel
58	Co	Cobalt
59	Fe	Iron
60	Cu	Copper
61	Ag	Silver
62	Au	Gold
63	Ru	Ruthenium
64	Rh	Rhodium
65	Pd	Palladium
66	Os	Osmium
67	Ir	Iridium
68	Pt	Platinum
69	Tc	Technetium[1]
70	Re	Rhenium
71	Np,Pu . . .	Transuranium Elements

$CrCl_2$

$ZnCrO_4$

Material presented under each Gmelin System Number includes all information concerning the element(s) listed for that number plus the compounds with elements of lower System Number.

For example, zinc (System Number 32) as well as all zinc compounds with elements numbered from 1 to 31 are classified under number 32.

[1] A Gmelin volume titled "Masurium" was published with this System Number in 1941.

A Periodic Table of the Elements with the Gmelin System Numbers is given on the Inside Front Cover